The Positive Mind

The Positive Mind

Its Development and Impact on Modernity and Postmodernity

Evaldas Nekrašas

Central European University Press
Budapest–New York

Published in 2016 by

Central European University Press

An imprint of the
Central European University Limited Liability Company
Nádor utca 11, H-1051 Budapest, Hungary
Tel: +36-1-327-3138 or 327-3000
Fax: +36-1-327-3183
E-mail: ceupress@ceu.hu
Website: www.ceupress.com

224 West 57th Street, New York NY 10019, USA
Tel: +1-732-763-8816
E-mail: meszarosa@ceu.hu

ISBN 978-963-386-081-6

Library of Congress Cataloging-in-Publication Data

Nekrašas, Evaldas.
 The positive mind : its development and impact on modernity and
postmodernity / Evaldas Nekrašas.
  pages cm
 Includes bibliographical references and index.
 ISBN 978-9633860816 (hardbound)
 1. Positivism--History. I. Title.

B831.N45 2015
146'.4--dc23

2015000626

Printed in Hungary by
Prime Rate Kft., Budapest

Contents

Preface

This book is an intellectual adventure story, a history of ideas, and a rigorous reappraisal of a major movement in philosophy, science, and culture that many have pronounced irrelevant, passé, even dead. Yet upon closer consideration, we may find that what we have come to call positivism has profoundly influenced our thought and practice in numerous ways. It would be difficult to find another philosophy that has been more influential, and at least by that measure, we may say that positivism not only survives, but thrives.

We aim here to investigate the positive mind, a term which astute readers would associate with the scientific mind, though it is often confused with the pop psychology of Norman Vincent Peale (1996 [1959]) presented in his perennial bestseller *The Power of Positive Thinking*. Here, we examine positivist philosophy, its significance and its legacy. We can treat it as a history of the ventures, achievements, and failures of the positive mind. The words "impact on modernity and postmodernity" in the subtitle indicate that, for many reasons discussed in this book, positivism, although no longer cultivated as a distinct current of philosophy, still continues to exert a profound influence on modern and postmodern developments inside and outside philosophy.

I embarked on writing this book many years ago. The project began in earnest while I was a Senior Fulbright Scholar at the Department of Philosophy at Harvard University. There I had an outstanding opportunity to discuss its main ideas and first sections with W. V. O. Quine and Hilary Putnam—philosophers who have significantly influenced the development of positivism and twentieth-century philosophy in general. We'll encounter their work in this book. I was encouraged to hear from Putnam that he had always wanted to have such a book at hand.

The Harvard experience was clearly useful for this book, although the idea to write it originated a few years earlier. While developing a course on positivist philosophy for my students at Vilnius University, I was unable to find a suitable book for their reading list. I needed a text in which different forms of positivism would be analyzed, its relations to other trends of philosophy scrutinized, and its influence beyond the realm of philosophy explained.

The situation has changed little in recent years. I think that the main reason for this gap in the available literature is the view, prevailing in the history of philosophy and the history of ideas, that there were only two positivisms—Comte's social positivism of the nineteenth century and the logical positivism of the twentieth century—and that they have almost nothing in common. The question why two philosophical trends that have been developed at different times are called the same name is rarely posed.

After having raised this question, I realized that once a more precise definition of positivism is given, we are compelled to recognize that more forms of positivism exist, and that the real progenitor of positivism is not Comte, as is usually supposed, but David Hume. Positivism is a philosophical trend that has existed for more than two centuries. It has taken various forms in different epochs, but it has always been founded on the same principles. Such diverse thinkers as David Hume, Jean le Rond d'Alembert, August Comte, John Stuart Mill, Ernst Mach, Moritz Schlick, Rudolf Carnap, Hans Reichenbach, and A. J. Ayer all adhered to these principles. They all share a common tradition.

In examining this tradition and its contemporary impact, I have several goals:

First, my aim is to define the notion of positivism more precisely and systematically.

Second, to describe positivism as a trend of thought concerned not only with the theory of knowledge and philosophy of science, but also with problems of ethics and political philosophy, and whose representatives usually thought that the latter cannot be solved without solving the former.

Third, to analyze the development of positivism as a movement which holds up a certain tradition and hence possesses some coherence, even though the forms of this movement changed in different historical circumstances: it was born in the eighteenth century during the Enlightenment, it took the form of social positivism in the nineteenth century, was transformed at the turn of the twentieth century with the emergence of empiriocriticism, and became logical positivism (or logical empiricism) in the twentieth century.

Fourth, to reveal the external factors and internal logic of this evolution, to determine not only the divergence of its various forms, but also the connections between them, to explain the differences between the views of the representatives of a particular branch of positivism, and to explore the reasons behind those differences.

Fifth, to explain the relation of positivism to other trends of philosophy, first of all to those that are akin, such as Marxism, pragmatism, critical rationalism, analytical philosophy, and the historical school of philosophy of science, and to reveal the character and scope of positivism's influence upon those trends. Also, to determine the *true* relation to positivism of thinkers and movements harshly critical of positivism, such as Nietzsche, Heidegger, or proponents of postmodernism. Having accomplished this task, the outline of a new, nonstandard image of the philosophical development of the last three centuries should emerge.

Sixth, to determine the influence the positive mind had not only upon philosophy, but upon other cultural phenomena: the natural and social sciences, law, politics, arts, religion, and everyday life.

Having listed the main tasks of this book, I realize that they may seem too ambitious. The reader will decide whether, and to what degree, I have succeeded. I did not intend to write an academic study confined to a limited set of problems solely for a scholarly audience, but to write a book that may be of interest to a broad circle of readers, that would include not only students, teachers, and researchers in diverse disciplines, but also all lovers of philosophy, a book in which positive thinking is presented in its multifarious facets and viewed from different angles, a book where the origins, problems, and dilemmas of the positive mind are discussed, where its challenges are revealed, and the entire drama of its development is presented. Today this drama is even more relevant due to harsh postmodernist attacks against the entire positivist project, which they take to be the worst incarnation of modernity and modernism. Yet those attacks did not destroy interest in positivism, especially in its latest form, namely that of logical empiricism. In the last two decades, the regard for logical empiricism is clearly ascending. This fact is attested by the rising number of articles and books devoted to the subject. In general they offer a view on logical empiricism which is more balanced than critical texts written earlier. From those later texts I would mention only a few: *The Cambridge Companion to Logical Empiricism* (Richardson and Uebel 2007), *Reconsidering Logical Positivism* (Friedman 1999), and *Origins of Logical Empiricism* (Giere and Richardson 1996).

What is the relation of this book to other texts that explore the development of entire positivism, and not just of one of its forms? I have not managed to find any works (at least in major languages) that raised all the questions I have listed above. The book closest to this one in its *chrono-logical* scope would be a relatively brief study, *Positivist Philosophy* by Leszek Kołakowski (1972 [1966]). I would also suggest the even shorter *Positivismus* by Reinhard Kamitz (1973). However, in these and other texts that I did not mention, including the most recent ones, only some of the questions I raise are solved, and even those are usually solved in different ways, or with a substantially narrower scope. Thus Michael Singer's (2006) book, with its promising title *The Legacy of Positivism*, is in fact a work devoted to Auguste Comte and *his* legacy.

This book would most likely belong to the genre of comparative philosophy. One cannot get away without comparative analysis while investigating the development of a *movement* that lasted for more than two centuries. It requires comparing positivism and other trends of philosophy, contrasting its different forms as well as the views of different representatives of a particular form. I have been mostly interested in the complicated relations between classical and logical positivism. Although they share the fundamental tenets of positivism, logical positivists have rejected many ideas of Comte, the founder of classical positivism. However, I was searching not only for the differences between these two branches of positivism, but also for their similarities. A quick example: the relation between Schlick and Carnap, the two key figures of logical positivism, resemble in many respects the relation between the two grandees of classical positivism—Comte and Mill. That is one of the reasons why two important chapters of this book have a bipolar structure and why rather different views of the philosophers in question are juxtaposed there.

Positive thinking is closely tied to scientific thinking. Often they are simply identified. But is such identification justified? I invite the reader to pursue the answer to this question together with me. However, I must note at once that during the last few centuries, science and technology have radically changed and continue to change not only the world around us, but also ourselves, our view of the world, and our place in it. Of all the philosophical trends, positivism undoubtedly exerted the strongest influence upon the progress of contemporary science. Thus, if we want to comprehend the nature of our contemporary technological civilization, and its (maybe baneful) course of development, we must understand the positive mind, its logic, and evaluate its successes and failures.

Before closing this preface I would like to draw the reader's attention to one more peculiarity of this book. We first look at positivism from within, and later, from without. In the first part of the book I devote considerable time to the scrutiny of the ideas of positivism, to the analysis of its development, and the examination of the relations among various forms of positivism, as well as the debate among its representatives. In the second part I will present the main arguments against positivism leveled by thinkers outside of positivism. There, I attempt to evaluate the fundamental ideas of positivism and examine positivism's relation to and interaction with other philosophical trends. Only after having explored the attitude towards positivism of such diverse philosophers as Nietzsche, Heidegger, Wittgenstein, Popper, Quine, Putnam, and Kuhn, and having discussed positivism's influence on them, can we attempt to determine the worth of the main ideas of positivism. But the reader should postpone his or her final verdict until he or she has pondered more over the impact of positivism upon various spheres of our public and private lives, our thinking and lifestyle, that is, until having turned the last page of this book.

Acknowledgments

I would like to express my deepest gratitude to all those who have helped me in polishing and refining the text of the book originally written in far from exemplary English. I profoundly appreciate the help and support shown by David Burgess and Jonas Dagys. They are my former students, teaching now at the University of Iowa and Vilnius University. My thanks also goes to Artūras Rožėnas, a former student of mine too, instructing now at New York University, who has helped me in improving the second chapter.

The Notion of Positivism

The main goal of this introduction is to define the notion of positivism. I have to say right away that this brief (as it has to be) definition is not a full and complete answer to the question *"what is positivism?"* My detailed answer (whether or not the reader agrees) will become apparent only after reading the whole book. The definition which I will unfold in no more than several pages has a different purpose—to provide a first explanation of what is the philosophical trend whose history and significance are examined in this book.

The simplest way to provide such an explanation would be to open some authoritative dictionary or handbook of philosophy and quote it. This, however, hardly serves our purpose. For example, in the well-known *Oxford Dictionary of Philosophy*, upon finding an entry for "positivism" we would read: "The philosophy of Comte, holding that the highest or only form of knowledge is the description of sensory phenomena" (Blackburn 1996: 294).

This definition is deficient for at least two reasons. First, although Comte really called his own philosophy "positive" (*philosophie positive* in French), the term "positivism" in the history of philosophy (and culture in general) is used with much broader meaning, by also considering other philosophers as positivists. Certain thinkers, who called themselves positivists, e.g., John Stuart Mill, and logical positivists even more so, completely disagreed with some of Comte's views.

Second, according to Comte, the aim of knowledge is a description not of sensory phenomena, but of things or events, and not so much their description, as their prediction. Comte was not a phenomenalist, that is, he did not claim that we only know our sensations (sensory phenomena), nor did he claim that things are merely a persistent possibility of sensations. The controversy between phenomenalism and antiphenomenalism had an important role in the history of positivism, and we are going to talk about that, but it would be an error to identify positivism with phenomenalism. Therefore, we have to declare that Blackburn's definition (although we've quoted only the main part of it) is false. But can definitions be considered true or false at all? I am going to provide a more precise definition of positivism and arguments in support of it, but its real value shall become clearer only when the reader becomes more familiar with the history of positivism and examines its influence.

I will define the notion of positivism by listing the key principles of positivist philosophy. *First of all, positivism maintains that only the empirical world is accessible to us and rejects all metaphysics which claims that we may know more. Second, it strives to create a scientific philosophy. Third, it has no doubt that scientific knowledge is the only genuine and reliable knowledge. Fourth, it holds the conviction that scientific knowledge of the world relies on experience—observation and experiment. Fifth, positivism believes in the unity of scientific knowledge, especially the unity of the natural and social sciences. Sixth, it strongly supports the strict distinction between facts and values, which also entails the thesis that scientific knowledge is value-free. And seventh, positivism holds that scientific knowledge, although value-free, is the main force of social progress.*

Now I will try to elaborate on these seven principles of positivist philosophy. It may seem slightly paradoxical that the major tenet shared by *positivists* is rather a *negative* one. However, after further consideration, it does not look so strange, since aversion to something—in politics as much as in other domains—usually unites people more than positive programs. The negative attitude I am speaking of is the hostility towards metaphysics. When speaking about metaphysics, positivists had in mind first of all transcendent metaphysics, which aspires to explain what is *behind, above, or beyond* experience. Positivists regarded questions about the essences of things, the nature of being, and the first (primary) and the ultimate (purposive) causes as typically metaphysical. Metaphysics has been often criticized in the course of the history of philosophy; however, positivists took not a merely critical, but a *radically* critical attitude towards metaphysics. They intended not to weaken metaphysics, but to *overcome* it, defeat it, and permanently remove it from the sphere of scientific thinking. Science cannot transcend the boundaries of direct or indirect experience.

The second constitutive principle of positivism is its intention to create scientific philosophy. Most historians of science have no doubts that various special sciences—from mathematics and physics to psychology and sociology—have essentially evolved from philosophy, although the issue of the genesis of specialized sciences is rather vexed and we will come back to this later.[1] The scientific revolution of the seventeenth century was of special importance to the separation of (specialized) sciences from philosophy as universal knowledge about humans and the world. During this revolution, science began to acquire the status of an autonomous intellectual force. The conditions emerged that allowed for the demarcation of scientific thinking, mostly supported by the principles of mathematized natural sciences, from the erstwhile philosophical, and especially metaphysical, thinking. At the same time the road was paved for the attempts to reform traditional philosophy, or, to be more exact, a part of traditional philosophy according to the example of (specialized) science.

The third constitutive principle of positivism is the thesis of scientific knowledge as the only genuine knowledge. Science is the prime object of interest for positivist philosophy. It is also the model of genuine knowledge. Knowledge acquired via methods that are not advanced in science, first of all in mathematics and natural sciences, cannot be considered knowledge, but is merely an illusion of knowledge. Insight, intuition, afflatus, and revelation are not paths to reliable knowledge. In effect, if someone claims that he knows something, he is obliged to clearly state where this knowledge comes from, that is, through which methods he acquired this knowledge, and plainly indicate its foundation and its sources. Knowledge deserves this name only if it is verifiable, intersubjective (i.e., public), understandable for all, and not based upon purportedly deep insights inaccessible to others. People who intentionally flout the methods of scientific thinking, in fact, propound what they themselves do not understand and what others cannot understand. Of course, we can quite sensibly talk about casual, everyday knowledge, but it also can only be called knowledge as much as it rests (at least ultimately) on experience.

We thereby reach positivism's fourth constitutive principle—the conviction that the basis of all scientific knowledge *of the world* is experience, that is, observation and experiment. In this respect positivism obviously relies on the tradition of empiricism. One could even say that posi-

---

[1] I will consider physics as a special science, although some authors would exclude it from the list of such sciences as a *supposedly* fundamental one. It is a special science at least in a sense—physics is *one* of the natural sciences which *evolved out* of philosophy. And it is impossible to reduce mathematics to physics. Thus physics cannot be regarded as *the* fundamental science. The question of the relation of physics to other special sciences will be dealt with below, especially in Chapter 4.

tivism grows out of the empiricist tradition, as it attempts to rethink it more rigorously. The fact that David Hume—the progenitor of positivism—is also the most prominent representative of empiricism, is quite revealing.

From the positivist point of view, logic and/or mathematics also play significant roles in acquiring knowledge. The views about their importance and nature varied among positivists, but they all tended to emphasize that those are but tools, not foundations, of knowledge. They allow us, on the basis of experiential facts, to determine the laws of nature that express regular relations among facts.

All positivists recognized that empirical and mathematical knowledge are the only kinds of knowledge. (One should note that while speaking about those two kinds of knowledge, positivists have clearly understood that empirical, i.e., nonmathematical, knowledge can have the form of a theory.) In his *Enquiry Concerning Human Understanding*, Hume propounded his conviction that only two sorts of knowledge exist in uncharacteristically strident language: "If we take in our hand any volume; of divinity or school metaphysics, for instance; let us ask, *Does it contain any abstract reasoning concerning quantity or number?* No. *Does it contain any experimental reasoning concerning matter of fact and existence?* No. Commit it then to the flames: For it can contain nothing but sophistry and illusion" (Hume 1981 [1748]: 165).[2]

The fifth tenet of positivism is the principle of unity of scientific knowledge. Positivists have been convinced that although various scientific disciplines explore different spheres or aspects of reality and analyze different phenomena, they all invoke the same or, at least, similar methods. Hence science is essentially unified. No insurmountable borders between scientific disciplines exist: nothing prevents scientists from achieving mutual understanding and agreement. The representatives of positivism were especially eager to emphasize that social sciences must be guided by the same methods as natural sciences.

---

[2] When exploring an intellectual tradition that lasted for more than two centuries, it is important to make known when a work quoted or mentioned was published for the first time. Therefore, a slightly modified Harvard system of quotation is used in this book— apart from the author, year of publication, and the page number(s) of the quotation, the parenthesis also indicates (in square brackets) the year of the first publication of a particular work. The works are listed in the bibliography in the same manner. The lack of square brackets means that, as far as the author of this book knows, the first edition of a work is cited. If a work is just mentioned but not cited, only the year of the first publication is indicated.

The sixth tenet is the strict dichotomy of facts and values. From the positivist point of view, what *is* and what *ought* to be are radically different matters. Science must confine its interest only to the investigation of what *is*, and to avoid any value judgments. The latter cannot be derived from facts. Normative statements have no cognitive content—they can only be formulated beyond science.

The seventh constitutive principle of positivism is the conviction that scientific knowledge, being value-free by itself, is the fundamental force of social progress. Progressivism was characteristic of all positivists, that is, they all shared the view that society and the social world are improving or at least may be improved, and that the progress of scientific knowledge is the basis of this improvement, of the amelioration of social relations, and of the growth of economic welfare.

Value-free science speaks about what the world is like, but in so doing, it does not affirm the social, political, or any other *status quo.* Scientific knowledge is instrumental and can be successfully used as a tool for the betterment of the social world. The statement that *knowledge is power*—Francis Bacon's famous dictum from before the birth of positivism—has always been a benchmark for positivists, who clearly comprehended their social mission. They felt responsibility for the future of the world and earnestly bore the burden of this responsibility. That is most apparent in the so-called classical positivism of the nineteenth century.

Another question that may deserve a brief discussion is the relation between the terms "positivist philosophy" and "positive philosophy." Positivists themselves, at least in the nineteenth century, used the term "positive philosophy," which expressed their belief that it is the only true, constructive philosophy that helps to solve the main problems of humanity, and that provides clear guidelines for so doing. Meanwhile, scholars and critics of positivism, who were unwilling to prematurely endorse this self-description of positivists, used the term "positivist philosophy." This latter, more neutral term became entrenched in the history of philosophy.

In general, far from all positivists liked being called philosophers—they often referred to themselves simply as scientists, scholars, or researchers. Some adherents of positivism disliked the term "positivism" as well, because positive philosophy was supposed to have overcome and risen above all "isms."

As I have mentioned, positivism has existed for more than two centuries. What were the main stages or phases of its development? Partly repeating myself, for I've mentioned it in the preface, I am inclined to take the early-eighteenth-century positivism to be the first stage of the development of positivism. Its most prominent representative was David Hume, and Jean le Rond d'Alembert should be mentioned among its lesser figures. The second phase was the classical nineteenth-century positivism (also called social positivism), whose key representatives were Auguste Comte and John Stuart Mill; sometimes Herbert Spencer is also mentioned next to them, although his connection to the main thread of the development of positivism is more tenuous. The third phase was a slightly specific, transitionary phase between classical, social positivism and the logical positivism of the twentieth century. Its most prominent representative was the empirio-criticist Ernst Mach, and we should certainly mention Henri Poincaré and Pierre Duhem among others. The last stage of the development of positivism is twentieth-century positivism—logical positivism, also called logical empiricism. It sprung out of the Vienna Circle, and its key figures were Moritz Schlick and Rudolf Carnap. Among other logical positivists, Hans Reichenbach should be duly mentioned, as well as A. J. Ayer, whose name is associated with logical positivism in the English-speaking world.

A question naturally arises: can thinkers so diverse, from various epochs and different countries, be taken to represent one single movement? In other words, aren't the boundaries of positivist philosophy excessively extended in this book? In answering this question I will repeat the fact mentioned in the preface that positivism took different shapes in different epochs. Those shapes were markedly different, and I will devote substantial attention to the analysis of those differences. Here, in the introduction, I would note that in the course of positivism's development, even the notion of scientific philosophy that its adherents intended to create has changed. Hume construed scientific philosophy as a general science of man that rests on the methods of the natural sciences. In Comte's view, however, human individuals are almost nonentities, and society is supreme. Thus, for him sociology is the supreme scientific philosophy, which, through analyzing social phenomena, explains the origins of various ways of thinking and various organizational forms of society. For Mach, scientific philosophy is mainly a psychology of science which explores the peculiarities of scientific thinking and is of a descriptive character. Meanwhile, logical positivists treated scientific philosophy as the logic of science, which they took to be normative, providing rigorous logical requirements that had to be satisfied by scientific statements and theories.

It seems all this is sufficient for putting in doubt the statement that positivist philosophy is an integral intellectual movement. One more argument against the view defended in this book could be advanced: although all positivists claim that scientific knowledge must rest on facts, they all interpret the notion of "fact" differently—some treat facts as external physical events, others view them as our sensations and perceptions. And there are yet more tensions within positivist philosophy.

In spite of all these and other possible counterarguments, I am prepared to persistently defend from criticism the notion of positivism provided above—the notion, not positivist philosophy itself, I have to stress. In this defense I rely on the observation that *all positivists of various epochs and different orientations that I've mentioned, even while disagreeing on some quite important issues, have always subscribed to the seven aforementioned fundamental principles of positivist philosophy.* In fact, it is the main reason why these tenets have been listed and why they are treated as constitutive principles of positivism. Precisely this core of shared beliefs allows us to speak about the development of *positivism* as a movement displaying certain unity, because *while its forms changed for intellectual and social reasons, its fundamental tenets held firm.*

The core of shared beliefs allows one to speak about the identity of positivism, which it has preserved through serious theoretical quarrels among its adherents, through severe criticism from its opponents, and through cultural and political changes. Of course, positivism changed together with the social world, upon whose evolution it also made a strong impact. The tenets of positivism received different interpretations and argumentations during different stages of its development. Some were considered more significant, others less. New problems were considered, different methods of analysis applied, new ideas and conceptions were formulated that enriched positivist philosophy, but despite all these transformations positivism had remained an important trend of philosophy for more than two centuries, exerting great influence upon other philosophical movements as well as various other forms of human culture. This influence is felt until now.

Until now in this introduction I have not said a word about the notion of *the positive mind*, which figures boldly in the title of the book. Now, after acquiring some preliminary knowledge about positivism, we may define the positive mind very simply: it is the mind shaped or inspired, directly or indirectly, by positivist philosophy, that is, the mind which takes for granted its main principles.

PART ONE

Development

CHAPTER 1

Early Positivism

The Divorce between Philosophy and Science

Positivism is closely related to science. In many ways, positivism originated from scientific progress. Hence, tracing the roots of positivism, it seems quite natural to start by analyzing the circumstances in which science begins to oppose and sever itself from philosophy, the discipline of which it was a part for centuries. This separation occurred in modern times and has impacted later relationships between philosophy and the sciences. However, before embarking on a discussion of how these relationships were changed by the first scientific revolution of the seventeenth century, one should perhaps evaluate an old belief that, allegedly, for many centuries, philosophy meant the same as what science means today. As many other received opinions, this one is neither entirely clear nor entirely correct. Let us begin with the problem of clarity.

The English word *science* originates from the Latin word *scientia*, which means knowledge, understanding, capability, or skill. A question arises whether term *philosophia* and *scientia* meant the same thing before the first scientific revolution of the seventeenth century. The answer appears to be negative: *scientia* meant not only theoretical knowledge but also practical knowledge acquired by experience. This seems to have been the reason why, in the aftermath of the first scientific revolution, when empirical methods started to dominate the natural sciences, these sciences were with increasing frequency referred to not as *philosophia* but as *scientia*. However, it should be noted that Isaac Newton's most important book that laid down the foundations for modern science, and foremost physics, was titled *Philosophia naturalis principia mathematica* (1686–87); that is, *Mathematical Principles of Natural Philosophy*.

The view that until at least the eighteenth century, philosophy and science meant the same thing could be also supported by an institutional argument: During the medieval era and at the start of modernity, European universities had four faculties or divisions—philosophy, theology, law, and medicine. All subjects, except the latter three, were taught and researched in faculties of philosophy. Even today, many universities award the degree of Doctor of Philosophy for dissertations that study various scientific fields, for instance, physics or mathematics. However, it is precisely in the seventeenth century that the paths of science and philosophy started to diverge.

Philosophy that avoids elevated questions, systematically uses experimentation, and employs mathematical language in formulating laws of nature differs greatly from the philosophy scholars were acquainted with before the scientific revolution. The name *natural philosophy* was retained and used very widely throughout the seventeenth and eighteenth centuries; yet the nature of the natural philosophy as practiced by Galileo, Newton, Euler, or Lavoisier was very distinct from the nature of the other branches of traditional philosophy, and especially that of metaphysics. In fact, the old name *natural philosophy* concealed the form of knowledge which did not have much to do with the first philosophy in the Aristotelian sense. Not all heroes of the scientific revolution were inclined to agree with this. Thus René Descartes still looked at science as a tree whose roots (metaphysics) nourish not only the trunk (physics) but, in addition, all of the other branches: mechanics, medicine, and so on. Yet doubts regarding the relation of physics (and other branches of science) and metaphysics began to loom large.

In the fifteenth and sixteenth centuries the fledging sciences sought to become more like metaphysics, which used to be regarded as the apex of knowledge. After the seventeenth-century, however, the authority of metaphysics began to decline. The success of physics and mathematics and lack of comparable progress in metaphysics reversed their relation. Now metaphysics began to seek to imitate them. In the eyes of many seventeenth-century scholars, the profundity of its questions mattered less than the precision, certainty, and efficacy of the exact sciences.

The latter statement needs to be explained and clarified. From the very start when Western philosophy originated in ancient Greece, it attempted, in contrast to the earlier Oriental philosophy, to separate itself from religion and mythology, although we may still find elements of mythological thought in some works of contemporary Western philosophy. However, as early as in the 6th century B.C., ancient Greek philosophy differed starkly from Indian or Chinese philosophy of the time. Greek philosophy assimilated some parts of the Egyptian and Babylonian culture, especially elements of mathematics and astronomy.

In a chronological sense, the first elements of scientific knowledge had preceded philosophy. However, Greek philosophers modified and developed those elements, transformed them into demonstrative propositions, and made them a part of philosophy. In this way, philosophy, as the discipline concerned with theoretical reasoning, subsumed all of the sciences. Aristotle divided theoretical philosophy into physics, mathematics, and the first philosophy, which later became known as metaphysics. Physics and mathematics remained to be considered as parts of philosophy up until the seventeenth century. Even though the peculiarity of physics and mathematics in the context of philosophy was always appreciated, their subordination to metaphysics as a discipline that was considered more profound and more general was, for quite a long time, unquestioned.

The scientific revolution of the seventeenth century not only witnessed important scientific discoveries in the fields of mathematics, astronomy, and physics. The revolution also initiated a long and painful process of divorce between philosophy and sciences. It was precisely in the midst of this revolution that science gained enough strength and confidence to start questioning its role as a branch of philosophy. The practitioners of sciences started to seek autonomy, intellectual self-rule, even sovereignty. They did not want to be subordinated to Aristotelian metaphysics. One of the most important developments of the first scientific revolution was that it discredited Aristotelian physics. Consequently, since Aristotelian physics was closely related to his metaphysics, the authority of the latter was also badly damaged. Having liberated themselves from the intellectual dependence on Aristotelian metaphysics that dominated scholarship for many centuries, physicists and other scientists did not want to be subjected to the authority of any kind of metaphysics. Most frequently, scientists questioned the attempts of metaphysicians to explain what is beyond experience and thereby acquire allegedly more profound and more fundamental knowledge than that attained by researching empirical phenomena. Science started to fight for its independence from metaphysics. Physicists and mathematicians objected to the treatment of their fields as divisions of philosophy. By overcoming a multitude of obstacles, they finally did succeed and established science as an independent intellectual force.

Positivism originates from these battles for independence of science. It is in these battles that the new notion of science's goals was forwarded, and a new scientific methodology was developed. Such ideas were soon identified with positivism and became the key elements of scientists' self-consciousness, as well as their Polaris, guiding them in their journeys of scientific exploration. Thus, many positivist ideas might be seen as tools used by researchers and scholars throughout the revolution. Those tools were no less important for achieving the aims of the revolution then the new devices used for observation and experimentation such as a microscope, a thermometer, or a telescope.

One of the most important developments related to the scientific revolution and growing independence of science from metaphysics was the transformation in the notion of the aims of knowledge. Francis Bacon's claim that *knowledge is power* expressed the essence of this transformation in a concise, seminal manner, and greatly aided the revolution itself.

The age of scientific revolution witnessed an important social transformation. Although this social transformation cannot be considered the cause of the changing notion of the aims of knowledge, we should discuss it briefly as well. In the sixteenth and seventeenth centuries, across many European countries, especially Italy, The Netherlands, and Great Britain, successful merchants, bankers, and industrialists rose to preeminent social positions. They were more interested in the discoveries and inventions with tangible benefits than in the "first causes" or the overarching harmony of the cosmos governed by the grace of God. Their worldview differed starkly from that of ancient philosophers or medieval monks. They were interested in the advance of experimental natural science, not metaphysics.

To be sure, the very idea that knowledge is power clearly precedes the scientific revolution. For example, the medieval alchemists appreciated applied and practical knowledge, which, they thought, would aid them in finding the philosopher's stone that can turn base metals into gold. We should note that alchemy, nowadays considered the earliest progenitor of chemistry, which, together with magic, also contributed to the development of the scientific method, was not considered a part of philosophy. "Real" philosophers preoccupied themselves with ideas and abstractions, not the philosopher's stone. Theoretical and intellectual objectives, not practical rules and methods that allow one substance to turn into another, captivated the minds of medieval philosophers. Their quest was to apprehend the world (or being), to discern fundamentals of things, and to grasp the principles of cosmic harmony. They were not interested in technological transformation of the world with the purpose of subjecting the material world to human purposes. It is somewhat paradoxical that alchemy and magic once had very scientific objectives, for now they are considered pseudosciences.

Of course, in addition to theoretical philosophy there was also practical philosophy. However, it did not concern practical matters in the conventional sense of the word but rather the moral nature of human behavior. Political philosophers since Aristotle discussed how to engineer political and constitutional systems, but they considered only social engineering, and not the technological one. Neither medieval, nor ancient thinkers reflected upon the relationship between social and technological engineering. The laws governing political life were being compared to the laws of nature instead of being perceived as mechanical principles designed to achieve certain objectives. Therefore, the main objective of politics, as a part of philosophy, was to determine the principles of just and perfect statesmanship and establish the foundations of political harmony. Little difference existed between determining the principles of social harmony and the principles of cosmic harmony.

In contrast to the ancient and medieval philosophers, the scientists of the seventeenth century were concerned with the technological value of scientific knowledge. In this regard, they have more in common with the alchemists and magicians than scholastic thinkers. A growing number of researchers and scholars started to presume that the *main aim* of scientific investigation is not to reveal the beauty or harmony of the world, but rather, to purposefully master the world through technological and theoretical prowess. Scientific truth ceased being seen as an end in itself and grew to be treated more and more often as an *instrument* of domination, used to control nature and society.

The growing notoriety of the instrumental notion of knowledge that started to dominate the classical notion of knowledge as an end in itself led to a new outlook as to what subjects of research are worthwhile to pursue. The changing attitude is most clearly reflected in the declaration by Galileo Galilei that he "prefers finding a single truth on a minor matter, to endlessly disputing the most elevated questions without achieving any truth" (Galileo 1890: 738).

Two things should be noted with respect to this statement. First, research results are ascribed higher value than its philosophical (or cosmic) importance. Second, Galileo is very clear that researchers ought to pursue only those problems that can potentially be solved. There is a conspicuous similarity between this attitude and that by later positivist thinkers like Moritz Schlick, and Rudolph Carnap. These thinkers also believed that, in a true science, all questions can be given answers, and the questions without possible answers are only meaningless combinations of words that only imitate questions. But how can one distinguish between solvable and unsolvable problems? Clearly, Galileo and his contemporaries relied on the notion that true scientific problems are those that can be pursued by a certain method. Science began to rely on methods that distinguish it considerably from metaphysical philosophy. This was the final straw that led to the eventual separation between philosophy and science.

What is the main difference between methods applied in sciences and those applied in philosophy? This is not an easy question mainly because there is no single prevailing method either in the natural sciences or in philosophy. And even those who claim that science relies on a definite method have quite different opinions concerning its nature. A rationalist will claim that reason is the source of knowledge while an empiricist will claim that true knowledge comes from experience. Scientific methodology, it seems, is not an issue of science but that of philosophy. Nonetheless, the scientific revolution of the seventeenth century revealed many features of the natural sciences that showed them to be very different from metaphysical philosophy. First, metaphysics is an utterly theoretical subject. On the contrary, natural sciences and most of all, physics, rely on observation and experimentation. The systematic use of experiential method led to growing understanding that physics is pursuing a very different path from metaphysics. One could perhaps argue that the distinction between the empirical methodology applied in physics and the purely theoretical approach applied in philosophy is somewhat artificial. In the end, Galileo himself extensively relied on thought experiments, which could be considered a form of theoretical speculation. On the other hand, metaphysical insights may be inspired by some experiences. However, the optical experiments described by Newton in his *Optics* required special equipment and instruments and were in general starkly different from speculations, in which metaphysical philosophers were and still are engaging. Moreover, physicists of the seventeenth century began to describe their observations and experimental results using quantitative mathematical language, not qualitative.

Until, as Aristotle had required, physics and geometry were separated, their difference from metaphysics did not seem very great. Aristotelian physics was describing phenomena in a qualitative fashion and did not involve experiments and empirical measurements. Aristotelian metaphysics was distinguished from Aristotelian physics not so much by its methodological approach as by its generality. Both were essentially speculative enterprises. Mathematics as well was not very different from metaphysics, as both were purely theoretical disciplines.

Aristotle was convinced that physics as well as other natural sciences seek to disclose properties of concrete objects of the sensible world. Mathematics, on the other hand, was thought to analyze abstract, idealized, and hence nonempirical objects like points, lines, planes, and numbers. Mathematical symbols did not, in Aristotle's view, denote empirical objects that exist in time and space. Therefore, Aristotle maintained that propositions of physics which concern empirical objects cannot be formulated in mathematical language. The physical realm is raw and qualitative. Hence, it cannot be expressed in the precise, quantitative notions of mathematics. Physics must disclose only the main features of phenomena under investigation (for example, to draw a distinction between the natural and forced movement of bodies) and to describe it qualitatively. Ideal mathematical forms do not exist in reality; thus, quantitative analysis of physical qualities is impossible. One can find such forms only in the heavens: the sun and the planets are moving around the earth in perfect circles. Thus mathematical astronomy describing eternal and perfect movement of heavenly bodies is possible. However, it is impossible to build a mathematical physics. When investigating the movement of earthly bodies which is neither perpetual nor flawless, physics must make use of empirical methods instead of geometrical (or mathematical) ones; it must use observation, not abstraction. For these reasons, Aristotle believed that mingling geometry and physics is fundamentally flawed.

During the time when physics and geometry were treated as incompatible subjects, it was relatively easy to support a view that they are two components of philosophy. However, perhaps the most crucial point in the scientific revolution of the seventeenth century came from the realization that physics and geometry can be combined. The Aristotelian program of scientific research was replaced by its predecessor, the Pythagorean program. Galileo astounded most of his contemporaries by declaring (in *The Assayer*) that the great book of nature which is always open before us is written in a mathematical language, and its signs are triangles, circles, and other geometrical figures. Mathematics, Galileo said, is inscribed in this book, and only those who apprehend its signs can read it. Galileo was convinced that mathematics not only can but also must be combined with physics. Mathematics can be applied not only to study the movement of the celestial but also terrestrial bodies. Led by the Pythagorean dictum that numbers govern the movement of all bodies, Galileo established the law of the free fall of bodies, $S=gt^2/2$, and opened the gateway for the advance of mathematical physics. Later, following the path laid by Galileo, Isaac Newton formulated his famous laws of mechanics, thus making obsolete the very Aristotelian distinction between terrestrial and celestial bodies.

Experimental-mathematical physics, which originates from combining physics and mathematics, has less in common with metaphysics than either Aristotelian physics or mathematics. However, there remained a danger of mixing physics with metaphysics. Newton was well aware of this. His recommendation was the following: physics, beware of metaphysics. Certain assumptions of the new, Newtonian, physics, like, for example, the view of the world as a mechanical system, *machina mundi*, which formed the basis of the classic natural sciences, *may* be treated as metaphysical. However, Newton and his followers hold different opinions on the issue. They treated this view of the world not as an assumption but as a result of scientific research.

The amalgamation of the mathematical and experimental method not only fostered the development of natural sciences but also prompted multiple breakthroughs in mathematics. These scientific advancements and their increasing application in the practical realm highly strengthened the reputation of the natural sciences. This was at the expense of metaphysics, whose reputation deteriorated. Mainly for this reason, in the seventeenth century the first attempts to remake metaphysics and all philosophy following the example of natural sciences took place. The rationalist metaphysics of the seventeenth century was the first to spare no effort to acquire scientific respectability by using methods borrowed from the exact sciences. It started the program of mathematization of philosophy: based on the example of geometry, it attempted to structure philosophy as a deductive system. Baruch Spinoza's *Ethics* is the best, though not really successful, witness to these efforts.

David Hume had a different idea as to how philosophy should be reformed. However, before we embark upon discussing Hume's reaction to the scientific revolution of the seventeenth century, from which science originated as a sphere of knowledge greatly distinct from the traditional philosophy, it is worthwhile to consider the following question: is it really the case that the concept of science as a type of knowledge distinct from philosophy appears *only* in the seventeenth century? I have already mentioned that the development of Western philosophy in ancient times included some scientific knowledge, especially, that acquired from Babylonian and Egyptian mathematics. However, the Oriental knowledge in the field of mathematics or astronomy was quite fragmentary and did not constitute a coherent and general system of propositions. Rather, it mostly consisted of algorithms to solve concrete problems. Ancient Greek philosophy, on the contrary, adopted a systematic method of proving philosophical statements. Consequently, mathematical knowledge was incrementally shaped into a new form by building theories, wherein separate propositions are connected via rigorous rules of logical inference. This is how mathematics and physics became *departments* of philosophy. I defend the view that these sciences became independent from metaphysics only in the seventeenth century. However, a different opinion on this matter traces the origins of the natural sciences not in the scientific revolution of the seventeenth century but in the scientific advancements of the Alexandrian sciences that flourished for half a millennium from 300 B.C. to A.D. 200. In his book *The History of Scientific Ideas from the Dawn of Man to the Twentieth Century* Charles Singer, a well-known historian of science, claims that in Alexandria, separate scientific disciplines were emerging. Eventually they lost ties with philosophy, of which they had been a part for a few centuries in Athens (cf. Singer 1959: 63).

When describing the development of independent sciences in Alexandria as well as Rhodes and Pergamon, which in those days were also important centers of scientific advancement, Singer prominently refers to Euclid and Aristarchus. However, as Singer points out, Euclid was perhaps a disciple of Plato's disciple and Aristarchus was relying on the idea advanced by one of the most famous Pythagorians, Philolaus, who is known to have claimed that the Earth is revolving around the central fire. Yet considering that Euclid's *Geometry* had an immense influence on the later development of mathematics and other sciences mostly because of its reliance on the Aristotelian deductive systematic approach to knowledge, it appears that Singer overestimates the independence of the sciences from philosophy in Alexandria. More importantly, no evidence in Alexandria shows social and intellectual changes, which were of vital importance for the scientific revolution of the seventeenth century.

The changes of the seventeenth century lead to a new concept of scientific knowledge, its method and objectives. Philosophy has always concerned itself with matters people may know and by what means they may acquire credible knowledge of those matters. Therefore these changes were of great importance to it. As I mentioned before, the first reaction to these changes came from rationalist metaphysics. The second reaction came in the form of positivist philosophy. In the seventeenth century, we see the first attempts to restructure metaphysics and all of philosophy in the image of the natural sciences. In principle, this was an attempt to create a new type of philosophy that was less philosophy in the old sense of the word and more like science in the new sense of the word. Spinoza's *Ethics*, written in 1675, is a superb example of how philosophers of that time applied methods used in natural sciences of the day to gain scientific respectability: it is written in a Euclidian manner as a system of definitions, axioms, and theorems. We must emphasize that the first attempts to create *scientific* philosophy were made in the seventeenth century and they have little in common with the Humean positivism of the eighteenth century. The only similarity that both Hume and Spinoza shared was in the belief that, in order to advance, philosophy must rely on a scientific method.

Hume's Positivism

After Comte's efforts, positivism acquired the character of a distinct philosophical movement, and its representatives often claimed that the most significant result of the seventeenth-century scientific revolution was the creation of a positive natural science (mainly physics) based on experiment and mathematics. Positivists regarded it as an extremely important achievement that this new science was independent of transcendental metaphysics and at once became its most formidable rival for claiming knowledge and truth. In this way, the intellectual field was fertilized for the rise of positivism. However, that certainly does not mean positivism as a scientific philosophy *originated* in the seventeenth century.

Granted, Bacon, Galileo, and other participants in the scientific revolution did make an array of assertions and proclamations which, being conducive for the emancipation of natural science from metaphysics, sounded quite positivist. Yet, abundant evidence exists which undermines efforts to present Bacon, Galileo, or Newton as positivists. In Newton's case, his interpretation of time and space in theological terms does not look at all positivistic. Newton's notion of the final aim of science as the means to meet the Creator through acquaintance with His creation also might seem to push the limits of positivist empiricism. And Bacon, who fiercely fought traditional scholastic, Aristotelian science, was heavily influenced by this tradition himself. It is enough to recall his use of the ancient notion of *form*, a concept most positivists would regard as metaphysical.

Contrary to Bacon, Galileo is almost free of the influence of the scholastic tradition, and he emphasizes the autonomy of science much more stridently than Newton. His way of thinking is quite close to the positivist one. Yet the aprioristic elements in his notion of scientific knowledge and, especially, his claim that we have nonempirical knowledge of necessary connections, does not allow him to be classified as a full-fledged positivist.

The first thinker whom we may without any reservation call a *positivist* is David Hume. One good reason to regard him as *the* founder of positivism is the fact that its representatives eagerly acknowledged their debt to Hume and his unique role in championing the movement. Auguste Comte valued him less than most positivists (it seems that Hume's *History of England* made a greater impression on him than his *Treatise* and *Inquiries*), but even he admitted Hume's significant role in the advance of positivist thinking. Of course, there are also more substantial reasons to treat him as the first positivist, and they will be explained below.

David Hume was born in 1711 to a relatively wealthy family in Edinburgh. After graduating from Edinburgh University, he began writing his greatest philosophical work, *Treatise on Human Nature*, which he finished in France. The *Treatise* was published in London in 1739–40. However, to Hume's great disappointment, it did not generate a lot of interest. Hume failed to become a professor of moral philosophy at Edinburgh University and he was rejected for a position as professor of logic at Glasgow University, arguably due to the skeptical view on religion that he presented in the *Treatise*. He worked as a diplomatic and war secretary, librarian, Undersecretary of State, Secretary to the British Ambassador to France, and was in contact with many French philosophers, including Jean-Jacques Rousseau. He published *Inquiry Concerning Human Understanding* (1748), *Inquiry into the Principles of Morals* (1751) reworking some topics of the *Treatise*, and six volumes of the *History of England* (1754–62), which was his most popular book in the eighteenth century, and a few other works on religion, morality, and politics, including *The Natural History of Religion* (1757). After quitting the civil service in 1769, he returned to Edinburgh. He was an acquaintance of Adam Smith. Hume died in 1776 (cf. Mossner 1980 [1954]).

Hume and Newton

As a *philosophical doctrine*, positivism is not the cause but the consequence of the scientific revolution of the seventeenth century. A great deal of time was needed to understand what really happened in this period and what impact it had for the later development of the sciences. This process was going on for about a century. To be sure, Hume did not analyze the dramatic changes brought about by the scientific revolution that he himself did not witness. However, he did witness its consequences, and they are very well reflected in his positivist philosophy. Conventionally, Hume's philosophy is considered in the context of the British empiricist tradition, not the development of the sciences. Yet British empiricism has been greatly influenced by the scientific revolution and the dominance of the empiricist tradition in British philosophy can be largely ascribed precisely to the scientific revolution.

Hume is a philosopher difficult to classify. He may be looked at from various angles and treated as an empiricist, phenomenalist, associationist, and skeptic among others. In this book I have no intention to examine the entirety of his many-sided, unpretentious, and brilliant, although sometimes not very coherent, philosophy. What I want to present now is a picture of Hume as a positivist. I mean by this the intention to expose those elements in Hume's thought which were especially important for the establishment and development of positivist philosophy, and to which later positivists were particularly indebted.

Of course, such a portrait cannot exhaust the richness of Hume's philosophy. Hume treated himself mainly as a representative of the British empiricist tradition. Yet the scientific revolution, by changing the intellectual climate of philosophy in its aftermath, not only nourished the entire British empiricist movement of the seventeenth and eighteenth centuries which enabled a remarkable flowering of new thought, but seemingly made the most indelible imprint on Hume himself. While being more remote in time to the revolution's beginning, he is nevertheless nearer to its spirit than his great empiricist predecessors, Locke and Berkeley.

While still in his teens, Hume got his education at the center of the Scottish Enlightenment, Edinburgh University, the campus ablaze in the early eighteenth century with new philosophical and scientific notions. Newtonian science was held in especially high esteem. Hume studied, it seems, both Newtonian optics and mechanics. Hume's instruction in mathematics at Edinburgh also gave him a powerful dose of Newtonianism.

Thus leaving university, and determined to pursue studies in the moral sciences or moral philosophy (at that time this notion covered a very wide range of subjects including what we presently call psychology, anthropology, ethics, politics, history, and social studies), Hume was acquainted, at least to a degree, both with modern mathematics and modern experimental science. "He was already provided with the scientific method which had been so brilliantly used by Newton in the realm of Natural Philosophy and was undoubtedly inspired—as were so many thinkers during the Age of Enlightenment—by Newton's own hint, dropped at the close of the *Optics*: 'If Natural Philosophy in all its Parts, by pursuing this Method, shall at length be perfected, the Bounds of Moral Philosophy will also be enlarged.' The examples set by Locke, Shaftesbury, Mandeville, Hutcheson, and Butler in the experiential approach to the study of human nature were ever before his mind" (Mossner 1980 [1954]: 73–4).

In a few postuniversity years crucial to his intellectual development, his enthusiasm for the experiential method applied to moral science was not diminished by the deeper studies of those philosophers who pursued this path. On the contrary, Hume decided to be more consequential, systematic, and even reckless in applying this method than his predecessors and contemporaries. His fondness for the new science and resolution to follow its methods, his audacity, and his uncompromising position made him most sympathetic to the scientific mode of thinking, the most radical of all empiricists, and mainly because of this, the first positivist.

Newton, whom Hume openly admired, perfected the natural science dealing with physical subjects. Hume regarded his task to be an advancement of moral science. Moral science is the science of man which concerns the human mind, life, and behavior. According to Hume, that science is extremely important because only after gaining knowledge about man may we decide what he may know about nature. Moral sciences include logic whose end is "to explain the principles and operations of our reasoning faculty, and the nature of our ideas," morals and criticism which "regard our tastes and sentiments," and politics which "consider men as united in society, and dependent on each other" (Hume 1967 [1739–40]: xix). For Hume, these sciences seem to encompass almost everything worth investigation and discussion.

Yet in spite of the preeminence of the moral sciences and all efforts of "some late philosophers in England" who have begun to build under them a new foundation—that of experience and observation—they still lag, according to Hume, behind physics. Newton's *Rules of Reasoning in Philosophy*, especially the rule regarding induction as a general method of justifying scientific generalizations, must be applied to the moral sciences fully and without reservation. Hume promises to draw no conclusion except that which he is authorized by experience.

The sanguine young Hume hopes to become the Newton of the moral sciences. He emphasizes his firm determination to follow the way of (a rather widely understood) methodological naturalism by giving to his *Treatise of Human Nature* a rather remarkable subtitle: "Being an Attempt to Introduce the Experimental Method of Reasoning into Moral Subjects." Yet when working on the *Treatise*, Hume became aware that more is needed for the creation of a new science of man than the simple expansion of the area of application of Newton's methods to include moral subjects. They must be adopted and modified. And no method can completely protect empirical science from error. As for the moral sciences, they are in an especially awkward position because their degree of certainty is substantially lower than that of the natural sciences. Thus, special rules for deciding what is *probable* are required. On the other hand, more than simply rules and methods are needed for the creation of reliable moral sciences. The "capital or center of these sciences" is, according to Hume, human nature of which an essential part is the mind. Hence, the establishment of moral sciences must begin from working out a theory of mind. In Hume's science of man, that theory plays a role similar to the one played by the theory of attraction in Newton's science of nature (cf. Passmore 1980 [1952]: 43).

Impressions, Ideas, and Metaphysics

The basic part of Hume's theory of mind is his explanation of the origin of our ideas. Hume's main thesis concerning this issue is as follows: "All the materials of thinking are derived either from our outward or inward sentiment: The mixture and composition of these belongs alone to the mind and will. Or, to express myself in philosophical language, all our ideas or more feeble perceptions are copies of our impressions or more lively ones" (Hume 1981 [1748]: 19).

This thesis is supported by two main arguments. First of all, "when we analyze our thoughts or ideas, however compounded or sublime, we always find, that they resolve themselves into such simple ideas as were copied from a precedent feeling or sentiment" (Hume 1981 [1748]: 19). Secondly, "if it happened, from defect of the organ, that a man is not susceptible of any species of sensation, we always found, that he is as little susceptible of the correspondent ideas" (ibid., 20).

Different ideas are connected together according to one or another principle of association. "To me," writes Hume, "there appear to be only three principles of connection among ideas, namely *Resemblance*, *Contiguity* in time or place, and *Cause* or *Effect*" (Hume 1981 [1748]: 24). Yet many later positivists, especially those of the camp of logical positivism who valued Hume very highly, seriously doubted if the methods of associative psychology, used by Hume, may be regarded as genuine scientific methods. From the perspective of the twenty-first century, it is rather clear that the experiential method applied in the moral sciences by British empiricists of seventeenth and eighteenth centuries does not have much to do with the methods of measurement applied in experimental psychology in the nineteenth century, or with empirical methods used in twentieth- and twenty-first-century political and social sciences. Hume's method of analyzing the content of the mind is based on introspection, which not only logical positivists regarded as an unscientific method. The methods Hume used in his theory of mind and human nature rather differ from those applied in the natural sciences.

Hume had ambitions to apply Newton's methods to the moral sciences. Yet the claim that the method Hume used is Newton's (cf. Mossner 1980 [1954]: 74) is, mildly speaking, an overstatement.

Being critical of Hume's associative psychology, later positivists were, however, in sympathy with his claim that complex ideas originate and thus are compounded of simple ones, and that simple ideas are copies of perceptions. They were especially happy with Hume's use of his doctrine of the origin of ideas for the demarcation of science and what he called "false metaphysics." Descartes still regarded metaphysics as the root of science. Hume's position on this issue is diametrically opposed. According to him, "the science of man is the only solid foundation for the other sciences" (Hume 1967 [1739]: xx). Hume is convinced that many metaphysical concepts and notions are in fact harmful for the development of both natural and moral sciences, and that the greater part of metaphysics is muddled and obscure. Thus, both in the *Treatise* and the *Enquiry Concerning Human Understanding* we find a declaration very characteristic for positivism of the intention to cultivate a "true" philosophy in order to destroy a "false" one. According to Hume, there is a plausible suspicion that the seemingly most profound philosophy is not properly a science. Many abstract ideas are dim and vague. Yet metaphysics permanently operates with terms referring (or supposedly referring) to such ideas. Which of them might be useful and which are only hindrances to the development of science? Experiential science might be helped only by those ideas which originate in experience because only they have meaning and value. Thus the only way to separate meaningless philosophical terms from those which do have meaning is to go down to the level of impressions from which meaningful ideas originate. The boundaries between impressions might be determined much more easily and more exactly, and this helps to separate the vague ideas from useful ones.

On the basis of such reasoning, Hume concludes the section "Of the Origin of Ideas" of his *Enquiry Concerning Human Understanding* with the famous and oft-quoted (especially by later positivists) words: "When we entertain, therefore, any suspicion, that a philosophical term is employed without any meaning or idea (as is but too frequent), we need but inquire, *from what impression is that supposed idea derived?* And if it be impossible to assign any, this will serve to confirm our suspicion. By bringing ideas into so clear a light, we may reasonably hope to remove all dispute, which may arise, concerning their nature and reality" (Hume 1981 [1748]: 22).

Almost two hundred years later, logical positivists proposed what they
called an *empirical criterion of meaning (or significance)* which had to
determine what sentences have cognitive (empirical) meaning and which
do not. Hume speaks about terms without meaning and not about meaningless sentences, yet the previous fragment surely can be interpreted as
the incipient version of the empirical criterion of meaning.

Although Hume sometimes uses the term "metaphysics" when speaking about the results of his own investigations, the requirement to look
into the origin of our ideas for determining their admissibility and legitimacy is the crucial point in the argument with metaphysics taken in a narrower sense to which later positivists were accustomed. Hume used various predicates to identify the sort of metaphysics which had no basis in
perception: false, adulterate, uncertain, chimerical, deceitful, and several
others. It was manifestly this kind of philosophy later positivists used to
call metaphysics *sensu stricto*. Thus having in mind all necessary stipula-
tions, we might claim that Hume was the first thinker who maintained that
metaphysics is meaningless. By this he stands out among various philosophers who may be called antimetaphysicians. Metaphysics and especially
transcendental metaphysics which tries to reveal what is above or beyond
possible experience was an object of philosophical criticism through the
ages. Beginning with the first sophists, many philosophers used to emphasize the impenetrability for the human mind of its objects, the inconclu-
siveness of the discussions on its subjects, its obscurity and fruitlessness.
Hume also castigates metaphysics with similar words. In this respect he
simply follows his antimetaphysical predecessors. Yet he goes farther than
them, and that makes him the first intellectual knight-errant crusading
against (transcendental) metaphysics under the banner of (a still to be
proclaimed) positivism.

Not only does he put forward a general criterion to be used as a
weapon in the confrontation with deception, he uses it in undermining
several specific and vital metaphysical notions: and not only that of *essence* at which earlier empiricists looked with suspicion, but that of *substance* also. By eliminating the notion of substance as the metaphysically
ultimate constituent of the world from his philosophical vocabulary, he
differentiates himself from his great predecessors—Locke and Berkeley—
and presents himself as a quite consequential positivist.

In both cases of *essence* and *substance* Hume simply did not find the
impressions from which these supposed ideas are derived. The elimination
from philosophy of these and other terms not founded on experience no
longer preserves a place for transcendent metaphysics which goes beyond

the boundaries of experience since it cannot do without them. In the aftermath of the scientific revolution, many metaphysicians still claimed that, notwithstanding all successes achieved by the natural sciences, they do not have access to deeper layers of reality by being limited to the investigation of phenomena. According to them, metaphysics, which is not bound by these limitations, provides a much more profound insight into the nature of reality. Hume now had a clear-cut answer to their allegations: any notion of a metaphysical reality behind empirical facts is meaningless.

Two Kinds of Knowledge

What sorts of things can we talk meaningfully about and have knowledge of? (Hume would certainly reject as unjustifiable the Kantian distinction between objects we can think about and those we can know.) Hume's answer to this question is straightforward:

> All the objects of human reason or enquiry may naturally be divided into two kinds, to wit, *Relations of Ideas*, and *Matters of Fact*. Of the first kind are the sciences of Geometry, Algebra, and Arithmetic; and in short, every affirmation, which is either intuitively or demonstratively certain. […] Propositions of this kind are discoverable by the mere operation of thought, without dependence on what is anywhere existent in the universe. Though there never was a circle or triangle in nature, the truths, demonstrated by Euclid, would ever retain their certainty and evidence.
>
> Matters of fact, which are the second objects of human reason, are not ascertained in the same manner; nor is our evidence of their truth, however great, of a like nature with the foregoing. The contrary of every matter of fact is still possible; because it can never imply a contradiction, and is conceived by the mind with the same facility and distinctness, as if ever so conformable to reality. (Hume 1981 [1748]: 25)

Early British empiricists, especially Francis Bacon, had been accused by rationalists for not appreciating the role of mathematics in science and giving no adequate explanation of its nature. It would be difficult to blame Hume for this shortcoming. Hume solves the problem of integrating mathematics into the empiricist tradition and creates a philosophy able to accommodate changes in the character of physical knowledge brought about by the scientific revolution. Drawing mainly on Leibniz, he presents an unelaborated yet clear-cut interpretation of mathematics as a system of analytic truths and thus sets up an exemplar for the later, mainly twentieth century positivism. But being compelled to draw a sharp dividing line between the two kinds of knowledge, Hume creates, of course, a new problem or, at least, a new tension in empiricism.

The unity of knowledge was always an ideal for philosophers and all other scholars. In this sense rationalism, and especially Leibniz's rationalism, had clear advantage over Hume's notion of knowledge. One of Kant's main goals, once he was roused by Hume from his "dogmatic slumber," was to restore the unity between mathematics and theoretical natural science, a unity undermined by Hume. Some French Encyclopedists, as well as nineteenth century positivists—most notably J. S. Mill— were also unhappy with Hume's solution and tried to restore the epistemic unity of science on a purely empirical basis. We will speak later in some detail about the role that the unity of science movement played in positivism. Hume, a staunch supporter of methodological naturalism, was no doubt himself a proponent of the unity of (empirical) science. Yet the price he (and his twentieth century positivist successors) had to pay for *such* a unity was rather high: mathematics, which through the centuries was regarded as an exemplary science, was pushed away from its place as the source of *real* knowledge about the world. The nature of mathematics was explained in a way formally consistent with the principles of (moderate) empiricism, yet a cleft in the main body of knowledge appeared.

Of course, it was possible to maintain that all knowledge about reality (matters of fact) is unified, and mathematics does not deal directly with real objects at all. Hume and those positivists who followed him did exactly this. However, we must not overlook the fact that, by proceeding in this manner, the *unity* of knowledge is achieved only through the *division* of sciences into those which deal with reality and those which do not, and by assigning to the latter a special, auxiliary function in the system of knowledge.

We will return to this problematic issue later. Irrespective of how later positivists treated the nature of mathematics, they all without reservation praised Hume for his firm conviction that mathematics, which concerns relations of ideas, and empirical science, which concerns matters of fact, are the *only* kinds of knowledge accessible to humans.

Critical Analysis of Causality

Hume did not see any pitfalls when explaining how we acquire knowledge concerning relations of mathematical ideas. He simply did not recognize the problems that later proponents of the analytic conception of mathematics faced. Hume failed to identify these problems simply because their formulation, let alone solution, required great advancement in formal logic.

He was much more interested in the nature of knowledge concerning matters of fact. The establishment of facts themselves presented no serious difficulties for Hume as the empiricist. Yet, being a phenomenalist and nominalist, Hume was convinced that only *particular* facts are given to us in experience. (Nominalists believe that only particular things, not classes of objects as a "tree" or "green" *really* exist.) But how can one establish generalizations beyond particular facts? For example, how can one claim that one phenomenon (say, a flame) is a cause of another phenomenon (heat)? Usually a statement "A causes B" is thought to mean that B is a result of A, that A has a *power* of putting B into existence. Thus, there is a necessary connection between A and B.

Hume struggled to find impressions from which ideas of necessary connection or power originate. Yet he did not find them. When a billiard ball is hit by another ball, it starts to move. One could observe this sequence of events repeatedly. However, one would never be able to see that the two events (the movement of the first ball and the movement of the second ball) are *connected*. One can observe a sequence of events but not the *power* of one event to cause another. Admittedly, the concept of power is void of meaning. "All events seem entirely loose and separate. One event follows another; but we never can observe any tie between them. They seem conjoined but never connected" (Hume 1981 [1748]: 74). If the relation between empirical facts (objects, events) cannot be described in terms of connections, all the more it cannot be necessary. Thus all such relations are merely contingent.

Hume's conclusion that events may be conjoined but not connected had deep implications on his treatment of causation because this conclusion is incompatible with the traditional view according to which causes produce effects "with necessity." Hume's analysis of the notion of causality is rather complicated and somewhat incoherent. In this respect it may be compared with his analysis of personal identity which also brought about many interpretations. The view according to which Hume considers causality as just our usual expectation that one event will be followed by another, is only one among many such interpretations. In accordance to this reading of Hume, his position is the following: Human experience is a chaotic wash of discrete sensations. But by picking out and comparing several of these separate sensations, we may perceive that one event (A) is repeatedly followed by another event (B). On that basis we may judge that the first one causes the second. "But in fact all that is known is that A and B have been regularly perceived in close association. The causal nexus itself has never been perceived, nor can it be said to exist outside of the human mind and its internal habits. Cause must be recognized as merely the accident of a repeated conjunction of events in the mind. It is the reification of psychological expectation apparently affirmed by experience but never genuinely substantiated" (Tarnas 1993 [1991]: 337–38).

This rather standard interpretation is not easily reconciled with Hume's claim that the Rule of Cause and Effect is one of the three principles of connection among ideas and with his conviction that all (or almost all) our inferences are based on causality. Yet specifically this interpretation was regarded by later positivists as most appealing: it was a step leading to the elimination from science of a suspiciously metaphysical notion of cause altogether. Stepping a little farther, it was possible to claim that science answers only the question *how* but not *why*, and to present the main goal of science as the *description* of phenomena but not of their *explanation*. This view was later fiercely defended by Ernst Mach. It also influenced Auguste Comte, who tried to avoid even mentioning causes of phenomena.

Certainty and Probability

Hume's analysis of causality has direct bearing on his treatment of the epistemic status of knowledge concerning matters of fact, or, more precisely, of its certainty. According to him, in reasoning about facts, we can only rely on what is directly given. Hence, we must examine the nature of causal relations.

Having carefully examined rationalist reasoning about causes and effects, Hume concluded that, since the cause and the effect are distinct, the latter cannot be contained in the former. On the other hand, according to Hume, our experience fails to suggest that distinct events we are inclined to treat as cause and effect are related in a physical, not merely logical sense. Hence in an attempt to establish a causal bond between events, we may rest only upon the fact that these events followed one after another in the past. Yet, this fact does not prove that one of them is the cause of another.

Although Hume examined only the relationship between cause and effect, his line of argument is relevant to any reasoning concerning the relationship of facts. From Hume's standpoint, reasoning which transgresses the limits of our actual experience cannot have sufficient ground. It would become justified if we could know "that instances, of which we have had no experience, must resemble those, of which we have had experience, and that the course of nature continues always uniformly the same" (Hume 1967 [1739]: 89). Yet proving this assumption, i.e., the principle of the uniformity of nature, is, according to Hume, impossible because we may perfectly imagine that "the course of nature" may suffer a change. The sun has always been rising every day but we *can imagine* that tomorrow it will fail to rise. Hence, it is impossible to justify extrapolations from observed facts to unobserved facts. Such extrapolations would necessarily rely on the principle of the uniformity of nature. Such inferences (we could say *inductive*, but Hume did not use this word) cannot be regarded as absolutely reliable and their conclusions as certain. And because our general knowledge concerning matters of fact is acquired by means of similar inferences, all of it is *uncertain* or *probable*, opposite to knowledge concerning relations of ideas, which is absolutely *certain*.

This is one of Hume's most important skeptical conclusions (another one is his doubt that our impressions are caused by things existing continuously and independently of the mind). Yet Hume's skepticism differs substantially from the radical, all-encompassing skepticism of Pyrrho of Ellis, and reminds us more of the moderate and discriminating skepticism of Karneades. I mean that Hume differentiates knowledge according to degrees of its certainty (or uncertainty).

In fact, he rejects the traditional distinction between *episteme* and *doxa*, or knowledge and opinion, on which almost all philosophy from ancient times is based. Although sometimes he still contrasts knowledge and probability as terms, more often he simply discusses two kinds of knowledge: one, derived from experience, which is not absolutely certain; and the second, mathematical knowledge, derived from comparing ideas, which is absolutely certain. In this respect he differs substantially not only from the rationalists Plato and Descartes, but also from earlier empiricists, especially from Bacon. We may recall that Bacon believed that the conclusions of inductive inference from empirical facts may be certain. Hume, however, perfectly understands that, firstly, we cannot prove this, and secondly, they may be, and sometimes are, false. Thus our knowledge concerning matters of fact is, in fact, merely probable.

"Is" and "Ought"

Hume was concerned not only with epistemology but also with moral, political, and social issues. But before we step into analyses of some of the latter questions, we must briefly sketch out the main presupposition which guides Hume in his moral and political inquiries. Hume made the distinction between factual and normative propositions, which was of very great importance not only for all subsequent positivists, but to all analytical moral and political philosophy as well.

The famous dichotomy is usually called the *fact and value distinction* or *Hume's law*. In the *Treatise*, it is stated as follows:

> In every system of morality, which I have hitherto met with, I have always remark'd, that the author proceeds for some time in the ordinary way of reasoning, and establishes being of God, or makes observations concerning human affairs; when of a sudden I am surpriz'd to find, that instead of the usual copulations of propositions, *is*, and *is not*, I meet with no proposition that is not connected with an *ought* and *ought not*. This change is imperceptible; but is, however, of the last consequence. For as this *ought*, or *ought not*, expresses some new relation or affirmation, 'tis necessary that it shou'd be observ'd and explain'd; and at the same time that a reason should be given, for what seems altogether inconceivable, how this new relation can be a deduction from others, which are entirely different from it. But as authors do not commonly use this precaution, I shall presume to recommend it to the readers; and I am persuaded, that this small attention wou'd subvert all the vulgar systems of morality, and let us see, that the distinction of vice and virtue is not founded merely on the relations of objects, nor is perceiv'd by reason. (Hume 1967 [1739]: 521)

Hume's main thesis is that normative propositions or judgments cannot be deduced from factual statements. The fact that snow is white does not yield any sort of normative statement. That is, we could not deduce from this fact that "snow should be white" or "snow should be black." A fact is an actual state of affairs. A value or norm is a desired, not actual, state of affairs. Accordingly, normative conclusions do not logically follow from factual statements. However, they also do not follow from analytic statements. That is the essence of "Hume's law."

According to Hume, normative propositions can be considered as "good" or "bad"; yet, they cannot be "true" or "false." Thus, normative propositions have no cognitive value. Although Hume did not elaborate on this insight, the consequences it had on later moral, political, and social theories are hard to overestimate. In was not until Hume made the fact/value distinction that an idea of a value-free social science could emerge.

Moral Principles and Social Progress

Hume was convinced that moral principles are based not on reason but on our natural dispositions. He believed that feelings and passions trump reason when it comes to human actions. "Reason is and ought to be the slave of passions, and can never pretend to any other office than to serve and obey them" (Hume 1967 [1739]: 415). The main function of reason is to inform us about the existence of a proper object of passion and to indicate the means of attaining that object.

Among various passions and feelings a special role is played by self-love, which is a powerful motive in human nature. On the other hand, it is quite clear that every individual depends on the community. Therefore, some philosophers have tried to deduce our concern for the public from the worry for our own happiness. Yet Hume repudiates any such attempt, claiming that there are instances in which private interest is separated from public, or even contradicts the latter, yet moral sentiment continues. That means that there are particular moral feelings or sentiments which do not coincide with selfish feelings. A man or woman is driven not only by personal pleasures and seeks not only personal happiness. According to Hume, "we must renounce the theory, which accounts for every moral sentiment by the principle of self-love. We must adopt a more public affection, and allow, that the interests of society are not, even on their own account, entirely indifferent to us. [...] If usefulness [...] be a source of moral sentiment, and if this usefulness be not always considered with a reference to self; it follows, that everything, which contributes to the happiness of society, recommends itself directly to our approbation and goodwill. Here is a principle, which accounts, in great part, for the origin of morality" (Hume 1966 [1751]: 219).

Hume agrees that particular feelings on which moral principles are based are by themselves weak in relation to self-love and strong emotions. Yet they are made stronger by the sentiment for humanity: sympathy writ large. According to Hume, we have a powerful sense of sympathy with our fellow humans. This sentiment is universal and extends to all mankind. The sentiments which arise from humanity reinforce one another. The sense of moral duty springs out of these sentiments and drives a person to pursue the common good, promotes the happiness of society, and opposes vice and disorder.

Hume quite often mentions traditions and habits as one of the sources of moral evaluation. Thus, he is not a strict consequentionalist. Yet, he principally speaks in a voice which is rather similar to that of later utilitarians (this is particularly evident in the *Enquiry Concerning the Principles of Morals*). Some scholars are even more outspoken. "It may fairly be said that Hume is a Utilitarian. It is certainly from Hume that the Utilitarian philosophy of Bentham is, through Helvetius and Beccaria, ultimately derived" (Morris 1952 [1931]: 151). They have little doubt that Hume was at least an indirect source of the utilitarian moral theory of the nineteenth century positivists—they, and especially J. S. Mill, draw heavily on Bentham.

To end our presentation of Hume's positivism, it is advisable to examine his position regarding the relation between scientific and social progress, an important issue for the positivists in the nineteenth and twentieth centuries. Frankly, this issue was not Hume's great concern. Yet, he touches upon some points of relevance in his *History of England*, "Essay on the Rise of Arts and Sciences," "Essay on the Idea of a Perfect Commonwealth," and elsewhere. It would be bizarre of the man who represents the epitome of the Scottish Enlightenment to pass over the subject in humble silence.

First of all, he admits that modern civilization outstrips ancient society. Important improvements in arts, sciences, and industry have taken place. The growth of knowledge stimulated the growth of commerce. Today a person feels more secure than in the past. At present, people in Europe enjoy more liberty than in ancient Greece or Rome. According to Hume, we have the right to make such assertions because in this case we rest upon historical facts, i.e., empirical evidence. Using this evidence we may scrutinize some statements regarding the past and confirm or reject them. We may discuss conditions conducive to the development of arts and sciences. Yet according to Hume, a general synthesis of history is unattainable. And we cannot predict the future of mankind; we cannot say with certainty that we are approaching some point of perfection. Moreover, it is quite possible that we are moving away from it. Of cause, at present we may testify the advance of arts and sciences, but history shows that a period of flourishing may be replaced by a prolonged period of decay. Besides, the perfection of arts and sciences in one country rarely influences their standing in other, even neighboring, nations.

Hume's skepticism regarding historical progress, and especially the possibility to predict, or all the more, to secure it, harmonizes well with his skepticism concerning human understanding, and does not surprise us. Yet there is one weighty additional reason why Hume was much more cautious regarding the prospect of the advance of humankind than the representatives of the French Enlightenment. This reason is related to Hume's political philosophy. He always emphasized the fact that political institutions are established by human beings. They cannot possibly function perfectly because man himself is an imperfect being.

We can safely conclude that Hume's views concerning a possible link between the scientific and social progress are consistently cautious and reserved. He clearly lacks enthusiasm regarding the advance of humankind, which was so characteristic of later positivists.

Even though Hume was a bit more moderate than the classical positivists of the nineteenth century, I believe I have provided enough evidence that Hume and not Auguste Comte should be considered as the first positivist. Let us briefly recall the concept of positivism presented in the beginning of this book. Positivism is an antimetaphysical branch of philosophy, which *seeks* to develop philosophy as a science. Admittedly, not every idea raised by a positivist was a scientific one in a strict sense of the word. However, it would be hard to deny that they *attempted* to develop philosophy as a scientific discipline. They have been consistently making these ideas known and they have been widely supported by scientists established in various scientific fields, including physics and mathematics. It is somewhat paradoxical that some thinkers question the scientific character of positivist philosophy, invoking precisely those standards of science that have been defined or codified by positivist philosophers themselves. Gradually those standards have been changing. This is a reason why some logical positivists were not satisfied with some earlier positivist ideas. However, one must bear in mind that the positivist focus on science and especially physics has never wavered. This focus should be considered as the main trait of the positivist philosophy irrespective of what we think about the scientific or general intellectual value of positivist ideas. In this sense, Hume and the later positivists are representatives of the scientific philosophy, a philosophy which preoccupied itself with sciences and sought to resemble them. Hume not only opposed transcendental metaphysics but he also aimed to practice philosophy as a science.

I have emphasized a few other important elements in the definition of positivist philosophy. Positivism focuses on the unity between social and scientific knowledge, it emphasizes the separation between facts and norms (values), and it considers positive (nonmetaphysical) cognition to be a value-free enterprise. All these features are present in Hume's philosophy. Even on the question of the link between scientific and social and progress where Hume's views, as I argued, are much less pronounced in comparison to later positivists, he takes a cautious but nonetheless essentially positivist stance. Therefore, we can consider Hume to be the first positivist, although the concept of positivism appeared only in the nineteenth century. This opinion about Hume is not a conventional one. However, it is also *not* unheard of. For example, John Passmore (1980 [1952]) and Leszek Kołakowski (1972 [1966]) also held similar views. The logical positivists also would not be surprised by this categorization of Hume, since they saw in Hume's work a great contribution to the development of their own ideas.

As I noticed before, classical positivists could point out that Hume's attitude towards human progress is rather guarded. This alleged deficiency in the positivism of the early Hume was, however, compensated more than sufficiently in the eighteenth century by the faith in the power of reason, the boundless enthusiasm, and social optimism so characteristic of the French Enlightenment. Not every philosopher of the French Enlightenment was conducive to the development of positivism. However, a few of them constituted an important and oft-ignored link between Hume and classical, or social, positivism.

The Idea of Progress in the French Enlightenment

Hume was the most famous philosopher of the Scottish Enlightenment. Classical positivism was grounded upon the idea of social progress developed mainly by representatives of the French Enlightenment. However, the relationship between philosophy of the Enlightenment and the era of social positivism is much closer than that. Positivism can be seen as an extension of the Enlightenment project or, more precisely, a version thereof.

The basis of the Enlightenment project is to free humanity from fear and prejudice, including the prejudice about man's own self, to reveal human nature, to investigate the human mind, and to determine how the power of reason could be used in organizing individual and social life. Philosophers of the Enlightenment prioritized reason over dogmas and freedom over traditions. They thought the education of society, and especially of its most influential members, was a necessary condition for a successful implementation of their project. The French philosophers of the Enlightenment strongly believed in the power of reason to free humanity from any blind faith. They thought that having learned their true nature, humans will naturally want to reform or entirely abolish the social relations and institutions inherited from the allegedly "dark" medieval ages. They believed that the main reforms will be implemented by enlightened monarchs. Many monarchs of the eighteenth century indeed had a genuine interest in the ideas of the Enlightenment.

The positive mind is certainly a form of the rational mind with which the Enlightenment thinkers were preoccupied. We shall return to the relationship between positivism and the Enlightenment in the fifth chapter when we discuss the positions of Max Horkheimer and Theodor Adorno. It will now suffice to mention that the concept of rational mind in the Enlightenment was never as well developed as it was in positivist philosophy. No single concept of rational mind was shared by all thinkers of the Enlightenment. Some of them, as for instance, Hume, were empiricists, while others, like Immanuel Kant, held apriorist views and attempted to reconcile empiricism and rationalism. Due to this variety, analyzing the concept of rational mind in Enlightenment philosophy would not be very rewarding. To be sure, this is not the main objective of this book. The concept of the positive mind (the first time this concept was used by Comte) is much more clear and precise than that of the rational mind. However, it is worth remembering that the concept of the positive mind originated from the attempts to give a clear meaning to a rather vague concept of rational human mind.

Of crucial interest to us will be those ideas of the French Enlightenment that were not elaborated in Hume's philosophy but were nonetheless important for the development of positivism. They concern the advancement of scientific knowledge and, in particular, its impact on social progress, the role of reason in transforming and perfecting society and state. Yet, representatives of the French Enlightenment were interested in a much wider spectrum of questions. They came from all quarters of the arts and sciences, and a few represented the medical and law professions or the state bureaucracy. Some of these distinguished men of letters were involved in research in mathematics and natural sciences, and their interest in this area was much deeper than that of Hume. Arguably the most important representative of the French Enlightenment coming from the circles of the natural sciences is Jean le Rond d'Alembert. He made important contributions to mechanics and astronomy, was one of the founders of mathematical physics and is often regarded as the most brilliant mathematical mind of the eighteenth century.

Together with Denis Diderot, d'Alembert edited what was likely the most impressive literary monument of the Enlightenment—the famous *Encyclopaedia, or a Systematic Dictionary of Sciences, Arts, and Crafts*, which appeared in 1751–72. A sign of the import of a new and more pragmatic, that is, positive thinking is the fact that for the first time in a major work of letters, Trade and Manufacturing are treated as equal in importance to Arts and Sciences. The main aim of the *Encyclopedia* was to present in a concise and integral form the results of the advance of human learning, to make them accessible to a wider public and enable their broad practical (including technological) application. D'Alembert was one of the *spiritus movens* of the Encyclopedia project and directly responsible for its mathematics and physics sections. This successful project prompted Comte in the nineteenth century to produce his famous encyclopedic six-volume *Course of Positive Philosophy* and impelled twentieth century positivists to engage in a project of creating *The Encyclopedia of Unified Science*.

Positivism's association with encyclopedic projects is not accidental. The task of projecting, shaping, and editing an encyclopedia imparts on an assiduous and diligent encyclopedist a posture rather typical of a positivist. He directly faces a vast array of knowledge he wants to order, systematize and understand. His effort is epistemological, historical, systematic, integrative, and, on occasion, polemical and therapeutic, because he wants to cure men from prejudices, fallacies, and misbeliefs. He seeks to grasp (1) what is the origin, foundation, and nature of knowledge; (2) how to separate knowledge from what only pretends to be knowledge but differs from it in its nature and must be excluded from the encyclopedia or included only on special terms; (3) what is the internal structure of disciplines comprising the system of knowledge and how do the parts of the system interrelate; (4) how this entire system and its parts developed historically; (5) whether this encyclopedia is a compendium of a more-or-less final state of the sciences, or whether it is a snapshot of a constantly progressing science which outlines areas where further exploration is necessary; (6) what practical benefits—both technological and social—are expected from this knowledge.

I do not claim that preparing an encyclopedia requires its authors and editors to address absolutely all these issues, but it does require an attempt to answer at least some of them. The approach of an Encyclopedist explains (at least partly) the range of the questions in which positivists were interested. The *Preliminary Discourse to the Encyclopedia of Diderot* written by d'Alembert and published originally in the first volume of the *Encyclopedia* in 1751 may be regarded as a paradigmatic positivist text in this sense. The *Preliminary Discourse* seeks to give a theoretical and methodological foundation to the ambitious project of the Encyclopedists. The starting point of d'Alembert when explaining the nature of knowledge follows Hume: an explanation of "the origin and generation of our ideas," because, according to an editor of the *Encyclopedia*, only on this basis it is possible to examine "the genealogy and the filiation of the parts of our knowledge, the causes that brought the various branches of our knowledge into being, and the characteristics that distinguish them" (d'Alembert 1963 [1751]: 5).

D'Alembert's claim that "infinitely varied branches of human knowledge" form "a truly unified system" is partly of a rationalist origin, although later positivists sought to justify it in empirical terms. Actually, d'Alembert did not overestimate the degree of unity achievable in science. The reduction of a large number of phenomena to a single one is a highly desirable step in the sciences, according to d'Alembert, yet he doubts that such a reduction is always possible: "the more useful such a discovery would be to the progress of science the more we have occasion to fear that it will elude our efforts" (d'Alembert 1963 [1751]: 23). Establishing the unity of scientific knowledge is, according to d'Alembert, more promising in the physical and mathematical sciences, as well as mechanics and astronomy. Yet, its prospects do not look encouraging in the other sciences. They are simply too diverse; their levels of development and areas of study are too disparate.

D'Alembert contends that the unity of science stems not so much from the close links between various scientific truths but from the single method applied in different sciences: "the single true method of philosophizing as a physical scientist consists either in the application of mathematical analysis to experiments, or in observation alone, enlightened by the spirit of method, aided sometimes by conjectures when they can furnish some insights, but rigidly dissociated from any arbitrary hypotheses" (d'Alembert 1963 [1751]: 25). In the *Discourse*, d'Alembert almost entirely ignores the problem of method applied in the human sciences, yet later he takes a virtually naturalist stance. He is less radical and less intellectually independent than Hume and often bows respectfully to established wisdom and philosophical tradition. This more traditional attitude, at least when compared with Hume, is clearly seen in d'Alembert's treatment of the problems concerning the status of metaphysics and the certainty of knowledge. Hume was known for his temper in attacking the "deceitful" metaphysics, whereas d'Alembert expresses much more measured opinions. To be sure, this moderate rhetoric can be partly justified by simple practical concerns: the Encyclopedic project was an ambitious one and it was rather critically received by the political and especially religious establishments. Therefore, while writing its introduction, an editor of the *Encyclopedia* was supposed to express more moderate opinion on the controversial questions of the day.

Yet it is necessary to add at once that while holding metaphysics in high esteem in the *Discourse*, d'Alembert nevertheless considerably changes the content of the term. According to him, sound metaphysics is nothing more than the experiential physics of the soul. Experimental methods are applicable not only in the investigation of the properties of bodies but also in the examination of the soul's properties. Complex ideas should be reduced to simpler ones and only by this procedure may we explain their origin. In the *Essay Concerning the Elements of Philosophy* (1759), d'Alembert takes a clearly critical stance towards those metaphysicians who pretended to reveal the nature of being and the first causes of things. He claims that science must hold facts, avoid hypotheses, and seek to rid itself of transcendent metaphysics, even including the materialist branch. According to d'Alembert, the seventeenth century asserted too much, and now the eighteenth century negates too much (for example, that there is no God and no soul). Negation may be no less dogmatic than affirmation; therefore, we must shun empirically irresolvable disputes. Our mind is too weak to adequately apprehend the infinitely complex universe. Yet the determination in advance of what we can and cannot know would itself be an entirely arbitrary and dogmatic procedure.

While in the *Essay* d'Alembert took a position concerning metaphysics that was rather similar to that of Hume, he unconventionally defined the tasks of "true philosophy." Hume claimed that his task was to create a science of man which may serve as the foundation of all moral (and partly of the natural) sciences. He did not separate accurately the tasks of philosophy from those of other moral sciences, especially psychology. Yet d'Alembert's position regarding the aims of philosophy is different.

According to d'Alembert, philosophy, like all other sciences, examines facts, yet the facts in its purview are second level, i.e., laws and principles of science. Thus, it becomes (or must become) the science of the principles of science. Such a philosophy will foster the reform of science necessary to achieve methodological unity, greater transparency of its results, and their wider practical applicability. D'Alembert is convinced that facts secure varying degrees of certainty of the knowledge which is based upon them. He does not try to make a detailed analysis of the certainty problems related to the generalizations from facts which Hume carried out in probabilistic terms. Yet the major difference between d'Alembert and Hume concerning facts as the foundation of our knowledge is not this one. More important is the following: d'Alembert is little interested in perceptions or impressions and believes that knowledge is based not upon internal but external facts. The facts he is talking about are facts concerning the physical properties of bodies and their movements.

In trying to establish a science of man and by focusing on the analysis of impressions, Hume actually makes psychology the foundation of all other sciences. In short, Hume hopes to achieve the unity of science on a psychological basis. D'Alembert's philosophy, on the contrary, displays antipsychologism and is based on a physicalist approach, according to which physics is a more appropriate foundation to other sciences. Here, in the eighteenth century, we encounter the first signs of a major controversy among positivists, to be continued in the nineteenth and the twentieth centuries. Then, Comte supported d'Alembert's antipsychological position, Mill and Mach were much closer to Hume, and Carnap, who initially was in Hume's camp, later changed his mind and sided with d'Alembert.

In general, d'Alembert is a less profound and original thinker than Hume. There is, however, one important innovation that d'Alembert makes which is absent in Hume's work and which becomes a momentous issue in classical positivism. I mean the historical treatment of mind. In the first part of the *Discourse*, d'Alembert still appears quite traditional. When explaining the relation between different sciences and tracing the origin of our ideas in the first part of the *Preliminary Discourse*, d'Alembert examines the operations of the "isolated mind" and inquires, as Descartes, Locke, and Hume did earlier, into the cognitive powers of a person. Yet in the second part of the *Discourse*, d'Alembert presents an inquiry of quite a different kind. In it, he aims at "a historical analysis of the order in which our knowledge was developed in successive steps" (d'Alembert 1963 [1751]: 60). What he in fact submits is a short history of "the renaissance of letters," beginning with erudition, continuing with belles lettres, and ending with philosophy. He examines the Latin and Greek scholarship of humanists, covers Corneille, Racine, Molière, La Fontaine, and developments in fine arts and theology, yet he pays the most attention to the philosophical works of "a few great men" who "silently [...] prepared from afar the light which gradually, by imperceptible degrees, would illuminate the world" (ibid., 74). As is well known, light was a sacred matter for Enlightenment thinkers. Here, d'Alembert has in mind the initiators and participants of the scientific revolution of the seventeenth century. The story d'Alembert tells is the story of the advance of scientific knowledge carried out by Bacon, Descartes, Newton, Locke, and others. In telling it, d'Alembert is guided by the principle he formulates himself at the beginning of his historical analysis: "When we consider the progress of mind since that memorable epoch [the Renaissance], we find that this progress was made in the sequence it should naturally have followed" (ibid., 60).

From the mid-eighteenth until the end of the nineteenth century, the inquiry of a progressive development of the human mind was a compulsory part of the positivist repertoire. It was precisely the historical analysis of the progression of the human mind (presented almost simultaneously with Turgot) which made d'Alembert an important figure for the history of positivism. As for his epistemology and methodology, they are of lesser importance. Yet, d'Alembert's philosophy nevertheless had one asset: it was more French in spirit, more Cartesian, and therefore was more easily assimilated by nineteenth century French positivism. The style of Comte resembled much more that of d'Alembert than that of Hume.

D'Alembert was more interested in the advancement of science than of society. Yet, he was firmly convinced that the former is a necessary precondition of the latter. Like all representatives of the French Enlightenment, he was optimistic concerning the possibilities of the progression of humankind in the direction of a more just, free, and prosperous society. Yet, a tint of skepticism characteristic of his thought bridled him from making overoptimistic predictions, let alone prophecies. The French Enlightenment certainly saw no shortage of more optimistic, and sometimes rather intemperate, thinkers, more than compensating for d'Alembert's reservations.

A convinced optimist of the day was the well-known economist and administrator Anne Robert Jacques Turgot. From 1774 to 1776 Turgot served as the comptroller general responsible for the finances of the state to King Louis XVI. Many representatives of the French Enlightenment had high hopes that an enlightened monarch could implement necessary social reforms. These expectations reached their peak when Turgot controlled all the finances of the kingdom and implemented progressive fiscal and economic policies.

Turgot's philosophy of progress was set forth in a very concise manner first and foremost in his lectures "Philosophical Review of the Successive Advances of the Human Mind" and "On Universal History" read at the Sorbonne in 1750. It seems that the demands of administrative duties precluded Turgot from implementing his plan to write the sizeable treatise he intended to title "Universal History." We therefore can only reconstruct his views from these outlines.

Turgot drew upon Montesquieu's *The Spirit of the Laws* (1748), and yet he was convinced that Montesquieu exaggerates the role of physical factors for the development of societies. Human reason, passions, and even errors are more important for human advance. According to Turgot, the growth of human knowledge and wisdom is inevitable, because each generation transmits past experience to the next. This growth was substantially accelerated by the invention of writing, and especially by that of the printing press. Of course, the advance of the human mind cannot be continuous. It depends on the general state of society and that of its institutions, both affected and afflicted by many factors, including baneful ones like war and natural disaster. They are responsible for short-term declines but cannot stop the progression of humankind forever.

In the initial stages of the development of the human mind, people believed that powerful but invisible actors are responsible for all occurrences taking place around them. Later abstractions, including those of essences and faculties, took the place of superhuman actors in explaining the course of nature and that of human life. Finally, the human mind reaches a stage at which it could relinquish efforts to find the ultimate causes of things, because it comprehends that observation, experiments, and mathematics give us the adequate picture of the real world. The scientific laws which create this new representation of the universe do not require a theological or metaphysical foundation.

Turgot was convinced that he lived on the brink of this new rosy stage. Scientific institutions such as academies may encourage and accelerate not only intellectual but also social progress by fostering the application of the results of scientific investigations. Social institutions, if properly administered, are able to carry out necessary reforms in education, system of law, economics, and elsewhere, thus quickening the pace of progress. These reforms must produce wealth, liberty, and justice. "[T]he interests of nations," wrote Turgot, "and the success of good government reduce themselves to a sacred respect for the liberty of persons and of labour, to the inviolable maintenance of the rights of property, to justice between all, from which conditions necessary result a greater production of things useful to man, the increase of wealth, and of enjoyments, and of enlightenment and all the means of happiness" (Turgot 2007 [1750]: 316).

No other thinker embodied the optimism of the French Enlightenment about the future of humanity as Marie Jean Antoine Condorcet, who was also mainly a mathematician, though not as famous as d'Alembert. He took a leading role in the French Revolution but was accused of treason by the Montagnards (a faction within Jacobins) led by Robespierre when they took the upper hand in the National Convention. Condorcet's overly optimistic views on human progress were laid out in his *Outlines of an Historical View of the Progress of the Human Mind* written at the time when he was hiding from arrest and had little motivation for a sanguine outlook on progress or happiness. Indeed, he died in prison in 1794. In *Outlines*, published posthumously (1795), Condorcet used some of the insights of his mentor and friend Turgot, which were developed a few decades earlier.

Condorcet's *Outlines* gives a more complicated explanation of the progress of human mind than that provided by Turgot. Of his planned great "Universal History" Turgot left us only notes. Condorcet had ambitions to achieve more than did his senior friend. The description both of the goals and of expected results given at the beginning of his *Outlines* looks as heady as it does impressive. As to the goals, Condorcet has to say that his picture of historical progress "should present the order of changes, reveal the influence which each moment exerts on the one that takes its place, and thus show, in the modifications which mankind has undergone, as it renews itself ceaselessly through the immensity of ages, the path which it has followed, the steps which it has taken toward truth or happiness" (Condorcet 1965 [1795]: 76).

It seems that Condorcet did not apprehend the immensity of this task. The main source of his optimism about the prospects of carrying out this gigantic project is not a cold appraisal by a scholar of the amount of historical material to be examined, but the faith of a visionary who before really starting already knows that the result of his work "will be to show by reason and by evidence that no limit has been set to the perfection of human faculties; that the perfectibility of man is really indefinite; that the progress of this perfectibility, henceforth independent of every power which might wish to stop it, has no limit other than the duration of the globe on which nature has cast us" (Condorcet 1965 [1795]: 77). This faith is reinforced, it seems, by the conviction Turgot did not lay stress on, namely that "this progress is subject to the [...] general laws," the same "which are observed in the individual development of our faculties, since it is the result of this development, viewed at the same time in a great number of individuals joined together in society" (ibid., 76). Condorcet does not present these laws *expressis verbis*, yet, it seems that he, like Hegel later, believed that the existence of laws of historical progress makes the ponderous labor of an historian less important.

The final criterion of progress is, according to Condorcet, the increase of happiness and the decrease of pain. The road of progress is, however, long and arduous. Many events have impacted the course of history: inventions, religious occurrences, geographical discoveries, and social and political changes. The advance may be divided in ten stages. The first three stages cover the primitive tribal way of life, the pastoral, and the agricultural ages, which end with the invention of alphabetic writing. The fourth stage is that of Greece with its characteristic advance of science, the climax of which is its division into discrete fields. The fifth stage accounts for Roman rule and the rise of Christianity. The sixth stage, which lasted to the time of the Crusades, brought about the decline of science in Europe, a decline which began already in the second half of the fifth stage. Yet Greek achievements were retained and transmitted by the Arabs, and in the seventh stage the gradual revival of science has begun. The invention of printing initiated the eighth stage and greatly fostered the development of science. The ninth stage is the era of the scientific revolution dominated, according to Condorcet, by Descartes. This stage is closed by the French Revolution and creation of the French Republic. The tenth stage remains just over the horizon.

The picture of the progress of humankind presented by Condorcet is much richer in detail and covers a broader spectrum of issues than that of Turgot. Yet he agrees with Turgot that, in the main, human progress is determined by the development of ideas used for the explanation of the world and man's place in it. He distinguishes four major types of explanation: (1) the anthropomorphic and theological; (2) the metaphysical; (3) the mechanistic-materialistic; and (4) mathematical-scientific (cf. Pollard 1968: 80). This distinction may be used to transform his ten-stage scheme into a much simpler scheme, comparable with that of Turgot.

The main difference between Condorcet's and Turgot's views is easily detected: Condorcet separates Turgot's final stage into two different stages, distinguished by two types of explanation and two ways of thinking—mechanistic-materialist and mathematical-scientific. Condorcet believed, it seems, that it is necessary to emphasize that the materialist explanation of the world based on its display as a kind of complicated mechanism was a necessary step *leading* to the present success of science. Yet it is clear that, for him, a consistent scientific thinking must avoid any reference to the substance of the world. In this respect Condorcet's position is fairly close to that of Hume.

The wide use of the mathematical-scientific mode of explanation in the last, tenth stage of human development, must ensure the efficacy of human actions based on this mode of thinking and the achievement of social and political goals conducive to humanity's manifest destiny of happiness and prosperity. By setting out these goals, Condorcet is a bold prophet of progress. "It is a measure of Condorcet's genius that in all essentials liberal-democratic Western Society still preserves long-term purposes laid down by him, and that his short-term demands have in fact been fulfilled in a remarkably large number of areas. Western society now accepts universal education, universal suffrage, equality before law, freedom of thought and expression, freedom and self determination for colonial peoples, national insurance and pensions, and equal rights for women; and a large section of it also holds greater equality of incomes to be desirable" (Pollard 1968: 84).

I would not go as far as to suggest that Turgot or Condorcet were positivists, although their ideas are very close to the spirit of the positive mind in the sense defined by Comte. However, in order to qualify them as positivists, we would need to take a closer look at their epistemological and methodological ideas, without confining ourselves to their views on human progress. Yet they did not write much on epistemology and methodology, thus this task would be nearly impossible to accomplish. And my goal was to expose and examine only those of their ideas that had significant effect on the later development of positivism.

One can conclude the following: the fundamental positivist ideas were formulated in the eighteenth century after the seventeenth century scientific revolution cleared the path for their appearance. Hume took a positivist standpoint regarding many issues of importance for this philosophy: the constitution of our ideas, foundations of knowledge, unity of science, meaninglessness of (transcendent) metaphysics. D'Alembert, Turgot, and Condorcet added to this the positivist treatment of the advance of the human mind and of the impulses it gives to the broader progress of humanity.

Thus the main questions that concerned positivism and some important answers to these questions were formulated before Comte. Therefore, in my opinion, one can speak of early positivism of the eighteenth century, best represented by Hume and quite clearly by d'Alembert. Yet positivist philosophy acquired a fully developed, integrated form only in the nineteenth century with Auguste Comte.

CHAPTER 2

Classical or Social Positivism

France after the Revolution

Whereas the emergence of early positivism is closely connected with the scientific revolution of the seventeenth century, and for this reason we began Chapter 1 on the first stage of the development of positivism by examining some of its features, the emergence of classical or social positivism was crucially influenced not by the scientific revolution, but by political revolution. The center of the scientific revolution of the seventeenth century, especially from the middle of the century, was Great Britain, although certain important scientific ideas were also advanced in other countries, first of all in Italy. British science prevailed in the world in the eighteenth century as well. Therefore it is not surprising that a Briton (or, more precisely, a Scotsman), Hume, became the pioneer of this nascent positivism.

The social and political revolution which we have in mind when talking about the circumstances of the emergence of classical positivism is the French Revolution. One cannot say that this social revolution bears no relation to the anterior scientific revolution. The widespread opinion is that the ideological foundation of the French Revolution was set by the French Enlightenment philosophers, who, as we have already seen when discussing the ideas of d'Alembert, Turgot, and Condorcet, considered the scientific revolution of the seventeenth century to be an especially important phase of the development of human reason. They, as well as other French Enlightenment proponents, sought such social and political reforms that would fit this new stage of the development of reason. Perfected human reason was to be transformed into an instrument for the creation of the society of reason and the state of reason.

At first sight, the claim that the philosophers of the Enlightenment, who are usually called (paying tribute to French) *philosophes*, should be praised or blamed for the intellectual preparations of the French Revolution, may seem to be an overstatement. The causes of the revolution were rather complex; most *philosophes* were far from being radical, and they preferred to address their projects on how to improve the social order and foster intellectual and social progress to monarchs and not to the people. They shunned violence and wanted progressive changes to be introduced cautiously and gradually. Yet there is no doubt that by criticizing many features of the *ancien régime* and emphasizing that the social and political order must be radically reformed in accordance with the requirements of Reason, they became accountable for changes aimed at replacing that order even if they disapproved the way by which these changes were introduced.

The Restoration, i.e., first of all the restoration of the monarchy, destroyed many of the innovations introduced during the revolution. Yet the Restoration of the Bourbons did not bring France back to its pre-revolutionary state. Nobody could simply ignore all the changes which occurred in the ruling class, French institutions, and Frenchmen's thinking. One of those transformations *in line with the Spirit of the Enlightenment* which the revolution catalyzed, Napoleon consolidated, and the Restoration had simply to accept by managing to introduce only slight modifications, concerned the educational system of France.

The requirement to change the obsolete eighteenth century educational system by restructuring it, broadening its base, and changing the content of learning was the most important practical postulate *all* philosophers of the French Enlightenment shared. They did succeed to an astonishing degree. Transformations in this area made France in the first half of the nineteenth century the most educated nation in the world and the leading nation in science. In accordance with the principle of universal education, a system of state elementary and secondary schools was created in the country. Much stronger emphasis was laid on mathematics and natural sciences. The number of schools and pupils increased greatly.

The universities were closed during the revolution, partly because they were regarded as centers of politically dangerous religious influence, partly because their curricula did not suit the needs of revolutionary reconstruction. Instead of these, specialized higher education establishments of a completely new kind were created, such as the Higher Normal School and the Polytechnic School. The Museum of Natural History and the National Institute of Sciences and Arts, predecessor of the present Institute of France, were founded. All four mentioned institutions became important centers of scientific research.

On all levels of education, schooling in subjects of practical, technical value prevailed over the study of classical philological and philosophical disciplines. The system of education created in France during the years of *political revolution* was the first in the world based on the principle of the seventeenth century *scientific revolution* that knowledge is power and must serve the implementation of practical aims and purposes. It was impregnated with positivist spirit.

The foundation of the new educational and research institutions created very favorable conditions for carrying out scientific investigation. Their results were impressive. Augustin Cauchy elaborated the theory of series and theory of limits, used it as a basis for differential and integral calculus, and also laid the foundations for the theory of functions of complex variables. Pierre Laplace systematized probability theory, refined methods for the calculation of planetary movement, and put forward a cosmological hypothesis on the origin of the solar system. Jean Fourier set the foundations of a theory of trigonometric series. Sadi Carnot developed thermodynamics, André Ampère—electrodynamics, Augustin Fresnel—the theory of diffraction and interference of light. Jean-Baptiste Lamarck created the first systematic theory of evolution (and introduced the very term *biology*). Many new elements were discovered by French chemists, among whom Claude Louis Berthollet was likely the best known. Important technical inventions were made at that period in France, from the turbine to the photographic camera.

The great achievements in mathematics and natural sciences substantially strengthened the prestige of science and the social position and influence of scientists. Some of them held for a longer or shorter time governmental posts and were very active in reshaping France's educational system. Compared with the remarkable accomplishments in sciences, French philosophy of the first decades of the nineteenth century looked especially bleak. The fame and recognition it enjoyed in the eighteenth century faded. In France, Reason was still in high regard. But now scientific, not philosophical reason was in vogue. The impact of only one of the French philosophers, who were active in this period, stretches far beyond his lifetime and his native country: Claude Saint-Simon. Both Auguste Comte and Karl Marx are much indebted to him, and with them, all positivism and Marxism.

A direct link is clear between the representatives of positivist thought in the French Enlightenment and Saint-Simon (we will speak about the direct link between Saint-Simon and Comte later). According to Saint-Simon, d'Alembert was his private tutor, and although there is no proof of that, the influence of the ideas of the Enlightenment upon Saint-Simon's conceptions is obvious. Yet by no means was he a simple follower of the philosophy of the Enlightenment. In a rather important sense he opposed this philosophy. According to Saint-Simon, the philosophy of the Enlightenment was critical and revolutionary, whereas the new philosophy conforming to the needs of the nineteenth century must be constructive and organizational. (One of his later works is even entitled *The Organization*.) Following Turgot and Condorcet, he emphasizes the role of ideas for the development of human society and contrasts the era of positive scientific knowledge with the previous ones dominated by theological and metaphysical speculation. Yet he has more to say about the significance of economic life for the advance of humanity. He treats his own times as an intermediary phase following the Middle Ages terminated only by the French Revolution. The latter created preconditions for establishing a new, industrial society.

In the treatise *On the Industrial System* (Saint-Simon 1821), he specifies social and political structures characteristic for that phase in the development of society and attempts to examine their mode of operation. He envisages the rule of an elite formed by scientists, engineers, and captains of industry. All those involved in productive activity, from bankers to workmen, form the industrial class, which is sharply contrasted by Saint-Simon with those who are not involved, directly or indirectly, in production, but hold present political power in their hands. In the rationally arranged industrial system, productive work would be compulsory, society would be transformed into a free association of industrialists in which everybody had the opportunity to use his skills and capabilities, and political power would be converted into an effective tool of planning and organizing production. Free associations of industrialists would gradually form a world-wide association in which national differences would tend to disappear. The achievement of solidarity characteristic for this highest stage of human development presupposes, however, not only a great advance of science, including the foundation of a science of society called by Saint-Simon in his later years *social physiology*, but also the creation of social and political structures which would conform to its ideas.

In the book *New Christianity* (Saint-Simon 1825), he emphasizes human brotherhood as the highest ideal man can seek and which may be established only on the condition of overcoming the selfishness and egoism of human creatures. With that end in view the new morality based on a Christian conception of fraternal love must be created and disseminated.

Saint-Simon and his followers presumed that the advance of humanity created many preconditions for the establishment of an industrial system.

They regarded this system as rational and realizable. Yet at the same time they saw, however, many obstacles on the way to it. One of the most important ones was the state of mind of his contemporaries. After drastic social and political changes, revolutionary terror, imperial wars, then the fall, miraculous return, and final defeat of Napoleon, followed by the restoration of the Bourbons and years of reaction, Frenchmen were disillusioned and frustrated. They were divided into royalists and republicans. Political life was turbulent, and some observers saw it as a permanent crisis bordering on anarchy. There was an element of overstatement in this characterization, yet it was not that exaggerated. And for sure, the intellectual and moral crisis in the years of the Restoration was even deeper than the political one.

The prevailing reaction of the French philosophers to the turbulence the country was going through was, however, that of indifference. With the exception of Saint-Simon and a few others (such as Charles Fourier, not to be confused with the mathematician Jean Fourier mentioned earlier), most French philosophers simply turned away from the social problems that overwhelmed the philosophers of the Enlightenment, and which became urgent once again. Disillusioned not only with the practical consequences of the philosophy of the Enlightenment but with this philosophy itself, they turned their attention from social and historical problems to metaphysical ones of which the Enlightenment was rather suspicious. The spiritualist metaphysics many philosophers of the Restoration practiced was second-rate, a far cry from the metaphysics of the seventeenth century, especially that of Descartes. The situation in epistemology and many other departments of French philosophy was no better. The level of inquiry into human knowledge and man's social and political dimension obviously did not match the high level of research in the area of natural sciences.

Auguste Comte

The philosopher who, already in the second decade of nineteenth century was extremely unsatisfied with this state of affairs—intellectual anarchy, the backward state of social studies, and especially, with the lack of clear prospect for France and all of humanity—was Auguste Comte. In fact, at that time the appellation of philosopher could be applied to him only with great reservations. He was then known neither to a wider public, nor to French philosophers. Born in 1798, he became a secretary to Saint-Simon in 1817, a year after he was expelled—for his republican sympathies—from the École Polytechnique (Polytechnical School) at which he studied mathematics and engineering.

He entered this prestigious school in 1814, at age 16. He studied at the École Polytechnique for two years only. Yet, it seems, studies and later work at this institution (although he held a humble post of an examiner) made an indelible impression on Comte's mind and influenced the contents of his project of the rational reformation of society. Established in 1894, the École Polytechnique soon became the most important center of studies and research in mathematics, natural sciences, and engineering in France, and, at least for a few decades, also in the world. The high level of studies and research was secured by an exceptionally modern curriculum and excellent professors.

The École Polytechnique became a prototype of Comte's model of the future state of society. Established in the revolutionary years, the school had, nevertheless, an elitist character. Yet the elite we are talking about was of a quite different kind than that characteristic of the *ancien régime*. It was an intellectual elite: both professors and students comprehended quite well that they formed an exclusive and important community. Many of them felt responsible for the progress not only of their own studies and research, but for the progress of the nation and of humankind as well. Primarily they thought about intellectual progress, but under the influence of the Encyclopedists and of revolutionary changes made in the name of Reason, many of them linked intellectual progress with social and political advancement. Because of the prevailing republican attitudes, the École Polytechnique was rather unfavorable to Napoleon in the time of empire, yet it supported him during the Hundred Days in 1815 as a lesser evil compared to the Bourbons, and even after Waterloo did not want to reconcile itself with his final defeat. Thus it is a small wonder that the Restoration authorities decided to reform the school and on this occasion to get rid of the most hostile students. When the École Polytechnique was reopened, Comte was one of those students who was not readmitted (cf. Gouhier 1965 [1933], vol. 1: 295–97).

After a short stay in Montpellier, his native city, he returned to Paris to give private lessons in mathematics, and produced a number of short mathematical articles. He read voraciously and became fascinated by the problems of the philosophy of science, and later, in the role of science in society (cf. Skarga 1977: 26–29). After becoming the secretary to Saint-Simon, Comte was strongly influenced by Saint-Simon, although after a few years Comte himself began to exert influence on his patron. They were having extensive discussions. It is quite probable that some of the already mentioned later works of Saint-Simon's contain Comte's ideas (cf. ibid., 35). In 1824 he left his job because of a dispute with Saint-Simon over the authorship of some texts written allegedly mainly by Comte and published by Saint-Simon as his own.

In 1826, he started a private course of lectures on philosophy attended among others by some distinguished scientists. Because of the breakdown caused by family and health problems, this course was interrupted after a few lectures and renewed only after two years. Six volumes of *The Course of Positive Philosophy* were published in 1830–42. Although favored by the rector, he had not become a professor of mathematics at the École Polytechnique, and had to leave the institution after the publication of the last volume of *The Course*, which contains a lot of social critique. Thereafter Comte was living on the financial support of his friends and followers. Comte's second great treatise in four volumes, *The System of Positive Polity*, appeared in 1851–54. Comte died in 1857.

Plan of Positive Labors

Comte's main works were devoted to the elaboration of the ideas he had presented in 1822 in his study, *Plan of Scientific Studies Necessary for the Reorganization of Society*. Comte begins this important study with a general characterization of his epoch. According to him, the old, "Catholic-feudal" system has already lost its power. And not only in France but in most civilized nations. This system brought society to "a profound moral and political anarchy which appears to menace it with a near and inevitable dissolution" (Comte 1975 [1822]: 9). The negative tendency, the movement of disorganization, coexists at present with the organic and progressive tendency, the movement of reorganization. Yet their clash only deepens "the grand crisis." The negative tendency still prevails and blocks the progress of civilization.

Comte was convinced that in order to guarantee the order and progress of society, one needs a unified system of ideas to guide society and strong government to unite separate segments of society and act so that human activities would be directed at a common goal. The Enlightenment's liberal principles of human rights and freedom were not favored by Comte. According to him, such principles are effective in destroying the outmoded feudal system, but they are useless for creating its alternative.

Comte found the earlier visions of reforms of society lacking in comprehension and depth. He had no doubts that before reorganizing society, a *theoretical* foundation of all necessary reforms must be prepared, the aims of society and the principles of coordination of social relations examined, and the plan of reorganization of society carefully developed. Acting in the political sphere and without a thorough understanding of the foundations of social life and the nature of the social progress, people mistakenly consider mere corrections of the old system on the basis of the constitution to be its reform. The thorough reorganization of society requires not lawyers, but theorists and scientists. Precisely because they are the greatest intellectual resource at the disposition of society, and only they can be relied upon when creating the new system.

Having established in the *Plan* that the most important task of scientific men at present is to establish a positive theory of society which he starts later to term *sociology*, Comte outlines the new discipline's theoretical foundation. The backbone of this theory is his notion of the three stages of the development of the human mind. This notion was elaborated substantially in the *Course of Positive Philosophy*.

The Theological, Metaphysical, and Positive Mind

In the *Plan* Comte faced the problem of determining the kind of facts on which social science must base itself and, accordingly, what character it must have. He was convinced that this science must examine collective phenomena of the human race and be clearly detached from investigation of phenomena related to an individual. While Comte uses various terms for this new science, including political science, political philosophy, social science, social physics, social physiology, and, finally, sociology (Comte is usually credited with coining this term), he is clear that it must be founded on historical evidence which may be used for tracking the development of civilization. When writing annals or chronicles, historians make use of historical facts as well. Yet the aim of social science is quite different—its main aim is to establish *generalizations* related to the progress of mankind. Because this progress is inseparable from the progress of the human mind, the determination of the law of its advance may furnish a solid basis for the construction of social science. Thus Comte begins his *Course* with the declaration that:

> From the study of the development of human intelligence, in all directions, and through all times, the discovery arises of a great fundamental law, to which it is necessarily subject, and which has a solid foundation of proof, both in the facts of our organization and in our historical experience. The law is this: that each of our leading

conceptions—each branch of our knowledge—passes successively through three different theoretical conditions: the Theological, or fictitious; the Metaphysical, or abstract; and the Scientific, or positive. In other words, the human mind, by its nature, employs in its progress three methods of philosophizing, the character of which is essentially different, and even radically opposed: namely, the theological method, the metaphysical, and the positive. Hence arise three philosophies, or general systems of conceptions on the aggregate of phenomena, each of which excludes the others. (Comte 1975 [1830/42]: 71)

This law is known as *the Law of Three Stages*. In the first volume of the *Course* (as well as in the *Plan*) Comte does not pretend that it was *he* who discovered this law. Later, however, he did not resist the temptation to change his mind. Although the examination of ideas of d'Alembert, Turgot, and Condorcet in the previous chapter reveals that Comte overestimates his merits in discovering this law, the widespread reference to it as *Comte's Law of Three Stages* is easy to explain. Comte's wording of this law was more suggestive than that of his predecessors and he was more explicit when defining the character of each stage, and through his own efforts the law became widely known.

Comte treats the Law of Three Stages as the main result of the historical examination of the human mind. Many philosophers were interested in the human mind. Yet Comte maintains that his (and, maybe, a few of his immediate predecessors') approach to the subject is quite different than that of earlier philosophers. They examined the human mind in an abstract way. They presupposed that a human being is in essence unchangeable. Thus they discussed the human mind as a part of constant human nature. Although Descartes and Hume were concerned with the individual mind, the individual they deal with is an abstract individual who may even be called the universal individual. No time coordinates are taken into account in the process of examining such a mind. It is treated as dissociated from time. And although Kant relinquishes the identification of the mind to be analyzed by philosophers with the mind of real individual and opposes the transcendental mind to the psychological one, he still treats it ahistorically.

Hegel's treatment of the human mind substantially differs from that of Descartes and Hume, and from that of Kant. The problem of the *development* of mind is of crucial importance for him. Yet his approach was distinct from Comte's. On the one hand the real aim of Hegel's efforts was to capture the nature of the World Mind (*Geist*). Hegel essentially identified the World Mind with the divine mind or God (the Absolute) himself, and he spoke of its development primarily in the logical, not a historical sense.

When Comte speaks about the development of the human mind, he does not mean the World Mind or the divine mind, but the mind of humankind. In fact, both notions are rather abstract constructs, but the differences between Hegel's three logical phases in the development of thought (thesis, antithesis, and synthesis) and Comte's three historical stages are far from negligible. Hegel's conclusions were founded (especially in *The Science of Logic*) mainly on *a priori* speculations, while Comte's historical analysis had much more empirical elements. Comte characterizes the development of the human mind as follows:

> In the theological state, the human mind, seeking the essential nature of beings, the first and final causes (the origin and purpose) of all effects—in short, Absolute knowledge—supposes all phenomena to be produced by the immediate action of supernatural beings.
>
> In the metaphysical state, which is only a modification of the first, the mind supposes, instead of supernatural beings, abstract forces, veritable entities (that is, personified abstractions) inherent in all beings, and capable of producing all phenomena. What is called the explanation of phenomena is, in this stage, a mere reference of each to its proper entity.
>
> In the final, the positive, state, the mind has given over the vain search after Absolute notions, the origin and destination of the universe, and the causes of phenomena, and applies itself to the study of their laws—that is, their invariable relations of succession and resemblance. Reasoning and observation, duly combined, are the means of this knowledge. What is now understood when we speak of an explanation of facts is simply the establishment of a connection between single phenomena and some general facts, the number of which continually diminishes with the progress of science. (Comte 1975 [1830/42]: 72)

These three states represent the different degrees of maturity of the human mind. The first is the starting point of intelligence, the third is its final state, while the second has a rather transitional character. In most respects it represents only a modification of the theological state.

Comte stresses that the human mind passes through these states in succession and this justifies their treatment as stages or phases of the process of development of the human mind. That still does not mean that different methods were not used in the same epoch. However, according to Comte, different methods and different philosophies *prevail* in different periods, and the shift in prevailing method and philosophy, the transition from one stage to another, is the clearest manifestation of the progress in human thinking. The succession of states of mind is not a simple contingency: the earlier stage is always a necessary precondition of the latter. Without the development of cognitive skills which took place in the theological and metaphysical stages, the human mind could not reach the positive state.

Comte emphasizes that the law of the three stages applies not only to the development of the collective mind of humankind, but to the development of the individual mind as well. The individual mind passes through the theological state characteristic of childhood, later through the metaphysical state, and reaches the positive state only in maturity.

Now we can examine Comte's views concerning each condition of the human mind and the causes of transition from one state to another a bit more thoroughly. We begin with the theological condition. It is, as Comte emphasizes, necessary and unavoidable. In its infancy the human mind may use only imagination when explaining the world around him and his place in this world. Imagination is an important human faculty. The process of cognition cannot do without it. Yet in order to achieve more important results, the imagination must be put under control of other cognitive faculties, namely argumentation and observation. The latter faculties being underdeveloped in the initial stage of the advance of the human mind, the imagination lacks any constraints. The theological mind is incapable of solving the most elementary scientific problems. It simply brushes them aside and concentrates on the questions which captivate the imagination and require nothing beyond imagination to solve them. It considers the origin and purpose of things and searches for beings concealed beyond phenomena.

The theological mind dominated throughout the greater part of the existence of civilization. Because of the very long period of its prevalence, it gradually changed its appearance, going through different phases. Comte singles out three forms of the theological mind: fetishism, polytheism, and monotheism. Fetishism is based on instinct and feeling, it animates and treats like lesser or greater gods trees, rivers, mountains, and stars. Polytheism deprives things of spirits and invents fictitious anthropomorphic or zoomorphic gods.

During the monotheistic phase, reason begins to limit the power of imagination and forms the notion of one God ruling all over the world. The monotheistic notion of God, according to Comte, is a precondition for the idea of all natural events being governed by unchangeable laws to spring up. One may say that the idea of manifestations of God's will is the precursor of the notion of the law of nature. During the theological phase the mind was developing its theoretical powers without relying on facts. Yet this phase was a necessary stage of its development, even though the theories created by sheer imagination were fictive and fanciful.

Comte does not renounce the most important empiricist principle, namely that theories must verified by relevant facts. However, by empha-

sizing the role theories play in the process of collection and systematization of facts, Comte overcomes the view of Bacon's and Hume's empiricism, that theoretical attitudes have no significance in selecting cognitively important facts. Comte's standpoint on the question of the relation
of facts to theories was taken over later by most leading positivists. Thus
widespread accusations of positivism as being a form of *naive* empiricism
are rather baseless.

To recall, the metaphysical condition is transitional. Comte emphasizes
that the metaphysical mind pursues absolute knowledge, and in this respect closely resembles the theological mind. The metaphysical mind
seeks to disclose the internal nature, origin, and purpose of things: the
essence of phenomena. Yet when searching for the first and final causes,
the metaphysical mind seeks not for spirits, gods, or God, but for abstract
entities. Comte calls them "personified abstractions," although the entities
Comte has in mind, such as Substance, Essence, Absolute, and The First
Cause, all differ from the notions of the theological mind by being deper-
sonified: they are much less similar to humans than gods.

In metaphysical philosophy, the imagination is put under control. Yet
observation is not regarded as the main tool of control: the only control-
ling instance is argumentation. Tormented by senseless doubts like those
concerning the reality of the external world, and endless controversies, the
metaphysical mind reminds Comte of an illness we must go through during the passage from childhood to maturity. Yet the wide use of speculative argumentation prepares the ground for scientific thinking. In this respect the metaphysical mind may be looked at as a more advanced form of
thinking than the theological one.

Only the positive mind is truly mature. After finally arriving at the
conclusion that both the theological and metaphysical minds hunt after
ghosts, man ceased to seek, according to Comte, absolute truths in which
he was so interested in childhood and during adolescence, and focused on
efforts to acquire a different kind of knowledge. Positive knowledge does
not stir so much discussion, does not pretend to be absolute truth, yet is
feasible and matches real human needs. Both its usefulness and attainabil-
ity follows from the fact that this knowledge is based on thorough and
careful observation.

Positive knowledge is not absolute, since the positive mind denounces
the Absolute and the search for essences and ultimate causes of phenomena, and it concerns itself only with the examination of phenomena. The
result of such examination is what Comte considers to be relative knowledge. But at the same time, Comte regards positive knowledge to be real,

reliable. He did not take this relative knowledge to be problematic or hypothetical, which is what logical positivists began to contend from the late 1930s.

The positive mind seeks to know in order to predict phenomena. Comte emphasizes that prediction is the most important task of cognition. This is the area where the theological and metaphysical mind fails. The positive mind excels in this and in other respects because (1) it explores what is real, i.e., within reach of the mind, and not indiscernible mysteries; (2) it deals with the useful, not fruitless; (3) it ensures certainty which replaces doubts and controversies; (4) it aims at precision, not indeterminacy; (5) it serves for organization, not destructive criticism. It is important to stress that Comte claimed that the positive mind which is examining phenomena should seek neither their primary, nor ultimate, nor proximate causes. The word "cause" according to him should be eliminated from the language of positive philosophy. By claiming that the investigation of phenomena and their regular relations does not require any speculation about their causes, he takes the same stance that was later adopted by Ernst Mach.

It seems difficult to defend Comte from the objection that his characterization of each of the three states of mind is unacceptably oversimplified. This is most clearly seen in his characterization of the metaphysical mind. Metaphysics is much richer and diverse than Comte portrays. Not all metaphysicians sought primary or ultimate causes, and even when they did, some of their ideas were later used in scientific theories. This happened, for instance, with the ideas of the Pythagoreans or the atomists. The boundary between the metaphysical and scientific thinking is perhaps not as sharp as Comte claimed. Most importantly, neither theological nor metaphysical philosophies were interested in the goals Comte suggests they *failed* to achieve. They were not concerned with the prediction of phenomena or technological applications of knowledge. Viewing them through the eyes of a pragmatic who is interested only in what is useful, Comte misinterprets their nature.

The Hierarchy of Sciences

Although Comte quite often terms a positive mind a scientific mind, he himself emphasizes that it is a rather superficial characterization. Science is comprised of different branches. Their development is uneven. Some of them reached the state of maturity already some time ago, others still are at the beginning stages of their development. Thus to claim that all science reached the positive state in the nineteenth century would be a glaring error. An even cruder mistake would be to maintain that all science was based on the positive method from the beginning.

Thus in order to understand the nature of the positive mind, the preconditions of its emergence and the gradual widening of its sphere of influence, it is necessary to examine the interdisciplinary structure of science and the relations between its different branches. This examination has to reveal the present condition of every main science and to give impulse for their advance, especially for the advance of those clearly underdeveloped. It may even disclose that humanity still lacks some science important for its own progress. Thus the evaluation and classification of the sciences is of crucial importance for the realization of the positivist program. It is impossible to present the positive knowledge needed for social reorganization without a clear delineation of the branches of science itself.

In devoting much attention to the problems of classifying science, Comte resembles both Francis Bacon and the Encyclopedists, and this is natural for a theoretician who himself embarks on devising—almost a century after Diderot and d'Alembert—the new Encyclopedia, the Great Book of Positive Knowledge that *The Course of the Positive Philosophy* had to be. Yet he is unhappy with the classification of the sciences produced both by Bacon and d'Alembert since they were based upon an "arbitrary division of the faculties of mind" and ignored the fact that in the process of scientific inquiry, all these faculties are often used simultaneously. The classification of sciences on which the exposition of the system of positive knowledge must be based, should take into account not the existence of different faculties of mind but the existence of different classes or orders of natural phenomena, and the mutual dependence of the sciences which results from the dependence of corresponding phenomena.

Proceeding to the construction of his own classification of sciences, Comte cautions that he does not aim at exhausting the allotment of all parts of our knowledge. His system covers only theoretical knowledge and does not include practical knowledge or the arts such as engineering or agriculture which depend upon theoretical knowledge. As for theoretical knowledge or "science properly so called," it may be divided into general or abstract, and particular or concrete science. General and abstract sciences "have for their object the discovery of laws that regulate phenomena in all conceivable cases," whereas the function of the particular or concrete sciences "is to apply these laws to the actual history of existing beings" (Comte 1975 [1830/42]: 90).

After narrowing his subject, Comte proceeds to the ascertainment of the "true order" of the fundamental, i.e., abstract and general sciences. When choosing the "true order" of such sciences, he follows the principle already mentioned: the dependence of one science upon another resulting from that of the phenomena under study. This order is determined by the degree of simplicity of their phenomena. Simpler phenomena are *ipso facto* more general ones. More complex phenomena depend on simpler ones, hence sciences which examine the former depend upon sciences which investigate the latter. Sciences which investigate more complex and concrete phenomena are based on the laws established by sciences examining more simple and general ones.

Guided by such principles, Comte classifies fundamental sciences in the following order: mathematics, astronomy, physics, chemistry, physiology (biology), and social physics (sociology). Comte emphasizes that mathematics plays a special role in the system of sciences. It has great value as a part of the system. However, it is even more valuable as "the most powerful instrument that the human mind can employ in the investigation of the laws of natural phenomena" (Comte 1975 [1830/42]: 100). Comte regards astronomy as a slightly more general science than it is seen today, because the law of universal gravity established, according to him, by astronomy and not by physics, is valid in the areas of phenomena investigated by all less general sciences. Astronomy is using mathematics most extensively. Chemistry is based on physics, biology on chemistry, and sociology on biology.

Comte regards the establishment of the true order of sciences as his great achievement, whose significance is not much less than that of the Law of the Three Stages. Besides giving us a clear and distinct picture of the main body of our knowledge, it has, according to him, great practical significance. It defines the true order of studies, both general and special, and lends guidelines to the reform of the educational system. The sequence of studies must correspond to the natural sequence of the sciences. Students can acquire a proper understanding of complex phenomena only after being introduced to simpler ones. And as for researchers, they, in order to proceed efficiently, must know perfectly the basic laws of sciences more general than their own because these laws also apply to the phenomena in which they are interested.

Comte's system of sciences bears the signs of his own epoch. He was aware of its imperfections and understood completely that it is simply impossible to produce an ideal classification. Yet its first and major part (slightly transformed by the partial marriage of astronomy and physics) has served well the cause of the advancement of scientific knowledge. Even the rise of the cognitive sciences does not discredit fully the basic order of the fundamental sciences Comte helped to establish—mathematics, physics, chemistry, and biology. This is arguably the most important part of his intellectual heritage from which we still benefit. He justly emphasized the importance of physics for chemistry and the significance of both to biology. Today we understand the import of ties linking these three sciences much more clearly than researchers understood it before Comte. The advance of molecular biology we expect so much from would be impossible without such an understanding. We have no doubts as to the importance of mathematics for all other sciences as well, although Kant stressed this before Comte.

The last part of Comte's classification appears much more questionable, and in fact, was criticized by many theoreticians, including representatives of positivism itself, especially, for not including in his scheme *psychology and economics*. We will consider this important issue later.

At first glance it looks even odder that a thinker who often characterized himself as a philosopher did not include philosophy in his system of fundamental sciences. Yet Comte identifies most often positive philosophy with positive thinking and with sciences which reached the positive stage in the process of their development. When Comte allots to the term "positive philosophy" more specific meaning, he dissolves it in fact in sociology. Comte regards the Law of Three Phases that we examined earlier as the basis of sociology, because the development of the human mind determines social changes, the transformation of social life. Granted, Comte emphasizes occasionally that an important task of a philosopher is to present a systemic synthesis of the results of various sciences. The establishment of the hierarchy of sciences is an important precondition of such synthesis. However, even in such a case, it would be unreasonable to claim that philosophy has its own specific object and therefore must have a special place in the system of sciences. By the way, even though some other positivists held quite different views from those of Comte on the character of positive philosophy, none was inclined to treat it as a *separate* science. And as to metaphysics, not a single positivist considered it a *science* at all.

When constructing his hierarchy of sciences, Comte was interested only in those sciences which employ (or, as in the case of sociology, may be expected to employ) the *positive method*. Yet the general characteristic of this method given by Comte in the context of explaining the differences between the theological, metaphysical, and positive mind is extremely concise. He says little about it more than indicating that it is based on observation, to which imagination and argumentation are subordinated. This is not a sign of a lack of precision and thoroughness, but a conscious position: according to Comte, we cannot grasp the positive method in the abstract. The only way to comprehend it is to study its application.

Of course, it is possible to investigate the impact science makes on the development of society and other similar, external issues without going into details of the relation of different sciences. Yet such an approach is impermissible when methodological questions are dealt with. According to Comte, method and doctrine are inseparable. Thus we must have some notion of physics before examining its method, and similarly, some notion of sociology or social physics before explaining its method. The method of a specific science can neither be laid down nor examined in any way other than through its use. No analogue of Mill's *Logic* exists in Comte's corpus. By the way, exactly for the same reason, his attitude to the very *science* of logic was rather skeptical. Yet Comte touches upon methodological issues when dealing with the problem of classification of sciences and when examining the "general spirit" of sciences and their relation to "anterior" and "posterior" (in the hierarchical system of sciences) disciplines.

The theory of the method of science plays an important role in positivist philosophy. Hume, John Stuart Mill, and especially the logical positivists devoted much attention to it. Yet because of the above-mentioned reasons Comte cannot be characterized, properly speaking, as a methodologist. Granted, he maintains that all science is based on observation, the most widely used in natural sciences. Thus with some minor reservations related to the fact that, according to Comte, different sciences use different modes of observation, he may be termed a methodological naturalist. But he is not a reductionist. He flatly rejects the possibility of the reduction of sociology to biology. In the same manner, he denies the possibility of reducing any one of the fundamental sciences to another. Terms of the less general sciences cannot be defined on the basis of the terms of more general sciences, and laws of the former cannot be deduced from the laws of the latter. The existence of the *natural categories* of phenomena prevent this. By exploring appropriate phenomena besides the procedures used by more abstract sciences, all sciences use some specific procedures and are partly based on the facts of their own spheres. Thus an all-embracing unified science is, according to Comte, simply impossible.

The most widely used scientific procedure is of course observation, which constitutes the kernel of the positive method. Astronomy is based almost exclusively on observation (and mathematics). Besides observation, physics widely uses experimentation. In addition to observation and experimentation, chemistry employs classification. Together with all these methods, biology uses comparison. Social science employs, additionally, an historical method.

Comte is eager to show that the Law of the Three Stages is applicable to every main branch of knowledge, and that each of the fundamental sciences passes through a theological, metaphysical, and positive stage. The speed of the advance of a science and its passing from one stage to another depends mainly on the degree of complexity of its object. Thus it is natural that astronomy reached the positive state earlier than biology. Yet even mathematics and astronomy, not to speak about more complicated sciences, were born in swaddling theological clothes. The Babylonians, who apparently first began to explore the properties of numbers, attributed to them special powers and regarded some numbers as bringing luck, others calamity. They regarded stars as gods and attributed to them special powers as well. And the Pythagorean claim that numbers are elements of things may be interpreted as a thesis of metaphysical philosophy.

The examination of the rise and development of sciences, specifically of the process of different sciences passing through theological, metaphysical, and positive stages, led Comte to important discoveries which have great worth irrespective of the value of his philosophy, and of the Law of the Three Stages in particular. These investigations made Comte one of the founders of a new discipline (which does not figure in his classification of sciences), namely that of the history of sciences. The first university department of the history of science in the world was established at the College de France in 1892 under the pressure of positivists, and its first chair was a leading representative of Comte's school, Pierre Laffitte. The history of science became a highly respectable academic discipline mainly thanks to Pierre Duhem, who also owes much to Comte.

Yet Comte is better known not as a leading historian of science and one of the originators of the discipline of intellectual history and the history of ideas, which has experienced rapid growth recently, but as the founder of a science *he* regarded as the most important. In his system of the sciences it is the last but not least. He used to call his system the hierarchy of the positive sciences. However, being sure of the great value of this science for man, for the reorganization of society, and even for the proper understanding of other sciences and guidance of their advance to the benefit of humanity, he had to state that the most important science has not yet reached the positive condition. Comte claims that metaphysical elements are present in many sciences. Yet in the most general ones they are already almost invisible. It is much easier to find them in chemistry. Comte emphasizes that in biology the metaphysical spirit still competes seriously with the positive one, and is reluctant to renounce *vis vitalis* and similar notions. The worst situation is, however, in social science. Here the non-positive spirit clearly prevails. Thus the most urgent task of the advance of human knowledge is the creation of social science with the aid of the positive method. Therefore, Comte devotes a great part of his *Course of the Positive Philosophy* to laying down positive principles of this discipline which must ensure the social order and progress badly needed for the contemporary world. An Encyclopedist realized that his set of books lacks the final and most precious one, and noticing no one appropriate to produce it, embarked on writing it himself.

Social Order and Social Progress

According to Comte, the importance of social science follows from the fact that it deals with the questions directly concerning man and is the theoretical basis for the reorganization of society. Moreover, in a sense it is the universal science: (a) other sciences may be treated as stages leading to it; (b) their significance in general and of specific problems they deal with is seen only in the context of sociology, only in retrospective.

At the same time this discipline concerns the most complex object of science, namely society. Therefore, it is rather natural that it is much less developed than mathematics, astronomy, and physics, which reached the positive stage in their advance, and are less developed than chemistry, which is nearing the positive stage, or biology, which is still dominated by metaphysical conceptions. According to Comte, social science remains in the theological stage. Thus the task which Comte sees laying ahead of him is really tremendous: to create a new fundamental positive science in the area which is still predominantly theological and only a little tinged with metaphysics.

Many social scientists and historians of the subject agree that in spite of the many shortcomings of Comte's sociology, he in fact made a great step forward in this area of knowledge. He is regarded quite often as *the* founder of social science. He introduced the term *sociology*, widely used as the name of the (positive) science of society, he exposed its significance much more convincingly than anybody else before him, and he put before social science many important problems, especially those concerning industrial society.

Comte's general notion of the place of sociology in the system of sciences was presented in the previous subsection. Yet because of the importance of sociology in this system and of the specific tasks Comte faces when dealing with this subject, it is worth examining the method of sociology and its relation to other sciences in a little more thorough way.

He claims that social science uses three methods, which natural sciences also use: observation, experiment, and comparison. As for observation, Comte is eager to emphasize that "no real observation of any kind of phenomena is possible, except insofar as it is first directed, and finally interpreted, by some theory. [...] Hence it is clear that, scientifically speaking, all isolated, empirical observation is idle and even radically uncertain; that science can use only those observations that are connected, at least hypothetically, with some law" (Comte 1975 [1830/42]: 241). Ascertaining the primacy of theory over observation makes Comte's stance on their relation quite different from that of Hume. Although many present-day philosophers of science would rather support Comte's position, it clearly opens the possibility of abusing theory, for which the later Comte is often blamed. In any case, Comte did not try to elaborate on the methods for empirically testing a theory which could make its primacy over observation a relative one.

As is to be expected, Comte regards experiment to be of inferior value for social science. Yet pathological cases are, according to him, true scientific equivalents of experiments, and they are quite common in revolutionary times and above all, as Comte notes, in his own.

Comparison, according to Comte, must prevail in all studies concerned with living organisms, both in biology and in sociology. Of special value for social science are, allegedly, historical comparisons of different periods of civilization. In his view, they are so important that Comte singles out the historical method as "a fourth and the final mode of the art of observing" to be used in social science alongside observation, properly so called, experiment, and comparison. Moreover, he claims that the present is, by itself, evidently misleading because when examining it, separating principal from secondary facts is impossible. Only examination of the past, of the successive stages of the development of humanity, may guide us in this enterprise.

Being the sixth fundamental science, sociology uses a greater variety of methods than any previous (anterior) science. The use of the historical method which other sciences do not use extends its cognitive possibilities. Yet Comte is quite aware that "this extension of means does not compensate for the increased imperfection arising from the intricacy" (Comte

1975 [1830/42]: 239–40). The complexity of social phenomena requires a special approach which differs from that characteristic of physics or chemistry. "[I]n the inorganic sciences, the elements are much better known to us than the whole that they constitute, thus we must proceed in them from the simple to the compound. But the reverse method is necessary in the study of man and of society, man and society as a whole being better known to us, and more accessible subjects of study, than the parts that constitute them" (ibid., 229). That means that social science cannot be separated out; its divisions cannot be studied apart. The most important consequence of this point of view is Comte's negation of economics as a distinct science. According to him, social science must cover all kinds of social phenomena to be treated in all their aspects and examined both historically and systematically. Of course, phenomena related to production and exchange must be covered by social science also. Comte, it seems, understood their role in the functioning of society, yet he clearly did not regard them independent enough to be dealt with separately. In this respect he substantially differs from Marx, who constructed his all-embracing social theory exactly on the basis of political economy. Interestingly, Comte denied psychology the status of a fundamental science for similar reasons as economics: the peculiarities of the behavior of humans as biological individuals must be investigated by biology; on the other hand, the examination of *social* behavior of human beings cannot be dissociated from the inquiry into other social phenomena.

Mill defended a quite different position on these issues by claiming that economics and psychology are self-sustaining subjects, and what is more, sociology can and ought to be based on their results. According to him, the starting point of sociology, like that of psychology and economics, must be an individual, not a collective. The disagreement among the two greatest positivists of the nineteenth century started in the social sciences the still unfinished debate between methodological collectivists who followed Comte and methodological individualists who took Mill's side.

In other methodological respects, however, social science differs little, according to Comte, from other sciences. It does not aim at absolute knowledge which all investigations into the nature of beings and their first and final causes pretend to offer. Like other sciences "it supposes a continuous progress of speculation subject to the gradual improvement of observation, without the precise reality being ever fully disclosed" (Comte 1975 [1830/42]: 220–21). And like other sciences, it must aim at the discovery of the laws of phenomena examined. These laws must enable the scientific prediction of phenomena which, according to Comte, is the test of true science. By the way, one of the important reasons Comte is against the separate consideration of social elements is the impossibility, he is convinced, of predicting social phenomena on the basis of such an approach. To achieve such a goal it is necessary to regard all elements as interdependent.

Until now, likely the most contentious issue in the methodology of social sciences is their value-freedom. Starting from Hume, positivists defend the thesis that social sciences ought to be value-free. Comte supports it and states his point of view rather clearly: "Without extolling or condemning political facts, science regards them as subjects of observation: it contemplates each phenomenon in its harmony with coexisting phenomena, and in its connection to the foregoing and the following state of human development; it endeavors to discover, from both points of view, the general relations that connect all social phenomena—and each of them is *explained*, in the scientific sense of the word, when it has been connected with the whole of the existing situation, and the whole of the preceding movement" (Comte 1975 [1830/42]: 239). Being value-free, social science favors, however, "social sentiment" and "enlightens political art."

According to Comte, the two main parts of sociology are *social statics* and *social dynamics*. The first one concerns social institutions, their way of operation, their interdependence, and most importantly, their role in the process of social integration. Social statics is the theory of the *Social Order*. It describes society in the most general way which does not take into account the stage society has reached in its development. Social statics defines—in the broadest sense of this word—the notion of society by examining its constituent parts, such as the individual and the family. Because it dissociates itself from the analysis of the specific forms these parts acquired in the process of historical development, social statics is quite abstract and not very informative. Comte himself regards it as much less interesting and important than the dynamic part of social science.

He lent much greater import to the second part of social theory, which he called *Social Dynamics, or the Theory of the Natural Progress of Human Society*. Before Darwin (and Spencer) but in the Lamarckian spirit, Comte maintained that our social evolution is only the final term of a progression of species. Although "the radical dispositions of our nature" are immutable, the highest of them, i.e., intellectual and moral, are continuously developing.

The rate of progress is affected by the natural increase of population, and especially, by its concentration upon a given space, which creates new needs and new difficulties and affords "a growing ascendancy to those intellectual and moral forces that are suppressed among a scanty population" (Comte 1975 [1830/42]: 283). Our social evolution depends upon various factors. All of them are connected and always act on each other. Yet one of them plays a crucial role—the human mind. Thus the "general history of human mind" must be regarded as "the natural guide" to the study of the historical development of humanity. The Law of the Three Stages governs not only the development of the human mind, but the whole advance of humanity. Therefore we must treat that law as the basis of social dynamics. Each of the three periods has its own social characteristics, and the task of social dynamics is, according to Comte, to disclose them, mainly by exposing the social functions that the theological, metaphysical, and positive mind play, and describing the ways and means by which each state of mind shapes social reality.

The theological mind fulfills its social function by bolstering man's moral courage. It lends belief that all phenomena are governed by a will much more powerful than man's, and therefore he may hope that religious rituals and prayers will help to acquire its favor and thus to reach tasks he undertakes. Theological philosophy provides a system of common opinions. Such a system of opinions is necessary for the cohesion of every society, but was especially important at the time when natural economy prevailed, and the links between individuals and families were tenuous. In those old days, this philosophy established a real intellectual communion. On the other hand, theological philosophy enabled the establishment of a corporation or a special class called by Comte a "speculative class" devoted to abstract, spiritual, or theoretic matters which comprised oracles, priests, and monks. Thus a permanent division between theory and practice was established. Without a class engaged neither in material production, nor in military activity, mental progress and intellectual advance would be halted.

Although the Church played an important role in the theological period by providing intellectual and moral guidance, the form of life which dominates at that stage is, according to Comte, military activity. The ascendancy of military spirit which instituted habits of regularity and discipline was indispensable for the consolidation of society. Military life is a natural tendency at the time when productivity is very low: the only way to acquire the wealth needed for social advance is through military expeditions. The leaders of successful military campaigns gained high positions in the secular hierarchy supported by the Church. No military system could persist without the support of the theological spirit. The sporadic rivalry between theological and military power must not conceal their essential attachment. The fact that they rose, increased, and declined together testifies that they were parts of the same regime.

Yet, according to Comte, the military regime fulfilled its functions with the creation of preconditions for the further advance of humanity. "This end was attained when the chief part of the civilized world was at length united under the same rule; that is, in regard to Europe, when Rome had completed its conquests. From that time forward, therefore, it declined, so as no longer to disguise that gradual rise of the industrial spirit, which had been preparing during the interval" (Comte 1975 [1832/42]: 294–95). This decline was closely related to the decline in belief, because both intellectual and social development gradually led to the expansion of the class of phenomena which have been supposedly subject to natural laws. The first sprouts of positive thinking finally emerged. The gradual replacement of supernatural causes by natural ones prompted the expansion of production, creating the preconditions for the substitution of industrial activity for military conquests.

Yet, according to Comte, "the industrial state was so radically different from the military as to require an intermediate term—and in the same way that, in the spiritual evolution, an intermediate term was required between the theological and the positive spirit" (Comte 1975 [1832/42]: 295). In this transitory phase the metaphysical mind replaced the theological. The transitory character of this period is testified by the fact that contrary to the theological mind, the metaphysical mind is not able to integrate society, because it does not produce a common set of general opinions. The metaphysical mind played an important role in undermining the foundations of the military society by vehemently attacking the theological views on which it was based. In this regard, its role was positive: it revealed new vistas for the advance of humanity. Yet by itself, the metaphysical mind is purely negative. It is useful for destruction but not for political construction.

In the metaphysical period, clergy and military nobility gradually lose their grip on society and are replaced by two classes which pretend to replace them as pillars of spiritual and temporal power, namely metaphysicians and lawyers. Metaphysicians, however, do not understand that neither the search for essences of things, nor for natural rights and liberties of individuals can lead to the formation of congruous opinions necessary for securing social cohesion and smooth and steady functioning of society. The lawyers err also. They are engaged in the frenetic activity of creation of laws which, allegedly, ideally regulate social life and solve all of society's ills. They do not see, however, that society is much more in need of positive goals and perspectives than of constitutions which give no direction to its activities. The domination of metaphysicians and lawyers produced not the order on which any permanent intellectual and social progress must be founded, but anarchy and frustration.

According to Comte, that means that the transitory period cannot linger long. The preconditions for the establishment of a new kind of social organization which could overcome the tendencies of social disintegration are already here. The most important of them is the positive mind, which has shown its potency and vigor in the area of science. It remains to establish a positive social organization of society which could match the state reached by the human mind.

Yet the positive organization of society is not an historical fact to be examined like other facts. It rests not in the past or present, but beckons brightly in the future. Although the positive social state is being prepared by all the development of humanity, and to bring it about it is necessary to take an active stance, because forces interested in preserving the *status quo* will not give up so easily. Hereby we must leave the domain of sociological theory based on empirical facts, and enter the domain of practical positive policy whose aim is the creation of new forms of social and political life and the new kind of intellectual and moral community.

Positive Polity and Positive Morality

It was mentioned already that Comte regarded the metaphysical period as superior to the theological in intellectual regard but as inferior in the social, political, and moral respects. Thus it is rather natural that when looking for the best way to overcome what he called the "anarchy" of modern society and to establish a firm and durable groundwork to secure intellectual and social progress, he looks for inspiration to a society which, albeit not very dynamic, had a social organization that ensured cohesion, high morality, discipline, social order, and at the same time did not utterly block progress. This was, of course, the theological society. The warrant of the stability and harmony of that society was a common body of beliefs, a well-established hierarchical social order in which everybody knew his role and functions, and a very wise political organization. These three elements, if adequately and carefully remodeled, may form the backbone of the positive organization of humankind and ensure its even greater harmony than that to which theological society might aspire.

The nucleus of the common body of beliefs must be, according to Comte, positivist philosophy. This universal doctrine has to check "erroneous action" and help to avoid "popular outbreaks." Yet Comte is perfectly aware that by itself a theoretical doctrine cannot change society. According to him, "there is another need equally manifest, the need of a spiritual power, without which it would be utterly impossible to bring our philosophy to bear upon practical life" (Comte 1975 [1851/54]: 335).

The fundamental principle of positive polity detaching it from the present one is, according to Comte, the permanent and strict separation of two different powers—temporal and spiritual. The stability of theological society has been based on this separation of powers, and it is a necessary condition of order and progress in general. These two kinds of power cannot be merged, because both their spheres of competence and their methods of ensuring authority are different. Temporal power is responsible for material affairs, and spiritual for the moral and intellectual sphere. Temporal power may use coercion, and spiritual power must restrict itself to persuasion. According to Comte, "the constitutionalist principle of separating the legislative from the executive is only an empirical imitation of the larger principle of separating temporal and spiritual power" (Comte 1975 [1851/54]: 346).

Who should rule society and what are the confines of their effective activity? For Comte the first question is very simple: only those who possess positive knowledge can rule. Ruling or governing is an activity which, like every other activity, must be based on real knowledge. Yet because the object of ruling is the most complicated of all—an entire society—the competence of rulers must be very high. Thus it would be silly to try to choose them in a democratic way by popular voice. In that case the least competent would judge who is the most competent.

This simple point is the foundation of the *technocratic* view on governance held until now. Accordingly, society must be governed by the *best* specialists. And since they cannot be chosen by the way of *popular* voice, Comte openly denounces democratic election and the very idea of democracy. Democracy would impede the progress because democratic institutions and democratic procedures are unable to guarantee that the best decisions (from the *scientific point* of view) will be made and effectively implemented. Modern technocrats, of course, are not so outspoken, because they are less upright than Comte, who openly claims that only the elite must be responsible for the selection of those who are worthy to enter their rank. And because of the necessity of separating the two kinds of power, the process of selection of two elites must go on separately as well. Temporal power or government must be concentrated in the hands of the managerial class, of the captains of industry and engineers. And the spiritual power must be, *of course*, in hands of positive thinkers and scientists who must withstand all temptations to become involved in politics. Otherwise they will not be able to forge doctrines which could unite and not divide society.

Both the temporal and spiritual powers must be organized hierarchically. Yet this does not mean that he bounds his hopes for a bright positivist future only with the representatives of the industrial and scientific elite. An important role in his political philosophy is assigned to the proletariat (Comte used exactly this term to refer to the working class).

Unlike the aristocracy and bourgeoisie (which, according to Comte's use of the word, does not includes managers and in which lawyers play the leading role), the proletariat has no interests related to maintaining the *status quo*. That class did not study metaphysics. Therefore it lacks metaphysical beliefs. That is its second advantage. Besides, the working class is the largest and least egoistic (Marx later claimed the same). Thus it is able to take a nonpartial position concerning public matters, check greediness and egoism incidental to some members of the industrial elite, and ensure progressive changes without which society cannot pass from the metaphysical stage to the positive.

The transition from the metaphysical to the positive stage must take place peacefully. Comte describes the social system or order to be established when the positive stage of historical development will be reached as *systematic socialism* (to be distinguished from the spontaneous socialism of workers which lacks theoretical foundations). According to socialists, humans must cooperate and not compete, thus Comte rebuts, of course, liberal individualism, which acclaims the social role of competition. Yet he was against the abolishment of private ownership and even against its limitation. He opposed any violent actions as a means of social transformation which the greatest devotee of the proletariat, Marx, later regarded as indispensable. Comte was convinced that positive organization of society must be established by means of spreading of ideas of positive philosophy and moral education.

According to him, after positive philosophy will gain wide-reaching acceptance, the *critical* examination of its doctrines will become as fruitless as free public criticism of astronomy or physics would be today. Besides, critical discussion of principles on which the social system is based would be dangerous both for social stability and social progress. Yet for the dissemination of their ideas during the transitional period, positivists must use instruments produced by metaphysical society. Moreover, they must aim at the perfection of these instruments, because refraining from violence means that they are the only means to be used in the process of transition. Hence, they must demand complete freedom of consciousness, freedom of the press, and freedom of association. They must always remember, however, that all political and economic changes are based on changes of a moral and intellectual nature. Thus all efforts must be concentrated on achieving the spiritual transformation of men.

The general task of both powers is to ensure social order and social progress. This is done by using both the traditional means of governing and new ones related to social planning and social engineering. A question must naturally arise: what are the limits of social engineering? To what degree may social phenomena, which, like all other phenomena, are governed by laws, be modified? This question is rather complicated, especially for a philosopher who believes that there are *laws governing the development of society*. Comte's answer, repeated later by Marx, boils down to the following. First of all, social phenomena are the most complex. Hence the laws formulated in sociology are less precise than those formulated in the natural sciences. By their nature, social phenomena are the most malleable of all. Admittedly, the order of the stages of historical development cannot be modified. Yet human actions, and especially actions taken by government, can speed up or slow down this development.

Comte was eager to emphasize the independence of spiritual power—which among other things is the highest intellectual authority—from temporal, economic, and political power. Yet he supposed that the temporal government would base its actions on positive knowledge and principles established by the intellectual elite. And this knowledge will ensure that the actions taken by temporal government will help to hasten the pace of history. In fact that would mean that the spiritual power may act *through* the temporal.

More important, however, are those functions of the spiritual power which it can fulfill by acting independently or almost independently from temporal power. In Comte's words, "[t]he principal function of the spiritual power is to direct the future of society by means of education" (Comte 1975 [1851/54]: 341). Education is the main means to disseminate positive knowledge and the positive way of thinking. To be able fulfill this role, the system of education must be radically reorganized. It must be more centralized and directly run by the spiritual power. Education must become universal and elements of positive sciences must be taught in their consecutive order. Yet the level and character of teaching must be adjusted to the specific job or functions of the individuals and may vary greatly. The system of education must have an elitist character because only such a system may effectively produce the elites needed for ruling society.

Yet intellectual guidance is only one of the functions of the spiritual power. No less, and maybe even more important is the moral guidance without which intellectual guidance will remain ineffective. In his late years Comte became much more aware of this. Accordingly, he came to the conclusion that the positive system of sciences must be supplemented by adding to the six fundamental sciences the seventh—the *science of morals*. In the *System of Positive Polity* he claims that the formation of a new system of universal morality is a major task of positive philosophy. Moreover, according to Comte, this system is "the ultimate object of all philosophy and the starting point of all polity" (Comte 1975 [1851/54]: 336). He claims that the moral value of positivism is inherent in its doctrine yet does not regard as very important the task of establishing a close connection between the science of morals and the rest of scientific knowledge. Thus Comte leaves unanswered some questions which modern ethics could raise regarding the foundations of this science and its epistemological status. He claims that moral rules are in principle "susceptible of scientific proof" (ibid., 340) not taking into account the fact that to produce such proof without renouncing the fact-value dichotomy is a formidable task. The practical value of positive morality and the ways of its social entrenchment is much more important for the later Comte than the rigorous scientific substantiation of its principles.

Like Hume, he emphasizes the close connection between morality and feelings. Moral education must be based on feelings and the main purpose of it is exactly the transformation of feelings. "To the positivist the object of morals is to make our sympathetic instincts to preponderate as far as possible over the selfish instincts, social feelings over personal feelings" (Comte 1975 [1851/54]: 337). In the social state, social feelings are constantly stimulated, and selfish propensities are checked. Yet the growth of social sympathies may be hastened by way of constant exercise. Thus morals are, in fact, less a science and more an art.

What are the principles of positive morality? The main one is altruism or the principle of universal love. Comte formulates it in the following way: *to live for others*. It is closely connected with another one according to which the *fulfillment of our duties* and not the exercise of our rights is our highest aim and vocation. Celebration of rights leads to anarchy, because those who pursue their rights find themselves in conflict with others pursuing *their* rights. Not less important is the fact that the very notion of the natural rights of an individual has a metaphysical character and is inconsistent with scientific findings concerning the relation of the individual to society. The individual is dependent upon society; his human features (different from those purely biological) are formed by society. He is a derivative of it, a product of its historical development. Owing it everything, he has social duties the fulfillment of which makes his life meaningful and ensures social harmony, which is a necessary precondition of social progress. The individual has only such rights society grants him, and he has no right to demand more. He can only require more of himself.

In the *System of Positive Polity* Comte comes to the conclusion that should not astonish an astute reader: the most effective way of practicing the art of moral education is through religion. At the theological stage, religion proved its social usefulness. It ensured social cohesion by establishing a common body of beliefs and a common cult practice. At the metaphysical stage, religion came under attack and gradually lost its integrative role and significance. Traditional religion, whose dogmas contradict scientific findings, cannot fulfill the same social role at the positive stage of historical development, including that of moral education that it played successfully in the theological stage. Yet for the establishment of the new moral order based on the principle of universal love, religion, with its ability to reach the human heart and affect feelings, is indispensable. Because the old theistic religion cannot be revived, a new religion called the Religion of Humanity must be established. Humanity must become the only object of religious sentiment and must be admired and worshipped as the Great Being.

Accordingly, in the last years of his life Comte is much more interested in the invention of rather curious forms of worship to be used in the Religion of Humanity and establishment of the positivist Church, with Comte himself as the high priest presiding, than in positivist philosophy. And because our subject is the development of the positivist philosophy, and not the teaching and life of Auguste Comte, we must stop here without delving into the details of his later ideas.

The System of Positive Polity, especially its concluding volumes, evidently lacks the unity, consistency, and clarity characteristic of the *Course*. At the end of *The System* not much is left of the positive spirit as it manifests itself in the *Course*. Thus every critical reader must confront the question: is it still possible to regard the *System*, and specifically, its political, moral, and religious notions, as a work of a positivist thinker? For Mill that issue was of crucial importance.

John Stuart Mill

John Stuart Mill is almost as important a figure of classical positivism as Comte. Though the foundations of it were laid by Comte, Mill overshadows Comte in some important respects. It seems that he had been read more widely than Comte in the nineteenth century and there is no doubt that Mill is much more popular today.

For Mill, Comte was both an important source of inspiration and a target of criticism. Like Comte, Mill got an excellent education albeit in a quite different way. At the very early age he found himself in the hands of the best British teachers. He described his extraordinary childhood and adolescence in his *Autobiography*, which fostered his fame probably not less than his philosophical works. The boy who was born in 1806 began to study Greek when he was at the age of three, Latin at eight, and read Plato and Hume before reaching twelve. His education was carefully planned and supervised by his father, the well-known historian, economist, and philosopher James Mill, who was a close associate of Jeremy Bentham. Inspired by some educational ideas of the Benthamites, he carried out a rare and rather positivist character experiment: breeding a genius in the shortest possible time. He himself taught his son languages, history, and mathematics. Mill's other teachers were no less famous—they included David Ricardo, who taught economics, and John Austin, who taught law. It is not strange that having such teachers Mill had no pressing reasons to enter any school whatsoever.

Mill's first publications appeared in 1822, the same year Comte published his *Plan of Scientific Operations*. In 1824 Mill began to cooperate with the newly established quarterly *Westminster Review* founded by Bentham and his father. Soon, articles published there and in other periodicals made him rather well-known. In 1823 Mill started to work at the East India Company. He climbed the ladder quite fast, yet the work related to the administration of India never absorbed Mill entirely. He had been publishing heavily and speaking occasionally on almost every subject of public interest: colonialism, liberty, the subjection of women, democracy, socialism, economics, and religion. As a well-known social reformer he was elected to Parliament in 1865 and represented Westminster in the Commons for three years. Mill spent the last years of his life in his beloved southern France and died at the age of 67 in Avignon.

Mill's major philosophical publications include *System of Logic* (1843), the famous *On Liberty* (1859), *Utilitarianism* (1863), and *An Examination of Sir William Hamilton's Philosophy* and *Auguste Comte and Positivism*, both of which appeared in 1865. *Considerations on Represen-*

tative Government (1861) is an important source for study of his political ideas. *Principles of Political Economy* (1848) acquired the status of a classical work on the subject already in Mill's lifetime.

Besides sharing an exemplary education and similar reading material (including Hume and Saint-Simon), as well as early acquired progressivist convictions, Mill and Comte share much more. Both were uncommon minds, both had very wide, nearly encyclopedic interests, and both were convinced that their vocation was to serve humankind. Both were social reformers. Yet these similarities cannot upstage their differences. The first acquaintance with their texts discloses their different styles of thinking. The moderate and slightly skeptical Briton clearly differs from the enthusiastic French prophet. Far from all readers of *On Liberty* appreciate that it is an essay written by a positivist. When reading Comte and Mill you may easily get the wrong impression that little connects them—so great are the differences not only in their styles of thinking but also in their views on a range of social issues. The tension between the two greatest representatives of classical positivism is really quite dramatic—they represent its two different faces.

Until now Mill is regarded as one of the most influential representatives of liberalism. Comte is surely not—he had no sympathies for liberalism at all. Mill, especially from the mid-fifties of the nineteenth century when Comte's *System of Positive Polity* was published, was rather keen to emphasize that he disapproved of many of Comte's political and social views. In his *Autobiography*, published shortly after Mill's death, in 1873, he went so far as to claim that Comte's political doctrine is the most complete system of despotism after St. Ignatius Loyola. Mill's theory of knowledge is more consequently based on the empiricist tradition, and the reforms of sciences and society he proposed were much more modest and undoubtedly more realistic. It is not surprising that unlike Comte's grandiose projects, many of Mill's proposals, especially those aimed at the improvement of the social and political status of women and workers, were implemented. Nevertheless, Mill regarded himself as a positivist and not only revered Comte and held him in high esteem, but also for some time corresponded with him and supported him financially.

Mill and Comte: Allies and Opponents

Mill's attitude towards Comte's early doctrines laid down in his *Course of Positive Philosophy* is in general rather sympathetic, sometimes even enthusiastic. He shared Comte's notion of "the fundamental doctrine of a true philosophy," which Mill renders in *Auguste Comte and Positivism* in the following words:

> We have no knowledge of anything but Phenomena; and our knowledge of phenomena is relative, not absolute. We know not the essence, nor the real mode of production, of any fact, but only its relation to other facts in the way of succession or of similitude. These relations are constant; that is always the same in the same circumstances. The constant resemblances which link phenomena together, and the constant sequences which unite them as antecedent and consequent, are termed their laws. The laws of phenomena are all we know respecting them. Their essential nature and their ultimate causes, either efficient or final, are unknown and inscrutable to us. (Mill 1993 [1865]: 6)

Laying down principles of positive philosophy, Mill emphasizes that knowledge of the succession and coexistence of phenomena is also the most necessary knowledge because it may be used to control the effects of events or to adapt our behavior to uncontrollable events. According to Mill, the doctrine of a true philosophy was doubtless correctly apprehended already by Newton, and "it was probably first conceived in its entire generality by Hume"; however, Comte "made the doctrine his own by his manner of treating it," especially by first formulating the Law of Three Stages—"the generalization which belongs to himself"—and then by proving it (cf. Mill 1993 [1865]: 7–12). Mill was convinced that this law was of extreme importance.

Yet unlike Comte, Mill claims that positive thinking is compatible with the conviction that the world was created by a supernatural being: "If the universe had a beginning, its beginning, by the very condition of the cause, was supernatural; the laws of nature cannot account for their own origin" (Mill 1993 [1865]: 14). Holding the view that believing is compatible with scientific knowledge (this view was held by many positivists later as well), Mill rejects Comte's project of the establishment of the new positive religion in place of the old one allegedly not in accord with positive thinking.

Mill accepts in general Comte's classification of sciences, yet has some important reservations. The first one concerns psychology. Mill claims that Comte misinterpreted its true position, and that in fact this discipline is not a derivative of physiology, but represents a separate branch of positive philosophy. According to Mill, Comte's arguments against self-reflection do not hold water because through the medium of memory we can study anything we were directly aware of earlier. Mill's second reservation concerns political economy, which Comte held in low esteem. Mill has no doubt whatever that this discipline represents "the only systematic attempt yet made by any body of thinkers, to constitute a science, not indeed of social phenomena generally, but of one great class or division of them" (Mill 1993 [1865]: 80). According to Mill, Comte seriously overestimates the value of the results of his own attempt to build a unified positive social science.

According to Mill, Comte's social theory has another serious flaw. Although his historical analysis of the development of humankind is mainly correct, his practical recommendations are scientifically unconnected with his theoretical explanations of the past progress of society. His "proposals are not, as we might expect, recommended as that toward which human society has been tending and working through the whole of history. [...] They rest as completely, each on its separate reasons of supposed utility, as with philosophers who, like Bentham, theorize on politics without any historical basis at all" (Mill 1993 [1865]: 118–19). And the very utility of Comte's political projects is, according to Mill, more than doubtful.

Turning towards Comte's later writings, most important of which is *System of Positive Policy*, Mill becomes even more critical. His opinion of the value of Comte's deliberations on politics, morality, and religion published after completing the *Course* is much lower than that of the doctrines laid down in the *Course*. Sometimes Mill's stance towards Comte's positive philosophy on the one hand, and towards his ideas concerning politics, morals, and religion on the other hand is schematized in the following way: he accepts with some reservations the early ("the first") Comte and rejects—with some reservations also—the late ("the second") Comte. Although this simple scheme has some merits, it is not exactly correct. Mill himself notes that in the *Course* we may find "the germ of perversion of his own philosophy which marked his later years" (Mill 1993 [1865]: 62). "Perverse" are first of all Comte's practical recommendations concerning the future positive organization of society and its activities irrespective of whether he put them forward in the *Course* or later. Mill rejects them on the ground of common sense, social experience, and/or his belief in liberal values. He repudiates the very principal aim Comte pursued—more and more persistently as time passed—namely, the establishment of complete harmony between individual and society, of total unity of personal and social existence.

Mill is convinced that if individuals themselves have the possibility to choose their own way of life, then the complete harmony and unity Comte is talking about cannot be achieved. Harmony, unity, and systematization as the principal aim of personal and social life demand absolute submission of the individual to society. They must act not according to their propensities and inclinations but according to a strict plan. The plan must, firstly, be carefully drawn and, secondly, be implemented fully and finally. That means that opposition of any kind either to the principles of its plan or to the methods of its implementation must be suppressed. This is the main source of Comte's antidemocratic tendencies. For Mill however, the notion of happiness for all to be achieved through sacrifice of each is simply a contradiction.

We find Mill even more horrified by another idea playing a crucial role in Comte's later writings, namely that the only legitimate object of the intellect must be the general good. Mill is firmly convinced that such a limitation imposed on the intellect must by necessity lead to the total suppression of all independent thought and, specifically, to the termination of all scientific research which does not provide utility to humankind. Mill emphasizes that nobody can know in advance what practical value the results of some studies will have in the future: who could predict that speculations of mathematicians in Alexandria investigating the properties of the ellipse would help explain two thousand years later the motion of planets and enable safe navigation?

Mill has no doubt whatsoever that Comte's later writings present a sad picture of intellectual debasement. He does not imply, however, that this degradation seriously undermines Comte's position as a great thinker. Mill ends his examination of Comte's intellectual legacy by concluding that Comte is as great as Descartes and Leibniz and hardly more extravagant (cf. Mill 1993 [1865]: 200). Mill regards Comte's principal doctrines laid down in the *Course* to be sound and important. He shares them and this makes him, in his own opinion, a positivist. Yet Mill's views differ in many respects from those of Comte, and not only on social matters but on the problems of philosophy of science as well.

Logic and Methodology of Science

A System of Logic, Ratiocinative and Inductive, Being a Connected View of the Principles of Evidence and the Methods of Scientific Investigation (1843) is Mill's main work in which he examines problems of the methodology and philosophy of science. He started to write it in 1830 and completed a voluminous book with intermissions at the end of 1841. In 1830 he had already some knowledge of Comte's early writings which, as Mill claimed later, made a great impression on him, but we must be aware of the fact that the major part of *A System of Logic* was written before the first volume of Comte's *Course of Positive Philosophy* appeared in 1840. Mill started to study *Course* thoroughly only after finishing his *System*.

In the nineteenth-century debates on the method of science, Mill's work was regarded as extremely important, the genuine *magnum opus*. It is devoted to the subject which, according to Comte, cannot be studied fruitfully outside the context of specific scientific theories. Mill's famous treatise is based on an opposite presupposition: the subject he was interested in—the way of acquiring and proving scientific knowledge—may be successfully studied separately from the study of its historical development. This is a crucial difference to note. Instead of studying the advance of sciences, Mill is aiming at the development of a theory of scientific method which diverges from the then widely read William Whewell's *History of Inductive Sciences* (1837) and *Philosophy of Inductive Sciences* (1840).

Yet the general purpose of Mill's methodological studies was essentially the same as that of Comte's studies concerning the system and development of the sciences. Namely, they were preparatory studies carried out with the far-reaching aim of broad social and political reforms in mind. Under the prevailing influence of his father, Mill had decided already in his young years that his aim was to become humanity's reformer. He believed that the only direct means to such reform is the reform of society's institutions.

And he did not doubt that anybody who wants to change them must fight any erroneous philosophy which supports existing institutions. He identified this philosophy mainly with what he calls an "intuitivist school," to criticism of the leading representative of which, William Hamilton, he devoted a special treatise. However, a much more important task, according to Mill, is the positive elaboration of the right, correct theories which are badly needed as the foundation of social and political reforms. Those reforms must be carried out on the basis of the laws established by social science, not on the basis of prejudices or the results of conflicts of interests on which much of political practice is based. For quite a long time Mill believed, like Comte, and rather naively that it is possible to form a system of convictions which will not require "like all former and present creeds, religious, ethical and political [...] to be periodically thrown off and replaced by others" (Mill 1961 [1873]: 101).

Mill did not pretend, however, to construct the whole texture of social science himself as Comte did. He was much more self-critical and understood well that it is not a task one man can accomplish. Yet he was convinced that he may greatly facilitate progress in the social sciences by elaborating their methods. Like Hume and Comte he was persuaded that the social sciences must model themselves after the natural sciences and use their method. Mill believed that the authority gained by the natural sciences is directly related to the unanimity existing among competent researchers as to the results of their investigation. And this unanimity is achieved because they employ the same criteria of determining what is proven in their area and what is not. The great task social science faces is to achieve the same. Mill supposed that to achieve this goal it is necessary, firstly, to expose the method used by the natural sciences and then to investigate what specific features, if any, discriminate the method of social science from that of natural science. Let's address at the outset the first question.

When describing what kind of issues he deals with in his *System of Logic*, Mill emphasizes its practical character. He treats logic as a theory of the art of reasoning. This art must be refined to improve our action. Yet our reasoning cannot be improved by elaborating metaphysical ideas.

Neither in *System of Logic*, nor in his later works does Mill attempt to give a clear-cut definition of metaphysics, although he indicates that it includes metaphysics proper (ontology), epistemology, and psychology. Accordingly, he does not say much about the opposition between science and metaphysics. Mill claimed in the *Autobiography* that Comte's doctrine "of the natural succession of three stages in every department of human knowledge [...] harmonized well with my existing notions to which it seemed to give a scientific shape" (Mill 1961 [1873]: 101). Mill suggests that like Comte he was inclined to juxtapose the scientific condition of the human mind with the metaphysical. However he was not very eager to condemn even metaphysics proper. In nineteenth century Britain, Mill was regarded as a radical, especially in politics. He was, however, a Victorian radical, rather prudent and circumspect both in moral and metaphysical matters. Mill's moderation concerning metaphysics makes him the least radical in this respect among all positivists. Hume among his predecessors, Comte among his contemporaries, and Carnap among his successors were all much more outspoken on the subject than Mill. We must take into account, however, that the notion of metaphysics he uses is very broad and encompasses psychology which neither Hume nor Carnap would like to exclude from a perfectly legitimate scientific discourse. In general, however, Mill shies away from transcendental metaphysics. In *System of Logic* he declares openly that he makes no research into efficient, ultimate, or ontological causes, and is not interested in the essences and inherent constitution of things (cf. Mill 1949 [1843]: 213).

The substantial part of his famous book is devoted to the analysis of induction, since Mill, like many of his contemporaries, regards induction as *the* method of science. He claims that all investigation of nature essentially consists in induction and that "all Inference, consequently all Proof, and all discovery of truths not self-evident, consists of inductions and the interpretation of inductions; that all our knowledge, not intuitive, comes to us exclusively from that source" (Mill 1949 [1843]: 185). This fact gives ample reason to term his methodology of science *inductivism*.

Inductivism is a traditional methodology associated with empiricism at least from Francis Bacon. Similarly and almost by necessity rationalists were deductivists. Indeed, the question of the source of our knowledge, and that of our methods of acquiring it, are simply two different facets of the same problem. If we are unable to show how, out of primary knowledge, secondary knowledge is inferred, we have no right to claim that either reason or experience is the source of all our knowledge. Rationalists must prove that all our secondary knowledge is being deduced from the primary truths of reason (since they are the most general, any inference must have the form of deduction), and empiricists must show how it is induced on the basis of particular experiential truths or at least how our hypotheses may be inductively proven on the basis of empirical facts.

Yet do we have the right to claim that results of inductive reasoning are certain? Hume showed that inductivism faces serious difficulties because it is impossible to prove indisputably that the results of inductive inference are sound. Later these difficulties—known as the problem of justification of induction—were regarded by proponents of deductivism as a major argument against inductivism. Mill, who was a great admirer of Hume (precisely due to his efforts a Scottish thinker not very well known in the eighteenth century became acknowledged at the end of the nineteenth century as the greatest British philosopher in history), was perfectly aware of this problem.

It seems that this problem is unsolvable. Until the twentieth century induction has been traditionally defined as inference from the particular to the general, inference of general conclusions from particular facts. The simplest standard example of inductive reasoning is as follows: the first (observed) swan is white, the second is white, the third is white, hence all swans are white. For quite a long time Europeans were convinced that this conclusion was true. Yet after black swans were met in Australia, it became evident that the conclusion was false.

This example attests that to prove that conclusions of inductive inference are *always* true when its premises are true is impossible. But how then may we justify induction at all? Mill's strategy when dealing with this issue is as follows: firstly, to present general arguments in favor of induction and, secondly, to propose more sophisticated rules of inductive reasoning which would reduce the possibility of falsity of conclusions of induction.

Mill presents the following reasons for induction: Firstly, there is an extremely wide agreement among scientists that induction is a valid inference and this wide agreement is the basic fact concerning the validity of induction. Secondly, this basic fact may be made understandable by another fact, namely that inductive inferences are in general successful. The successfulness of induction, i.e., the truthfulness of the conclusions of inductive inferences, justifies induction, although this kind of justification of induction does not mean proving it is a perfect method which *always* leads to Truth. Thirdly, inductive inference is based on the principle of uniformity of nature according to which "that what happens once, will, under a sufficient degree of similarity of circumstances, happen again, and not only again, but as often as the same circumstances recur" (Mill 1949 [1843]: 200–1). The only problem is to determine precisely those circumstances.

The first reason is, according to Mill, the most important. The second one enables us to understand the first, and the third one lays down "the fundamental principle, or general axiom, of Induction." It is important to emphasize that, as the author acknowledges, those reasons do not prove that induction is a *perfect* method.

Many logicians and philosophers of science—both in the nineteenth and twentieth centuries—claimed that Mill's argument is a circular one: on the one hand, the principle of uniformity of nature is regarded as a basis of inductive inferences, and on the other hand, this principle may be justified only by making use of such inferences. Notwithstanding the fact that Mill himself was clearly aware of the perplexities, they maintained that Mill did not take seriously Hume's problem of induction. One of these authors was Hans Reichenbach, a leading representative of logical positivism.

Hume demonstrated that the premises of our inductive inferences are perfectly compatible with the negation of conclusions of those inferences. Thus it may be legitimately claimed that he not only posited the problem of (the justification of) induction but also solved it. However, at the same time, Hume claimed that the conclusions of many of our inductive inferences (of the kind "the sun will rise tomorrow") are practically certain and there is no doubt that we in fact rely on them when conducting our everyday activities. Being of the same opinion, Mill aims solely at explaining *why* we are using induction. He regards inductive inference as a tool whose previous usefulness is an important reason to use it in the future. There is little doubt that not only Reichenbach but Mill also perfectly understood that it is impossible to prove *logically* that this tool will always remain an efficient instrument. Thus, like Reichenbach's, Mill's justification of induction might be regarded as a *pragmatic*, not a logical one. His arguments concerning the problem of induction may be interpreted as a rather fruitful effort to explain on an empirical basis why we rely on induction.

Mill is aware that even if the principle of uniformity of nature would be proven, this would not guarantee the truthfulness of all conclusions of inductive inferences. He has little doubt that "the universe, so far as known to us, is so constituted, that whatever is true in any one case, is true in all cases of a certain description; the only difficulty is, to find what description" (Mill 1949 [1843]: 201). The simplest way to find such descriptions is to ascribe the character of general truths to all propositions which are true in every instance that we happen to know of, i.e., to use a procedure known as *enumerative induction*. Francis Bacon, among others, has claimed that enumerative induction is a childish thing and that its conclusions are uncertain. Although, contrary to Bacon, Mill held the belief that enumerative induction is the basis of any inductive reasoning he, like Bacon, wanted to improve methods of inductive inference and was convinced that science must go beyond simple enumerative induction, and use methods of eliminative induction. He sought to draw precise rules enabling us to eliminate and exclude from the set of phenomena or circumstances (let's call them A_1, A_2, ... A_n) preceding the phenomenon in which we are interested (let's call it B) those phenomena which are not causally connected with B. This elimination leads to identification of that single A_k which is the cause of B. On the presupposition that the cause of B is one of the members of a given set of circumstances A_i, eliminative induction becomes infallible inference (cf. Szaniawski 1994: 148–62). Regrettably, we do not usually know if this presupposition is true or false.

The rules of eliminative induction are known as Mill's methods or canons. They form the nucleus of his inductive logic. According to Mill, the simplest modes of experimental inquiry are those termed by him the *Method of Agreement* and the *Method of Difference*. The first one enables the laws of phenomena to be established by comparing different circumstances in which the phenomenon B occurs. In Mill's words: "If two or more instances of the phenomenon under investigation have only one circumstance in common, the circumstance in which alone all the instances agree is the cause (or effect) of the given phenomenon" (Mill 1949 [1842]: 255).

The *Method of Difference* is expressed by Mill as follows: "If an instance in which the phenomenon under investigation occurs, and an instance in which it does not occur, have every circumstance in common save one, that one occurring only in the former; the circumstance in which alone the two instances differ is the effect, or the cause, or an indispensable part of the cause, of the phenomenon" (Mill 1949 [1842]: 256).

Mill's canons are rules for establishing causes of phenomena. He was convinced that establishing causes (as well as effects) is the most important part of scientific inquiry. He was interested in the possibilities of practical use of scientific knowledge and believed that if we want some events in nature or society to ensue, we must know their causes because only in this case we will be able to bring about desired effects.

Mill's emphasis on the significance of the knowledge of causes sets him apart from Comte. When critically appraising the theological mind, Comte exposes its orientation towards first and final causes. And he persistently avoids using the word *cause* when describing the tasks of the positive mind. Mill notes this fact and regards it as extremely unfortunate. However, if one takes into account the consequent development of positivist philosophy and, especially, the stance towards the study of causes taken by Ernst Mach, it becomes clear that Comte does not use the word *cause* deliberately. Comte, it seems, held the view that the difference between first causes and immediate (proximate) causes is only that of degree, and that if we admit that the search for causes of phenomena is legitimate, this will soon lead to the vicious search for causes of causes *ad infinitum*, thus invoking the very metaphysical demons science was supposed to exorcise. Mill had no such fears. He clearly was not obsessed with the idea shared by more radical positivists that crossing the line between positive science and metaphysics is deadly for science.

According to Mill, causes are established with the aid of inductive procedures. As his work on induction proceeds, it becomes clear, however, that he uses the term induction in such a way that it comprises also what he calls the hypothetical method and what we would call today the hypothetico-deductive method. In this respect Mill's notion of induction looks quite modern, at least much more modern than Popper's: he does not doubt that the acceptance of hypotheses on the basis of positive results of testing their deductive consequences is an inductive procedure. Thus for Mill it would be perfectly legitimate to claim that the law of gravitation has been established by induction, although it is a result of neither enumerative nor eliminative induction.

When talking about induction as the method of science, Mill, however, does not draw a clear distinction between simple inferences covered by methods of enumerative and eliminative induction and the hypothetical method. In his polemic with Whewell concerning the significance of hypotheses for science, Mill consented to the opinion that hypotheses may explain known facts and that they may be tested by making deductive inferences and comparing results with observable facts. Nevertheless, he insisted that the method of hypothesis cannot be regarded as the main method of science. He emphasized that it is always possible to figure two or more different hypotheses explaining the same facts; thus any hypothesis, even a carefully tested hypothesis, cannot be regarded as certain as are results of induction "properly so called." And certainty this was for Mill an extremely important aim of scientific investigation: only by establishing methods of inquiry which lead to maximally certain results could the ground be cleared for the uniform opinions in social science which so preoccupied Mill.

Mill was convinced—and the present state of social science does not supply evidence that he was wrong—that if social scientists are allowed to put forward any hypotheses they fancy to explain social occurrences, while not bothering to infer the posited laws of these occurrences from the facts according to strict inductive rules, unanimity among social scientists will forever remain a futile hope. Hence the possibility of using universally accepted scientific results as a theoretical basis for far-reaching social reforms carried out by representative government will simply disappear.

Mill's theory of induction is partly based on the ideas advanced by the British astronomer John Herschel in his methodological treaty *A Preliminary Discourse on the Study of Natural Philosophy* (1831). Yet Mill pays little attention to an important distinction which Herschel makes but which Mill does not draw consequently. It is the distinction between the problem (or context) of the *discovery* of laws and that of their *justification*. This distinction became standard in the philosophy of science of logical positivism. Mill indicates that discovering laws is not the same thing as proving them; however, he does not make the point as clearly as Herschel does. The reason is simple: Mill considers that the certainty of science depends not only on the way its statements are tested but also on the way they are discovered. Accordingly, his inductive logic is rather a logic of discovery of new knowledge than a logic of justification or appraisal of an already available hypothetical knowledge. In this respect, Mill's standpoint is different from that of De Morgan, Whewell, and Jevons, who view inductive logic as logic of testing and appraising hypotheses.

When comparing Mill's understanding of the role of induction and hypotheses in science to the views of these authors, not to speak about those of Carnap, one may form the impression that Mill does not differ much from Bacon and represents a rather naive inductivist point of view according to which in science we proceed initially from facts to lowest generalizations, later to the middle principles (*axiomata media*, in Bacon's terms), and then to the most general laws, invariably using standard rules of inductive inference. This impression is faulty. According to Mill, Bacon is wrong. Many sciences, including mechanics and astronomy, must be regarded as deductive, and in those sciences "it is evident that the higher and the middle principles are by no means derived from the lowest, but the reverse. Thus most general truths of these sciences are those earliest arrived at" (Mill 1949 [1843]: 568). As we will see later, even social sciences have, according to Mill, a predominantly deductive character. Thus notwithstanding the fact that in *A System of Logic* Mill pays much attention to the analysis of relatively simple forms of inductive inference, he understands rather well the role deduction plays in scientific knowledge.

Mill is convinced that the improvement of our methods of observation and systematization of facts and, especially, better methods of inductive inference make our knowledge of facts and their causes more certain than Hume believed. This knowledge is, of course, fallible, and in this respect the views of Hume and Mill coincide. Mill, however, rejected Hume's claim that mathematical knowledge is infallible. According to Mill, mathematics, like natural science, is based on experiment, concerns facts, and is synthetic, not analytic—*a posteriori*, not *a priori*. He maintains that the traditional juxtaposition of necessary knowledge with experiential knowledge is unfounded. Truths of mathematics are in a sense necessary. However, they are not *a priori*. They are necessary only as necessary logical consequences of mathematical axioms, postulates, and definitions.

The foundations of mathematics are, however, not necessary statements by themselves. According to Mill, there are no objects whatsoever which exactly correspond to their mathematical definitions. Euclid speaks about points having no size and lines having no width. Where can we find such objects? Mill does not doubt that they do not exist in the physical world. He rejects also the widespread opinion that they exist in our minds. We cannot imagine a one-dimensional line or a point without size. He maintains that when examining mathematical objects we always think about real physical things although we abstract ourselves from their qualities which are unimportant from a mathematical point of view. Thus definitions are in fact nothing more than generalizations. The geometrical definition of a circle according to which all its points are equidistant from the center is faultless. However, it corresponds only roughly to physical—let's say drawn on paper—circles. Hence, concludes Mill, the extraordinary exactness we impute to the foundations of geometry is simply an illusion. He points out that the real foundation of geometry is, in fact, not a system of axioms, postulates, and definitions, but a supposition that things corresponding to the statements comprising that system do indeed exist. He is convinced that *definitions* are only approximately true: in fact they are hypotheses. Contrary to them, *axioms* of geometry are not hypotheses. However, even they originate in experience and cannot be treated as absolutely certain *a priori* truths. Mathematical truths differ neither in their origin nor in their epistemic status from the statements of the natural sciences and are experiential truths. In Mill's view, the unity of science derives not only from the unity of methods used in natural and social sciences, but also from the fact that the relation of mathematics to experience does not differ from that of other sciences.

In an epoch during which the philosophy of mathematics was dominated by Kant's ideas, Mill's empiricist interpretation of mathematics looked rather novel. Yet it was not entirely new in the European tradition. It was advanced by representatives of the French Enlightenment, mainly by Pierre Louis Moreau de Maupertuis, already in the mid-eighteenth century. This interpretation had been supported—even before Mill put arguments in favor of it—by some leading mathematicians of the nineteenth century as well. Notably, almost all founders of non-Euclidean geometry: Gauss, Lobachevski, and Rieman regarded geometry as an empirical, synthetic theory. Being aware that the sum of the interior angles of a triangle differs in non-Euclidean geometry from 180 degrees and that the difference increases with the increase of the area of a triangle, both Gauss and Lobachevski believed that if we managed to measure the sum of the inner angles of a sufficiently large triangle formed by beams of light, we would empirically ascertain which geometry—Euclidean or non-Euclidean—is true. Rieman's celebrated lecture—a landmark in the history of mathematics—in which he presented a general theory of geometry covering geometries advanced by Gauss, Lobachevski, and Rieman himself, was dubbed by him, characteristically, "On the Hypotheses on which Geometry is Based." Admittedly, among philosophers, the interpretation of mathematics as a science with an empirical content long had few supporters. The situation changed only in the 1960s when new arguments for the empiricist interpretation of mathematics were advanced by Imre Lakatos, who, going even further than Mill, claimed that mathematical truths can be criticized, reformulated, and, in specific circumstances, rejected.

By presenting arguments to the effect that the nature of mathematics does not differ essentially from that of the natural sciences and that both have an empirical and inductive character, Mill opposed not only Hume, but also Comte, who, it seems, believed that because of the special role mathematics plays in the system of sciences, an attempt to interpret it along the same lines as physics or biology would be rather misguided. Mill and Comte are at variance also over the relation of social sciences to natural sciences.

Social and Natural Sciences

The nineteenth century positivism represented by Comte and Mill is justly called social positivism. The main aim of social positivists was to create social science which at the time quite often was called moral science. Their principal views as to how that science should be advanced were similar. Like his predecessors Hume and Comte, Mill believed that social science or sciences must make use of the methods of natural sciences. Mill supported and continued the tradition of methodological naturalism. Yet Mill's and Comte's opinions both as to the possibility of building a positive social science *promptly* and as to its structure and foundations were far from identical.

Mill was confident that all positive sciences are parts of science, which exhibits a considerable degree of unity. Thus, from his point of view, the differences of various sciences that Comte paid a lot of attention to are not so very important. Mill was not convinced by Comte that social science must start with the inquiry of the whole (society, humanity) and only later pass to the inquiry of its parts, the individuals. He was sure that it is easier to cognize individuals than all of humanity. As far as the status of the historical method in social science is concerned, he bows to the opinion that it is simply a variety of the method of observation and comparison used in the natural sciences. And as to the possibility of subdividing social science, Mill's opinion is exactly the opposite to that of Comte: disciplines concerning specific kinds of social phenomena have the full right to exist; moreover, in fact they already exist and at least one of them—political economy—has achieved successes not yet achieved by general social science (sociology), launched only recently.

Mill also has no doubt that other social sciences can and will be established. Human actions are subject to laws no less than natural events. The principal predictability of human conduct does not imply, however, that social science may be exact in the sense that astronomy is an exact science. Agencies which determine human character are so numerous, and circumstances of their actions so diversified, that even if we had known the laws of human nature perfectly, we rarely would be able to predict with certainty the thoughts, feelings, and actions of an individual. However, what is only probable when asserted of an individual may become almost certain when asserted of society. Regardless, though, social laws must be based on the laws of the nature of individual human beings—in this respect Mill's opinion clearly differs from that of Comte.

Mill believes that Comte's claim that social science must be based on physiology (biology) is a great error. Human actions depend upon thoughts and feelings, and uniformities of succession of states of mind can be ascertained by observation. Even if states of mind are produced by states of body, uniformities of succession of states of mind exist and are revealed by psychology, which is in a much more advanced state than physiology.

As was already mentioned, Comte even did not include psychology in his list of fundamental sciences. Mill was of the opposite opinion concerning its status. He was convinced that psychology is the very basis of social sciences. In his words, "[t]he laws of the phenomena of society are and can be nothing but the laws of actions and passions of human beings united together in the social state. Men, however, in a state of society, are still men; their actions and passions are obedient to the laws of individual human nature. [...] Human beings in society have no properties but those which are derived from, and may be resolved into, the laws of the nature of individual man" (Mill 1949 [1843]: 573). Those laws are primarily the laws of psychology. In regarding the laws of psychology as the basis of the laws of social science, Mill is more similar of course to Hume than Comte. Opposing Comte's *methodological collectivism* he clearly posits himself as a *methodological individualist*.

Because the laws of the phenomena of society depend upon the laws concerning individual human beings, Comte's demand to base all social science on specific social experience is, according to Mill, unfounded. A purely experimental method which proceeds "as if the nature of man as an individual were not concerned at all, or were concerned in a very inferior degree, in the operations of human beings in society" (Mill 1949 [1843]: 573) overlooks this dependence. The right method of social science has a deductive character: it treats every effect as a result of many causes and is dependent upon the variety of laws of human nature.

Mill emphasizes that the laws of the most complex social phenomena are reducible to the laws of their elements. Social science is a deductive science exactly for this reason. Like physics, it takes into account not a single law but all causes which together determine the effect. Yet because of the multiplicity and variety of acting causes, in social science we cannot hope to compute their exact joint effect. In any case we must seek to verify the consequences of deductive reasoning by comparing them with concrete phenomena or with their empirical laws—if we can establish them.

All general propositions of social science are, according to Mill, hypothetical propositions. They indicate what effects this, or other causes, would have under certain circumstances if no other factors had influenced the effect. When we try to determine the effects of the increasing number of jointly acting causes, the uncertainty of the results of our deductions grows considerably. Thus it is quite understandable that we may try to limit ourselves to the investigation of those social phenomena which are influenced directly by only a few social factors. Different kinds of social phenomena are influenced mainly by different kinds of causes. Thus *different social sciences* may exist and, moreover, may be cultivated with good effects.

Another difference between Mill and Comte concerns the appraisal of the actual state and the nearest perspectives of social science. Although Mill was rather optimistic as to the prospects of developing not only a special social science like political economy, but also of a general social science, he foresaw many roadblocks to a scientific theory of society and did not expect it to mature soon. It seems that being more cautious about the prospects of social science reaching the level of development of natural science he was more close to reality than Comte.

On the other hand, Mill draws the line between facts and norms or values much more carefully and precisely than does Comte. He lays special emphasis on the substantial difference between propositions enjoining or recommending that something should be, and propositions of science asserting matters of fact. A proposition of which the predicate is expressed by the words *ought* or *should be* is generically different from one which is expressed by *is* or *will be*. Mill bends to the *emotivist* interpretation of the former: they imply only one fact, namely that conduct recommended "excites in the speaker's mind the feeling of approbation" (Mill 1949 [1843]: 620).

The substantial difference between two kinds of propositions means that no normative propositions can be inferred from the propositions of social science. The province of ends and purposes is quite different from the domain of facts and laws. That leads Mill to an important conclusion Comte did not make: norms may be inferred only from the first principles of human conduct. According to Mill, "there are not only first principles of Knowledge, but first principles of Conduct. There must be some standard by which to determine the goodness or badness, absolute and comparative, of ends or objects of desire" (Mill 1949 [1843]: 620–21). Moreover, there can be but one ultimate principle of conduct, because in the opposite case it would be possible to approve some way of conduct according to one criterion and to condemn it according to another. Thus some principle would be needed anyway to decide which side engaged in an argument is right.

Mill himself is certain that such an ultimate, most noble standard is the promotion of happiness of humankind. According to Mill, it serves "quite as well for the ultimate principle of Morality, as for that of Prudence, Policy or Taste" (Mill 1949 [1843]: 621). The principle of the promotion or increase of general happiness in the specific form it acquired in the British philosophical tradition is known as the *principle of utility*. The promotion of happiness of humankind was the most important motive of Mill's work, and the main aim of all social reforms he proposed. Comte, as we remember, was more eager to promote the harmony of humanity. However, in a very wide sense it may be regarded as a kind of utility also.

Utility and Liberty

Ideas of utilitarian ethics dominating in Britain in Mill's youthful years made a great impact on him. Mill was inspired by Bentham's credo: "The greatest happiness for the greatest number." Although in 1826 his attitude towards Bentham became more critical (in the *Autobiography* Mill characterizes that period as a spiritual crisis), he did not renounce that credo. He revised some precepts of the Benthamites but never ceased to regard himself as a utilitarian. In 1861, he presented his own version of that ethical theory in *Fraser's Magazine*, then very popular in Britain. In 1863, his *Utilitarianism* was published in book form.

Mill, unlike Comte, was a proponent of personal liberty, but both of them were concerned with the common good. That is very characteristic of all classical positivism. Mill, like Comte, strived to serve mankind. Hence it is rather natural that Mill gives priority not to individual but to social happiness. Like Comte, he was a proponent of altruism, not of egoism.

Mill explicates the principle of Utility (or of the Greatest Happiness) on which utilitarian theory is based in the following way: this principle "holds that actions are right in proportion as they tend to promote happiness, wrong as they tend to produce the reverse of happiness. By happiness is intended pleasure, and absence of pain; by unhappiness, pain and the privation of pleasure" (Mill 1972 [1861]: 7).

He is perfectly aware that this principle concerning the *ultimate end* of our actions is not amenable to direct proof in the strict sense of the word. That does not mean, however, that no arguments might be presented in its favor. He emphasizes that there is "a larger meaning of the word proof, in which this question is as amenable to it as any other of the disputed questions of philosophy. [...] Considerations may be presented capable of determining the intellect either to give or withhold its assent to the doctrine" (Mill 1972 [1861]: 5). Mill's stance on the issue is different from one most logical positivists took later on—contrary to them he believes that rational argumentation concerning the problems of moral choice is possible. Looking for arguments to support this principle he turns to the theory of life and the empirical evidence, if it can be so called, of human desire which claims that "pleasure and freedom from pain, are the only things desirable as ends; and that all desirable things (which are as numerous in the utilitarian as in any other scheme) are desirable either for the pleasure inherent in themselves, or as means to the promotion of pleasure and the prevention of pain" (ibid., 4).

Critics of utilitarianism emphasize, however, that both the status of and the truthfulness of the claims that the first, the supreme principal that the moral decisions of human beings should be based on utility, and second, that human beings desire pleasure and only pleasure, are both uncertain. They suspect that this claim is tautological, not an empirical statement: people seek in fact different aims, but if you define pleasure as that which people seek, you make the claim automatically true. Secondly, even if the claim that people desire happiness is accepted as an unproblematic empirical statement, its link to the claim that it is *right* to desire happiness is problematic. Isn't maintaining that a normative statement is grounded in some sense or another on a statement of fact, or somehow supported by it, actually a typical naturalistic fallacy of which Mill himself warns in his *System of Logic*: are things that humans actually desire worth desiring?

In *Utilitarianism*, where he addresses a wider and less reflective public, Mill seems, however, more concerned not about dissipating these reasonable doubts but in explaining what the relation is between personal happiness and general happiness. Mill does not doubt that the main motive of most people's actions is their own well-being. And what is more, he emphasizes that the good of the world is composed of the benefits of individuals. In most cases people attend to their own happiness or interests, and the happiness of a few people. Yet, according to Mill, universal happiness not only derives from individual happiness, it is also a condition of it. Although not everybody pursuing their own interests sees the link between both kinds of happiness clearly, laws and social arrangements are capable "to establish in the mind of every individual an indissoluble association between his own happiness and the good of the whole" (Mill 1972 [1861]: 18).

Let us admit that this is in fact the case. The establishment of an ultimate criterion of human conduct does not solve, however, all the problems a utilitarian faces. When applying the principle of the greatest happiness, it is necessary to compare the increase in (or decrease of) happiness of different people. Bentham believed that this is purely a *quantitative* exercise: every person's happiness has the same value as everybody else's. Mill was convinced that this is not the case. "It is quite compatible with the principles of utility," claims Mill, "to recognize the fact, that some kinds of pleasure are more desirable and more valuable than others. It would be absurd that while, in estimating all other things, quality is considered as well as quantity, the estimation of pleasures should be supposed to depend on quantity alone" (Mill 1972 [1861]: 8). By introducing into utilitarian ethics considerations related to the *quality* of pleasures, Mill moves away from leveling egalitarianism which is characteristic of the Benthamites. They were convinced that happiness of every person must be taken into account on even terms when deciding what is good and what is bad. In a famous quote, Bentham claims that the children's game of pushpin is as good as poetry considering the amount of pleasure they produce. According to him, there are only quantitative differences between the degrees of happiness. Mill, however, makes people and their pleasures qualitatively different. He believes that in calculating the degree of general happiness, more attention must be paid to the educated minority able to enjoy higher kinds of pleasure than to the uneducated majority incapable of having them. This elitist judgment had not only moral but also political implications. Like Comte, Mill was convinced that the educated minority must play a greater role than others in the process of political decision making. Mill's rejection of the political principle "one person, one vote" is a direct consequence of his ethical views.

By introducing considerations as to the quality of pleasures, Mill considerably modifies the central precept of earlier utilitarianism. Bentham and James Mill would stress that only the consequences of human actions matter, not the actors themselves. John Stuart Mill holds, however, that not only the actions but human beings themselves may be appraised in moral terms. Because of the capability of the educated and intellectually gifted to indulge themselves in higher pleasures and thus enjoy happiness unattainable to others, education and intellect itself acquire moral value.

Mill's political theory, unlike Comte's, is based on the principle of representation. In *Considerations on Representative Government* he deliberates extensively on forms of government, functions of representative bodies, modes of voting, requirements for members of parliament, and other issues related to representation. It is regarded a classical work in political theory. When writing *Considerations*, Mill was inspired rather by Locke and later liberals, not by Comte. Yet as could be expected from a positivist concerned with social progress, Mill does not subscribe to Locke's notion of minimal state, according to which the functions of the state must be limited to ensuring physical security of the individual and his property. According to Mill, those functions are much more extensive: the state must take care of education and social security and provide other public services that the free market is unable (or unwilling) to provide. Mill also did not doubt that the state must protect interests not only of individuals but also of their groups, especially of those notably exposed, like women and workers.

Thus Mill certainly cannot be regarded as an adherent of the night-watchman notion of the state. He is one of the first proponents of social liberalism, which opens the door to the welfare state. Radical liberals, i.e., libertarians like Murray Rothbard or Robert Nozick, claimed that by substantially widening the functions of state Mill betrayed the ideals of liberalism. It seems, though, that the most important factor which determined Mill's receding from Locke was precisely his positivism and progressivism.

Mill claimed to be a resolute proponent not only of representative government but also of democracy. From a contemporary point of view, the foundation of democracy is the political equality of all citizens. Yet Mill, who insistently fought for the equal rights of men and women and is regarded today as one of the founders of feminism, in no way assumed that all citizens must have equal political rights. He was confident that those who lack educational credentials cannot take part in elections. Comte, we should remember, was against any political elections whatsoever. Mill's political views are only a little less antiegalitarian than those of Comte. What makes the political views of both positivists substantially different is

Mill's admiration of liberty and individualism, which was completely alien to Comte, who clearly preferred discipline and collectivism. Mill's concern with liberty distinguishes him not only from Comte, but also from the earlier utilitarians.

Much more than Bentham and rather like Alexis de Tocqueville, whose *Democracy in America* influenced his views profoundly, Mill was aware that even a representative government founded upon and governed with the best democratic principles may suppress individuals. "The will of the people," wrote Mill in *On Liberty*, "practically means the will of the most numerous or the most active part of the people. [...] The limitation, therefore, of the power of government over individuals loses none of its importance when the holders of power are regularly accountable to the community, that it to the strongest party therein" (Mill 1972 [1859]: 72–73).

Mill, however, does not claim that an individual has a *natural right* to be free and enjoy liberty. Like Comte, he rejects the very idea of natural rights—both of them regarded it as rather metaphysical. "It is proper to state," says Mill, "that I forego any advantage which could be derived to my argument from the idea of abstract right as a thing independent of utility. I regard utility as the ultimate appeal on all ethical questions; but it must be utility in the largest sense, grounded on the permanent interest of man as a progressive being" (Mill 1972 [1859]: 81). That does not mean that he appreciated liberty less than founders of the theory of natural rights. In *Subjection of Women* he clearly states that after essentials like food and clothes, liberty is the most important human need (cf. Mill 1991 [1869]: 576).

Mill regarded liberty not only as a basic human need but also as a necessary factor of any social progress. Progress for Mill means first of all an increase in general happiness. In this respect he differs from Comte, who prioritized an increase of harmony between individual and society. Both of them regarded intellectual progress as a means to social progress. But while Comte stressed the role of discipline as a precondition of rapid social development, Mill emphasized the role of individuality. Mill was convinced that the development of humankind is fostered by the diversity of human beings, not, as Comte presupposed, by their uniformity.

According to Mill, individualism is a means to ensure progress— political, economic, cultural, and moral. But no individualism exists without liberty. Thus liberty and progress are closely related. Despotism is incompatible with the "good condition of human affairs." Liberty as protection against despotism is a means to guarantee their good condition. On the other hand, liberty by itself is an essential element of happiness of those who value it. Freedom is the precondition of happiness, at least of the happiness of the most creative people, and because of their importance for society and its advance, of general happiness also.

The appropriate domain of liberty is, according to Mill rather wide: it comprises liberty of conscience, liberty of expressing and publishing opinions, liberty of tastes and pursuits, "of framing the plan of our life to suit our own character" and freedom "to unite for any purpose not involving harm to others." He is firmly convinced that it is necessary to defend all this domain from the encroachment of society which always sought "to compel people to conform to its notions of personal as of social excellence" (Mill 1972 [1859]: 81) and diminish the power of the individual. Mill finds it especially regrettable that some modern social reformers, and first of all Comte, seek to strengthen society's power over the individual even more by aiming "at establishing (though by moral more than legal appliances) a despotism of society over the individual" (ibid., 82).

In his praise of liberty Mill repeats some well-known arguments. In other cases, however, he advances his own. One of them was often used by Popper afterward, in the twentieth century. Mill in his later years came to the conclusion that an important factor of intellectual progress is a free discussion in which parties agree that even the most improbable opinion may possibly be true and a popular opinion false. He anticipates Popper by linking rationality and fallibility directly. He explains the preponderance among humankind of rational opinions—of which he has no doubt because in the opposite case human affairs were much worse off—by the capacity of human beings to correct their errors. He does not make this point in *A System of Logic*. However, in *On Liberty* Mill sounds like an exemplary fallibilist. He asks: "In the case of any person whose judgment is really deserving of confidence, how has it become so" and answers immediately: "Because he has kept his mind open to criticism of his opinions and conduct. Because it has been his practice to listen to all that could be said against him; to profit by as much of it as was just, and to expound to himself, and upon occasion to others, the fallacy of what was fallacious" (Mill 1972 [1859]: 88). According to Mill, the best way to safeguard our beliefs is to invite everybody to prove them wrong. Because we are fallible beings, it would be unwise to silence any opinion. Thus diversity of opinions is highly desirable. To ensure progress, diversity of actions, of "different experiments of living" and different modes of life must be guaranteed as well. In this environment of liberty, the fittest ideas survive.

By emphasizing the value of individualism and diversity, and the harm of conformism and unification, Mill championed social and political ideals very different from those of Comte. Mill was convinced that only by making choices between different opinions and modes of behavior may we exercise and develop our intellectual and moral powers and search for what is best. Instead of progress, the unity of opinion, discipline, and strict collectivism advocated by Comte would lead to stagnation. For Mill, a Europe in which all people would hold the same opinions (even positive ones) and behave in the same way would slow into stagnancy as did China's empire. Mill was really horrified by attempts undertaken by some well-known reformers, including Comte, to push Europe in this direction.

Coming to the end of our examination of the views of both architects of classical positivism it is worth saying a few additional words on their comparative merits and demerits. From a liberal point of view, Mill's critique of Comte's political project is just and well-founded. Yet, in support of Comte, we must concede that even though his political ideas do not look very attractive today, he was a more resolute, audacious, and consistent thinker than Mill. He comprehended well that the positive social order as he understood it cannot exist without the sacrifice and self-devotion of the individual to society. An individual must always prioritize the common good over a personal one. He must be brought up or reeducated (positivists would help) in such a way that he would think constantly not about himself but about mankind and its progress. Who has the right to determine what serves the progress of society and what doesn't? Comte's answer is clear: the poorly educated segment of society who lacks positive knowledge of the laws of the development of humanity simply cannot judge, and therefore should not be granted the right to judge. Basic questions of social life must be decided by a scientific and industrial elite which is dutifully imbued with positive spirit. And what is more, Comte understands well that decisions taken by the elite for the sake of humanity may not be willingly enacted by people, who do not always behave as perfect altruists. Therefore he emphasizes the importance of social discipline.

Mill's position is not so clear and consistent. On the one hand, as a utilitarian professing the principle of the greatest happiness of the greatest number, he holds society and the collective over the individual. On the other hand, he emphasizes that every individual has the right to seek personal good. According to him, "[t]he only freedom which deserves the name, is that of pursuing our own good in our own way, so long as we do not attempt to deprive others of theirs, or impede their efforts to obtain it. Each is the proper guardian of his own health, whether bodily, *or* mental and spiritual. Mankind are greater gainers by suffering each other to live as seems good to themselves, than by compelling each to live as seems good to the rest" (Mill 1972 [1859]: 81).

An awkward question looms large: if everybody is free to pursue their own good in their own way, why must anyone obey the principle of Utility and bother about mankind, society, or community? It is evident that Mill's liberal individualism conflicts with his utilitarian and positivist collectivism.

The Positivist Movement in the Nineteenth Century

Positivists and especially Comte were convinced that they have an extremely important mission: to help humanity pass from the metaphysical stage to the positive. The historical tendencies of the advance of humankind were ascertained and the theoretical foundation for social reforms was laid down, but it was clear for the founder of social positivism that one man cannot transform the mode of thinking of the people, spread a new morality, and initiate necessary political transformations. The coordinated activity of people devoted to positivist ideas and doctrines and able to carry them out was needed. Especially later in life, Comte knew that a school of devoted pupils was indispensable to his positivist mission.

Comte's circle of followers, quite small in the late 1820s, gradually grew. Until he taught and published his *Course*, he was content with informal ties with his disciples. Yet in the late 1840s, Comte decided that a formal organization was needed to propagate his positivist ideas and prepare the necessary moral, political, and religious reforms. In 1848, Comte founded the Positivist Society. The next year he established the Universal Church of the Religion of Humanity. The Positivist Committee was also created to direct and supervise the activities of the positivist movement. At that time, Comte's school hardly resembled most of the known philosophical schools, such as those of Plato and Aristotle or, say, the Vienna Circle and the Frankfurt School. Subordination, discipline, and unconditional acceptance of the leader's authority on all issues made this school a bit like a monastic order. Earlier, the Pythagorean School was likely of a similar nature. And later only the Marxist movement was akin to positivism in this respect.

Among his most faithful disciples, Comte clearly preferred Pierre Lafitte. He became the head of the Positivist Committee after Comte's death, holding the title of the Director of Positivism. Yet the most brilliant among close collaborators was, without doubt, Émile Littré, who, in fact, was most instrumental in making positivist ideas known to the wider public. He was the first who reacted to the *Course* in the press by writing in 1844 six articles on it, and there is no doubt that he made the greatest contribution in popularizing the principles of positivism after Comte's death. Yet after Comte supported Louis Bonaparte's *coup d'état* in 1851, after which the latter took the title of Emperor Napoleon III (Comte, like philosophers of the Enlightenment, cherished the hope that the monarch would be instrumental in carrying on progressive reforms), Littré's relations with Comte dramatically deteriorated and he left the Positivist Society, although he continued to propagate positivist ideas. He became very famous by publishing the voluminous *Dictionary of French Language*, on which he worked for thirty years, and which was the best dictionary of French in the nineteenth century used until now. It came to be called simply *Littré*. In 1875 he was elected a senator of the French republic *ad vitam*.

Comte cherished hopes that spreading the positive way of thinking and establishing a positive social and political order in Western Europe would enable five of its "most progressive" nations—Frenchmen, Italians, the Spanish, the British, and Germans—to create a united *Republic of the West*. It was one of the first projects of the peaceful integration of Europe. The Positivist Committees had to direct the work of the positivist missions abroad. Partly through the work of the Spanish mission, and partly by direct impact, Comte's ideas became very popular in Latin America, especially in Mexico and Brazil.

An important role in the dissemination of positivist ideas was played by positivist periodicals, the *Revue occidentale philosophique, sociale et politique* and *La Philosophie positive*. The first one (officially termed *organe du positivisme*) was published by the Positivist Society and expressed the views of orthodox followers of Comte who gathered after Comte's death around Lafitte. They were convinced that their teacher had all the right answers to every main question he raised and held that the Religion of Humanity was an extremely important element of positivism. Taking almost everything Comte said, especially in his late years, as a kind of revelation and final truth, they saw their missionary task in expounding and spreading his teaching, not in deepening or revising it.

Philosophie positive was edited by Littré (together with Grigorii Vyrubov). He also held Comte in high esteem yet was much more critical of him in general, and of his late works especially. The notion of positivism characteristic for the circle around Littré was broader, slightly closer to that of Mill and, in fact, more influential.

Outside of these two circles were located thinkers connected with Comte more loosely than Laffitte or Littré. In this group Mill must be named primarily. Some other thinkers whose link with Comte and his ideas is weaker than that of Mill ought also be mentioned, first of all Herbert Spencer in Britain and such important French figures as Claude Bernard (physiologist and methodologist), Joseph Ernest Renan (historian of religion and orientalist), Hippolyte Adolphe Taine (historian and art critic), and Émile Durkheim (sociologist and anthropologist).

The second half of the nineteenth century was dominated by positivist thought. First of all the spirit of positivism corresponded to the spirit of the epoch, which was fascinated with great achievements in fundamental science and research and development, and which believed in a bright future. The idea of progress enthralled positivists, and harmonized very well with the hopes of a wide public in Europe and the Americas.

The greater part of the nineteenth century was one of the most peaceful periods in the history of the Western world. This circumstance was conducive to the growth of optimism enhanced by other factors as well. The nineteenth century saw prolific technical inventions, the spread of education, progress in health care, and a rise of living standards, especially brisk in its last decades. Positivism used these facts to support its own ideas concerning the advance of humanity. Moreover, its proponents used the opportunity to claim that their ideas facilitated this progress. And there is no doubt that, say, Mill's activities had tangible effects on social and political conditions in Britain. Even the great reconstruction of Paris commenced in the 1853 and directed until 1870 by Georges-Eugène Haussmann may be attributed partly to a positivist influence. (The work on his project continued until 1927.)

In the middle and second half of the nineteenth century, positivism had few rivals able to capture the imagination of the wide public. With Hegel's death in 1831, it seemed the age of metaphysics had ended for good. Metaphysics simply fell out of fashion. Many felt that it was incompatible with the spirit of the epoch. Positivism, by claiming exactly this, i.e., that the metaphysical era came to an end, and the new, positive epoch was dawning, gained much sympathy and scored many points among the intellectuals and general public.

Today not a few philosophers would claim that Marx and Nietzsche are greater thinkers than Comte and Mill. Nevertheless, in the nineteenth century, a very different opinion prevailed. Comte and Mill were famous and influential, and relatively few knew Marx and Nietzsche, much less Kierkegaard. Positivist ideas and attitudes influenced developments in politics, the arts, and sciences. The positivist movement had a surprisingly wholesale, sweeping, and comprehensive character. Thus the fact that the nineteenth century sometimes is termed, especially by historians of culture, the "positivist century" should not be a great surprise for us.

CHAPTER 3

From Classical to Modern Positivism

Reappraisal of Positivism at the End of the Nineteenth Century

Classical or social positivism was at the peak of its glory in the 1860s and 1870s. Its influence remained quite strong until the very beginning of the twentieth century, but from the mid-eighties it began slowly to diminish. Besides the evident fact that intellectual trends share the fate of all trends and thus cannot persist indefinitely, there are other, more definite reasons to indicate when explaining the demise of classical positivism at the end of Victorian era.

From the mid-nineteenth century, positivism was quite popular among different social and professional groups, counting among its supporters part of the political establishment and managerial class as well as workers. It is fairly natural that a philosophy revering scientific knowledge and claiming that its advance is the main cause of social progress had many supporters in scientific circles also. Both Comte and Mill and their less famous followers, such as Lafitte and Littré, were highly regarded by many scientists and had a number of friends among them. However, it was precisely the scientific elite which turned out to be the first important professional group to become increasingly disappointed with positivism, especially as it was developed by Comte. In the 1890s classical positivism lost its grip on the scientific elite.

When praising scientific progress, classical positivism was at the same time too dependent on it. We do not mean that at the end of nineteenth century scientific progress stopped or slowed and that this undermined the credibility of the principles of classical positivism. What happened was quite the opposite: it was namely the rapid scientific progress which undermined many of Comte's theses, and indirectly weakened the positions of Mill's positivism.

It seems that when writing the *Course of Positive Philosophy*—his great encyclopedia—Comte did not reflect on the fact that it cannot be anything but an encyclopedia of *present-day* knowledge, and that the very advance of science he so praised will inevitably destroy his great synthesis. All encyclopedias become outdated sooner or later. Comte's encyclopedia shared their fate: it had not a few shortcomings at the very moment the last volume of it was published, but fifty years later, in the last decade of the century, it had become obsolete. The social science he was very critical of—economics—flourished. The social sciences in general made a huge leap forward. Psychology, which began to widely use experimental methods, prospered as few other human sciences.

Important changes occurred in mathematics and the natural sciences. Non-Euclidean geometries sprung up, mathematical logic was born. Maxwell's electrodynamics opened new vistas for research in physics. Darwin's theory of the origin of species changed the character of biology. In his *Course* Comte did not shy away from sweeping statements about all the sciences. I did not find it necessary to examine most of them in this book, but by the end of the nineteenth century, it was clear to most scientists that he had erred often and gravely. Especially annoying for the members of the scientific community at the end of the century was the affirmative, oftentimes arrogant, manner in which Comte presented the state of knowledge of his times as almost final and his unwillingness—noted already by Mill—to leave at least some questions open or even to hedge his bets.

The staleness of Comte's scientific picture of nature and society corresponded to the outdatedness of his political views. The gradual development of the leading Western nations towards liberal democracy evident in the rather quiet decades of the end of the nineteenth and the very beginning of the twentieth century made many of Comte's political ideas look quite obsolete. The specific sentimental and at the same time categorical manner in which the later Comte presented them made them even less digestible at the time when the Victorian epoch was drawing nearer to its end.

Mill's version of positivism was more in line with the tendencies characteristic for the end of the nineteenth century. Positivism in Mill's version had a more analytic than synthetic character and developments in science did not have such a devastating impact on it. He was able to predict the growing role of psychology among the social sciences. Mill's notion of social science as a system of different sciences with economics (political economy) serving a very important position had won out in the competition with Comte's idea of a single, unified social science. And most evidently, Mill's political ideas were very much more in line with the developments in Western politics than Comte's.

In spite of this, at the very end of the nineteenth century, Mill's positivism also began to lose supporters. Partly it was a consequence of the collapse of Comte's positivism. Notwithstanding all the efforts of the later Mill to emphasize the differences between his and Comte's positions, Mill's positivism was too closely associated in the public eye with Comte's standpoint and the breakdown of the latter damaged the former. His ideas in ethics and especially in political philosophy attract attention even today, but in the last years of the nineteenth century, they were looked at already as simply *Mill's* and not *Mill the positivist's* ideas. This interpretation prevailed also in the twentieth century.

At the end of the nineteenth century, the attention of the wider public (and especially of members of the scientific community) was attracted to the works and ideas of theoreticians who—being influenced more or less by Comte and Mill (and maybe even more by Hume)—concentrated much more on methodological and epistemological, and not on social or political, issues. They are of interest to us, because, as we see today, they were intermediaries between nineteenth century social and twentieth century logical positivism. Most of them were representatives of exact sciences.

Many of their ideas, which inspired later logical positivists, originated in the process of reflection on the conceptual apparatus and epistemological foundations of science. They were most interested in the sciences they were involved in, i.e., in physics and mathematics. At the end of the nineteenth century these sciences made great progress. Some of their representatives understood rather well that the results these sciences achieved, and on the other hand, the problems they were facing, have wider implications and are important not only to the physicists and mathematicians themselves. They had a well-developed methodological conscience and at the same time shared Comte's and Mill's conviction that science has a great significance both for understanding the world we live in and for our way of life. They were progressivists in the broad sense of the word: convinced that the advance of science leads to progress in other areas, that the use of new technologies and techniques raises living standards, that education improves manners, customs, and a person's economic prospects, and that promotion of better laws and methods of governing ameliorates the human condition. Yet most of them evidently did not have a calling to draw elaborate projects of social and political reforms. They felt more at ease in explaining to laymen the value of science, its aims, foundations, concepts, and methods. Even the most socially and politically involved of these scientists—Ernst Mach—who advocated social reforms of the liberal-socialist sort and for some time served as a member of the Austrian parliament, did not develop anything approaching a consistent social philosophy.

In general they did not think that Comte or Mill (or Littré or Spencer) succeeded in explaining the nature of science. Representatives of classical positivism looked at the natural sciences mainly from the outside, not from the inside, and misunderstood many facets of science. Specifically, the inductivism of which Mill was the main advocate was, according to them, a fallacious philosophy of science. Mathematics and hypotheses are much more important in science than simple inductions of the kind Mill considered. And mathematics itself is not an inductive enterprise. When explaining its nature, Mill makes still greater mistakes than when interpreting the nature of the natural sciences.

According to the leading scholars and researchers active at the end of the nineteenth and the beginning of the twentieth centuries, an even more important shortcoming of the positivist philosophy of science was its uncritical attitude towards science. Classical positivism simply continued the devotional apology of science so widespread in the Enlightenment. The value of science is great, but that does not mean that nothing remains to be changed in the foundations of science. Representatives of classical positivism claimed that mathematics, astronomy, and physics have already reached the positive stage. But does it mean that there is nothing to be improved in *these* sciences? Comte and Mill came to exactly this conclusion. They were mistaken. The condition of these sciences is better than that of others but it is far from perfect. Comte and Mill simply did not take care to examine thoroughly the foundations of these sciences and of science in general. If they were more careful, they would understand that even the notion of fact they use so often is far from being obvious: not everything we hold as fact is a fact indeed. Their views concerning the nature of scientific claims are in need of reexamination as well.

These reasons prompted some prominent scholars and researchers to reject the old positivist theory of science and to start a serious search for alternatives. A few new options were of special importance to the subsequent development of positivism: empiriocriticism, conventionalism, and hypothetism. Let's examine them briefly.

Mach's Empiriocriticism

Ernst Mach (1838–1916) was a famous Austrian physicist, especially renowned for his investigations concerning shockwaves (some of concepts used in this field bear Mach's name), as well as for his studies on mass, inertia, and inertial systems of reference. In his *Mechanics* (1883) and numerous scientific articles, he examined Newton's notions of absolute space, absolute time, and absolute movement, rejected them as having no empirical content, and thus built much of the necessary foundation for Einstein's theory of relativity. Many results of Mach's investigations, including those just mentioned and those devoted to the analysis of problems located in the border region between natural science, psychology, and epistemology, especially that of perception, had import both for physics and philosophy.

For almost thirty years starting from 1867, Mach was a professor of physics at the German Charles-Ferdinand University in Prague and served for some time as its rector. From 1895 until 1901 he was professor of philosophy at the University of Vienna. It is not an overstatement to credit his lasting influence for creating the favorable conditions for the flowering of logical positivism in Vienna.

Most of his philosophical ideas Mach presented in two very influential, widely read and translated books: *The Analysis of Sensations* (the first edition was published in 1886 under the title *Contributions to the Analysis of Sensations*; later Mach expanded the book substantially and slightly changed its title) and *Knowledge and Error* (1905). These ideas bear resemblance to the concepts of Swiss philosopher Richard Avenarius outlined in his books *The Critique of Pure Experience* (2 vols., 1888–90) and *The Human Concept of the World* (1891). At the turn of the twentieth century, both were regarded as the major representatives of *empiriocriticism*. Logical positivists drew, however, almost exclusively on Mach, who sounded much more scientific and positive than Avenarius, and in retrospect it is clear that Mach's impact exceeded that of Avenarius, even if the latter must be credited with anticipating (in his *Habilitationschrift* in 1876) many of Mach's ideas. Since we are interested much more in the process of transition from social to logical positivism than in empiriocriticism itself, we will limit ourselves to the presentation of some of Mach's key points without trying to define to what extent they are original.

Likely the most important empiriocriticist notion is the idea that the precondition both of the advance of science and of adequate understanding of the nature and foundations of scientific knowledge is the critical examination of experience, and specifically, of the concept of fact. By stressing the importance of examining experience, empiriocriticism more closely resembles Hume's and not Comte's positivism. It tries, however, to refine Hume's analysis. And at least Mach claimed that his adjustments were based on the results of the most modern—mainly his own—results of scientific investigations. Many doubts still remain if there is such a thing as a scientific *foundation* of philosophical ideas at all (although there is no doubt that philosophical ideas have a scientific *context*). However, the status of the latest word in science greatly facilitated the spread of Mach's views among the educated public in various countries from the Americas to Russia and Japan.

Mach begins where Hume ends. In examining experience, Hume concludes that it is impossible to decide whether or not anything independent from experience causes our sensations. He is convinced that there are no empirical means to assure ourselves that our sensations or perceptions depict objectively existing things. By the way, it is worth noting that Hume does not claim that the question regarding the existence of things causing our perceptions is the question of a "false" or meaningless philosophy. Moreover, when he examines other issues, say the problem of causality, he sounds quite often as if he supposes that our perceptions are caused by objectively existing things.

Mach thinks that the very question Hume and other philosophers consider—are our perceptions caused by things whose ontological status is quite different from that of perceptions and which are neither perceptions, nor complexes or constructions out of perceptions—is put wrongly. According to him, experience gives us no pretext to think that there are two different categories of objects: things or physical phenomena on the one hand, and perceptions, mental pictures, or psychical phenomena on the other. Only one sort of entity is given to us in experience. It is *we* who *divide* this sort of homogeneous entities into two: into things and their mental pictures. We believe that physical things cannot exist in our mind. Thus we postulate that they exist outside it and that inside it we have only their pictures. In fact, however, only one kind of phenomena exists. Yet, according to Mach (and later phenomenologists), we are free to treat them either as a physical or psychical phenomena. The way we conceive them depends upon the context, upon issues in which we are interested. If we are interested in the relation between color and light, and pose a question about the length of electromagnetic waves which corresponds to the color yellow, we are treating color as a property of physical bodies, thus as a physical phenomenon. If, however, we are interested in perception and in the eye's role in the process of perception, then we treat color as a sensation, thus as a psychical phenomenon. We must treat the world as an agglomeration of elements which are neutral in the sense that they are neither physical, nor psychical. When examining different problems we, however, convert, transform, or, better to say, interpret these elements as either physical or psychical.

Mach himself regarded this position as a third way in philosophy different both from materialism and idealism, because it gives preferences neither to material nor to ideal (mental) entities. At least some of his opponents were not persuaded that this position is a new one and neutral in the argument between supporters of materialism and those of (immanent) idealism. It reminded them of Berkeley's and Hume's phenomenalist ideas. However, it seems that they underestimated Mach's originality and significance. It is clear now that Mach is an extremely important figure not only in the history of positivism but also in that of phenomenology, especially because of his direct impact on Husserl and through Husserl, Heidegger: two major trends based on very different philosophical perspectives crossed paths at one point only to branch out to become adversaries after a few decades.

When speaking about neutral elements of which consist all facts of the world, Mach quite often used the term "sensations." The title of his main epistemological work was *Analysis of Sensations* (*Die Analyse der Empfindungen*) and his epistemological position may be, without doubt, characterized as phenomenalism, even if it bears some resemblance to that of phenomenology. In fact, Mach was not only epistemological but also an ontological phenomenalist. As an epistemological phenomenalist he claimed that we could know only sensations, and as an ontological phenomenalist he maintained that the world consists of sensations. Moreover, his conception of the purpose of science may be called phenomenalistic as well: the aim of science is to describe sensations.

In his celebrated—at least in Marxist quarters—book, *Materialism and Empiriocriticism*, Lenin regarded Mach's position as bluntly idealist, because he was convinced that it was only a slight variation of Berkeley's well-known position. It seems that the difference between Mach's and Berkeley's points of view is in fact bigger than Lenin claimed (for example, the notion of "soul" plays a crucial role in Berkeley's philosophy but not in Mach's).

But in the context of mapping the direction of positivist thought, a different question is of greater interest to us: is Mach's claim that the world is composed of sensations as neutral elements (that is, the thesis of ontological phenomenalism) a metaphysical claim or not? Hume would regard this claim in the best case as problematic. He would likely emphasize that there is no way to substantiate it. But there is little doubt that Comte's attitude towards Mach's speculations would be definitely negative. Comte (and most logical positivists beginning from the late 1930s) would say that positivist philosophy cannot enter into the examination of *epistemological* questions as far as they transcend the methodological issues, and venture into metaphysics. In case the process of perception becomes an object of philosophical examination, questions which cannot be decided by scientific methods inevitably arise. No strict boundary exists between epistemology and *metaphysics*, and Mach's postulating the existence of neutral elements of the world gives evidence to this. In fact, he treats these elements as the ultimate constituents of which the world is composed. Most of the earlier and later positivists were convinced, however, that the issue of the ultimate constituents of the world—regardless of what they are called—is clearly a metaphysical problem alien to the scientific, positive mind.

However, another important issue Mach raised, namely the question concerning the nature of facts on which our knowledge is founded, likely would have been regarded as perfectly legitimate by earlier positivists (it was regarded as legitimate by logical positivists), although it involves some epistemological questioning also. Inquiry into the nature of facts is probably the best example of what empiriocriticists meant by the *critique of experience*. From their point of view, facts are relatively stable groups of elements. The neutrality of elements in an above-mentioned sense is transferred to facts as well: they may be treated either as physical or mental structures. Yet notwithstanding how we treat facts, we must not err in determining what facts are. Facts must be described by notions which are defined in terms of sensations (which is simply another name for elements). The latter form the real and *absolutely certain* foundation of knowledge. Accordingly, facts must also be *given* to us. "For me," Mach wrote, "every scientific work is lost which is not *solidly grounded* in the immediately given" (Mach 1976 [1905]: 13). Facts may be looked at also as propositions. In that case they must be self-evident: "Science has always required self-evident propositions as a safe foundation upon which to build" (Mach 1996 [1886]: 56).

However, according to Mach, quite often we consider as a fact something which evidently is neither immediately given, nor self-evident. We often bring in, add, or attach to the given experience or sensation something which is produced or generated by ourselves. Thus if we really want to know facts and base all our knowledge of the world on facts, we must *purify* our experience. From the alleged facts we must eliminate everything which is not given. We must reduce our experience (or what we *hold to be* our experience) into pure experience which contains only *really* given facts.

What elements contaminate experience? They are of three major kinds. Firstly, they are appraisals. Pure facts by themselves are neither good nor bad, neither beautiful nor ugly, and do not contain or imply any evaluations. Facts must be clearly separated from values, and factual statements from value judgments. Secondly, experience must be purified from anthropomorphic, often simply mythological elements, from the notion that things act like persons and move under the influence of desires and passions. Thirdly, it must be freed of all kinds of metaphysical entities like things-in-itself, matter, forces, and causes. They are also of anthropomorphic origin but these elements have a more clearly expressed intellectual and formal character. From Kant's point of view, *a priori* categories (say, that of causality) are *necessary conditions* of experience, and they give to it form; for empiriocriticists they must be removed as an arbitrary addendum.

The last kind of metaphysical entity of which experience must be purified deserves special commentary. Mach's position concerning causation substantially differs from that of Mill, who was convinced that the main task of science is to determine the causes of the phenomena under investigation. Mach, on the contrary, believed that the main task of science is not to explain *why* phenomena happen, but to describe *how* they happen. According to him, modern science, whose degree of mathematization constantly increases, is interested mainly in establishing functional links between different variables. The scientific laws at which scientists aim are not causal statements as Mill believed. They have the character of mathematical equations in which one or more (dependent) variables are treated as functions of some other (independent) variables. Maxwell's equations do not contain any references to causes. Moreover, even Galileo's venerable law of the free fall of bodies having the simple form $S=gt^2/2$ does not contain any reference to the causes of the phenomena. It simply describes the process of the free fall by defining distance S covered by a falling body as a function of the time t that the body falls. Description of this type is the main aim of science, its primary function. Facts must be described in notions which are defined in terms of sensations. The latter form the real foundation of knowledge. Other functions, say prediction, have a derivative character. Thus the notion of cause must be substituted by the notion of mathematical function, which, contrary to the notion of cause, harbors no elements of anthropomorphic ideas.

Scientific laws having the character of mathematical descriptions help us to deal with an enormous amount of facts. We cannot embrace all facts, and oftentimes cannot remember them. Thus we invented laws as shortened descriptions of whole classes of facts, serving as a kind of their abbreviation. We must act by sparing our physical efforts because our physical capacities are limited, and we must think economically because our mental capacities are not boundless. The search for causes and explanations is epistemologically suspicious because it involves postulating dubious (in the best case) entities not given directly in experience. Yet even more important is another circumstance. Such a search is pragmatically idle, moreover harmful, because it violates the *principle of the economy of thought*. In Mach's words, "No knowledge worthy of the name can be gathered up in a single human mind limited to the span of human life and gifted only with finite powers, except by the most exquisite economy of thought and by the careful amassment of the economically ordered experience of thousands of co-workers" (Mach 1996 [1886]: 198). The search for causes and explanations is incompatible with the requirement of "the most exquisite economy of thought" and would mean wasting our intellectual capabilities.

Although science is based on the use of our intellectual capacities, its primary and essential function is practical, not theoretical: science is an instrument of survival. Here Mach takes the same stance as Spencer and the pragmatists. In evolutionary terms, science is a result, and at the same time the means of progressive adaptation of an organism to his environment. Because of this, the description of phenomena enabling us to predict and control them is everything we need of science.

In fact, Mach distinguishes between science's *internal* and *external* purposes (cf. Blackmore 1972: 28). The internal purpose of science is the simplest description and relation of sensations, the most economical abstract expression of facts. Its external purpose is to help satisfy our biological needs and aid the survival and welfare of the species. The internal purpose is subordinated to the external purpose and justifiable in terms of it. By describing facts economically, science provides us with as perfect a means as possible of orienting ourselves in the world. Scientific laws are concise forms of descriptions. They tell us what is happening. Yet at the same time they may be looked at as *limitations* concerning the area of possible facts. They restrict the eventuality of some states of affairs, that is to say "forbid" some facts from happening, and thus help us to prepare for the future and make it manageable.

Mach's principle of economy of thought plays a very important role in his philosophy and methodology. Even his theory of "neutral elements" of the world may be justified in its terms. By rejecting the traditional distinction between appearance and reality, "unnecessary doubled existence" is abolished, and in the best traditions of Ockham, ontological economy is attained. Yet no less important is the fact that the principle optimizes scientific endeavor: it orients scientists towards successfully achieving the external purposes of science and gaining the greatest possible amount of knowledge with the least possible labor and in the shortest possible time. And the most important means to attain this, according to Mach, is to use mathematically simple abstract descriptions. For laymen this may not look simple at all, but for a scientist, description by a system of differential equations often is the most economical way of expressing facts.

Limitation of the internal goals of science to mathematical expression of facts helps to achieve another aim Mach and many other positivists regarded as critical: to unite the sciences. It seems that Mach did not regard sensations to be qualitatively different, thus he saw no uncrossable boundaries dividing different sciences. His own experience of research in the boundary region of physics, physiology, and psychology confirmed his confidence in the essential unity of science.

When specifying the internal aim of science, Mach emphasizes, especially in the *Analysis of Sensations*, that it is the *description* of *observable* facts. Some of Mach's pronouncements give the impression that he is convinced that science simply describes sensations (using, preferably, mathematical language), and these descriptions are never *inferred*. In that case neither induction nor hypothesis would play any role whatever in science.

In fact, Mach is not so radical, although his use of the term "description" is really ambivalent: sometimes he opposes it to prescription, sometimes to explanation, and sometimes to any kind of inference indeed. Yet he does not flatly deny the role of induction and hypotheses. He tries to downplay, however, the role of *both* induction and hypotheses, and in this respect takes a position substantially different from that of most positivists. Thus we saw earlier that Mill—to whom Mach was close in many respects (including not only phenomenalism and psychologism but, let us add, the empiricist interpretation of mathematics)—downplayed the role of hypotheses in science also. Instead, Mill acclaimed induction. It may seem that an empiricist—and Mach was one of them—faces a choice between praising either hypotheses *or* induction. When rejecting hypotheses and granting even to theories only provisional value, Mach, however, did not consider it his duty to commend induction. From his point of view, induction cannot be regarded as the main or even a very important method of science. According to Mach,

> Syllogism and induction do not create new knowledge, but merely make sure that there is no contradiction between our various insights and show clearly how these are connected, and lead our attention to different sides of some particular insight teaching to recognize it in different forms. Obviously, then, the genuine source from which the inquirer gains knowledge must lie elsewhere. In view of this it is rather strange that most inquirers who have dealt with the methods of inquiry nevertheless denote induction as the principal means of inquiry, as though natural sciences had nothing to do but directly classify individual facts that lie openly about. Not that we wish to deny that to be important to, but it does not exhaust the inquirer's task: we must above all find the relevant characteristics and their connections, which is much harder than classifying what is already known. The name "inductive sciences" for the natural sciences is therefore not justified. (Mach 1976 [1905]: 231)

We should consider an important fact here: Mach is interested much more in the procedures of inquiry as they are seen from a psychological point of view than in the logic of inquiry. According to him, the main function of the latter is coordinating discoveries already made and presenting them in a nice form. Let's remind the reader that Mach's *Knowledge and Error* bears a really remarkable subtitle, especially when looking from the vantage point of Carnap: *Sketches on the Psychology of Inquiry*. Mach's orientation in examining scientific knowledge differs essentially from that characteristic for logical positivists, who drew on Mach sometimes and referred to him quite often but were interested in quite a different thing—not the *psychology* but the *logic* of science. Trying to *describe* how scientists find "relevant characteristics" of facts and how they establish connections of facts, he produces a statement which logical positivists would rather reject because it not only emphasizes the psychology of inquiry and not the logic of science (that would be bearable), but indirectly *subordinates* the latter to the former.

> Logic provides no new knowledge. Whence, therefore, does it come? Always from observation, which can be "external" through the senses or "internal" through ideas. Depending which way our attention is attuned, it will emphasize now one connection of elements, now another, fixing it in concepts: if this turn out to be tenable and stands the test vis-à-vis other findings it constitutes knowledge; if not, error. The basis of all knowledge is thus intuition, which may relate both to sense perception and to intuitive ideas, as well as to what is potentially intuitive and conceptual. Logical knowledge is only a special case, concerned purely with findings of agreement or contradiction, inoperative unless there are sense perceptions or ideas from findings previously fixed. (Mach 1976 [1905]: 233)

What role is played in science by hypotheses? According to Mach, their role boils down to the completion in thought of observable facts. "A provisional and tentative assumption that cannot yet be established but helps us to understand a range of facts, we call," says Mach, "a hypothesis" (Mach 1976 [1905]: 173). Hypotheses make facts intelligible and quite often trigger new experiences which confirm, refute, or modify our surmise and so widen experience.

Thus, according to Mach, Newton's claim that he does not invent hypotheses cannot be taken at face value. Newton used the term "hypothesis" in its old sense, as an arbitrary and empirically unprovable assumption. In conformity with modern usage, Mach treats hypotheses as conjectures which may be empirically proved, confirmed, or disconfirmed. Such hypotheses function as vivid pictures of the facts which greatly facilitate the discovery of new facts. Mach agrees that hypotheses can contain elements which cannot be derived from facts themselves. Yet he is convinced that they must remain as close to facts as possible. Their fate depends upon the discovery of new facts. "It lies in the nature of hypotheses to be changed in the course of inquiry, becoming adapted to new experience or even dropped and replaced by a new one" (Mach 1976 [1905]: 178). It is very characteristic of Mach's treatment of hypotheses, however, that according to him, they may be replaced by a complete knowledge of facts. Precisely because of this they play only a *tentative* role. Mach's views on the role of hypotheses in science differ substantially from those of his contemporaries Poincaré, and, especially, Duhem, which will be examined below. According to Mach, fruitful hypotheses are *self-destroying* because they lead to descriptions directly representing the facts. In "complete parts of science [...] there is no room for hypotheses that have a beneficial role only in growing areas" (ibid., 182).

The main means by which hypotheses are framed is, according to Mach, analogy. Induction also plays some role in widening our knowledge, yet it is fallible like a hypothesis, and, therefore, its conclusions must be carefully tested, corrected, and, if necessary, rejected. In general, Mach is eager to emphasize that knowledge is gained "on very tortuous paths." The single steps depend on prior ones but "there can be no widely effective instructions for inquiry by formula." Accidental physical and mental circumstances, imagination, and intuition play, according to him, a much more important role in scientific inquiry than logical rules.

Mach's claim that science is the description of the observable requires additional commentary. The issue of *unobservable* entities was much discussed in the positivist philosophy since Hume's times. They were regarded as needless, unwanted, superfluous, nonexistent, or simply meaningless. Yet the question of how to define unobservable entities, of what is the difference between the directly and indirectly observable or between in principle unobservable metaphysical entities and physical objects unobservable merely for technical reasons, was fiercely discussed not only between positivists and their opponents (who were not so ready to apply Ockham's razor and sever all superfluous entities) but also among positivists themselves. On the basis of a criterion which can be called the *degree of observability*, various notions (or their referents) may be attributed to different classes. The (philosophical) notion of *substance* was rejected by all positivists without exception. *Essences*, *occult qualities*, and *substantial forms* shared the fate of *substance*. These notions are metaphysical *par excellence* and form the first class. Characteristic members of the second class are notions of *matter*, *force*, and *cause*. Many positivists claimed that these notions are unwanted in science. Yet others disagreed. *Atoms* belong to the third class. Mach was convinced that this notion hampers the reduction of science into pure description and must be eliminated from it together with the notions of absolute space and absolute time. Yet according to all known definitions of that notion, atoms have at least some qualities of observable physical things, first of all extension (however small). Today we are convinced that we *know* mass of different atoms and *know* that even parts of atoms may be indirectly observed. Moreover, we do not doubt that atomic theory allows us not only to *understand* some known (observable) phenomena but also to *predict* some unknown phenomena. Thus *for us* Mach's stubborn struggle against atomism at the time when it gained more and more support may seem strange and incomprehensible. Yet Mach did not know much of that which we know now. And it is worth adding that many other famous scientists of the epoch shared his opinion on the matter. Logical positivists disapproved, however, his tenacious efforts to remove from science all directly unobservable entities.

Poincaré's Conventionalism

Ernst Mach was one of the leading physicists of his epoch. His contemporary Henri Poincaré (1854–1912) was one of the leading mathematicians, maybe even more prominent in this field than Mach in his own. Poincaré's scientific interests were certainly no less vast than Mach's and they partly overlapped. After studies at the École Polytechnique—the same

Comte attended—and at the Mining School, Poincaré in 1886 became a professor at the University of Paris and won fame with his works in the theory of differential equations, topology, and mathematical physics. Independently of Einstein he examined the mathematical consequences of the postulate of relativity.

Poincaré was one of the leaders of the new science emerging at the turn of the twentieth century. He was deeply interested in its nature and the possibilities of practical application of its results. He wanted to remain essentially a scientist looking at science, and the titles of his non-mathematical books seem, at the first glance, philosophically neutral: *Science and Hypothesis* (1902), *The Value of Science* (1905), *Science and Method* (1909), and *Mathematics and Science: Last Essays* (1913). Yet those works reveal not only the domain of Poincaré's general interests but also his philosophical position. This position became known as conventionalism. Conventionalism was one of the important trends which shaped twentieth century scientific self-consciousness, and by this influenced modern positivism. It led to a much more instrumentalistic interpretation of science than was characteristic of previous epochs. In this respect Poincaré goes further than Mach, although it is not difficult to see the similarity between Mach's notion that the leading principle of science is *economy of thought* and Poincaré's idea that science is in fact a *convenient* way of dealing with the world. The words *economic description* and *convenient portrayal* express in two not-so-different ways the same idea that science is a *tool*, an *instrument*. Yet Mach was clearly more concerned with truth than Poincaré. The latter did much in undermining the very idea that science is Truth and nothing more than Truth.

He started with the rejection of the idea that geometry is Truth about space. He presented suggestive reasons supporting this point of view. Before tracing out his thought, it is important, however, to make a more general point by noting that a great deal of modern positivism was built on some or another interpretation of mathematics which quite often was related to new mathematical developments. In this respect not much has changed from Plato's times. Positivism influenced mostly the advance of *social* sciences; it recommended that they use methods of *natural* sciences, first and foremost those of *physics*, but it was built in great measure on the interpretation of *mathematics*.

Let's begin, as Plato recommended, with geometry. Yet because we are interested in the trajectory of positive thought, let's not go back further than Mill's empiricist interpretation of geometry, according to which geometrical propositions are based on experience, and in technical terms, are synthetic *a posteriori* truths. We recall that when the first non-Euclidean geometries were produced, most of their creators were convinced—as Mill was—that geometry is a theory of space with empirical content and/or of empirical origin. Yet unlike Mill who, it seems, knew little about non-Euclidean geometries, they confronted a new problem: how to determine which geometry is empirically sound. Their answer to this question was as follows: to measure, for example, the sum of inner angles of a large triangle formed by light beams. This answer was based on the supposition that light beams are, or at least rightly represent, straight lines. For those who believed that geometry is of empirical origin it was quite a natural stance.

Developments in mathematical logic and David Hilbert's formalization of (Euclidean) geometry changed the situation substantially. Earlier mathematicians were convinced that geometrical terms defined by Euclid such as point or line indicate or represent real objects. From this perspective, by proving theorems, a geometer is determining properties of objects already known to him. Yet at the end of the nineteenth century, logicians concluded that what Euclid called definitions do not conform to the stricter requirements of definitions established by the new logic. On the other hand, geometers concluded that explicit definitions are not needed at all for proving geometrical theorems. In fact, geometry is a deductive system whose foundation is not allegedly explicit definitions, but axioms determining the relations between different geometrical objects. Although Poincaré's and Hilbert's positions as to the nature of mathematics were far from identical, they assented the view that axioms of geometry are in fact disguised, implicit definitions of abstract geometrical objects and their relations. Previously, geometers thought that geometry describes properties of some *given* object or structure. Thus when non-Euclidean geometries were constructed and it became clear that axioms of one geometry are incompatible with axioms of other geometry, it was natural for them to search for ways of determining which one of these geometries is a truthful description of that *given* object or structure. The empirical measurement of angles of a triangle was one of the answers to this question, because in different geometries, the sum of the angles of a triangle is not the same and may depend on its size.

Poincaré's position was different. According to him, geometry is not a theory of some *given* object or structure called space. Geometry is an intellectual construction, an intellectual tool to be used primarily in physics. The character, usefulness, and convenience of this instrument depend on our selection of axioms of geometry. They are disguised (implicit) definitions of geometrical objects and we are free to choose them. By this we *construct* an object or structure called space. And because there are different systems of geometrical axioms, they define different spaces. To ask which of them is real is nonsense. These systems of axioms have the same right to exist. The only thing that really matters is the logical compatibility of axioms of the same system. And mathematicians proved that if Euclidean geometry is noncontradictory (nobody seriously doubted that), then non-Euclidean geometries are noncontradictory systems as well.

Therefore, we may look at axioms of geometry as *conventions*. In this respect different systems of geometrical axioms are like different systems of coordinates or systems of weights and measures: we are free to choose any of them. The main criterion is pragmatic convenience. True, this criterion does not always give unequivocal results. We decide what is most convenient for us. By the way, Poincaré was convinced that we shall continue to prefer the Euclidean system because we are accustomed to it.

Poincaré's view of axioms as disguised definitions directly influenced logical positivism. Although most logical positivists later adopted a slightly different interpretation of mathematical propositions, their notion of mathematics was much closer to that of Poincaré (and especially to his notion of geometry) than that of Mach. They agreed with Poincaré that observation and experiment has nothing to do with mathematics and rejected Mach's empiricist interpretation of it.

Some of Poincaré's contemporaries, and first of all his compatriot Édouard Le Roy, tried to extend Poincaré's point of view concerning the conventionality of axioms of geometry to *all* statements of science claiming that they are pure conventions. Poincaré, however, was much more cautious. Strongly emphasizing the conventionality of axioms of geometry and the constructive character of its objects, Poincaré was of different opinion as to the character of objects of arithmetic. According to him, conventions are uncharacteristic of arithmetic. We have intuitions of number, and axioms of arithmetic express these intuitions. We are not free to choose them. According to Poincaré, arithmetic, unlike geometry, is a fundamental mathematical theory, thus it is quite natural that its nature is different from that of geometry.

Poincaré was convinced that most propositions of natural science are not conventions either. They must correspond to experience, which is the last judge in deciding their fate. In general, propositions of the natural sciences are hypotheses. Yet in his words "there are various kinds of hypotheses; [...] some are verifiable and, when once confirmed by experiment, become truths of great fertility; [...] others, without being able to lead us into error, become useful to us in fixing our ideas, and [...] others, finally, are hypotheses in appearance only and reduce to definitions or conventions in disguise" (Poincaré 2001 [1902]: 4). A typical example of the latter are fundamental principles of mechanics. According to Poincaré, it is by definition (and therefore beyond the reach of any experiment) that force is equal to the product of mass and acceleration. No experience can falsify this principle.

Other propositions of the natural sciences that are of a high degree of generality are not conventions. Yet because they are quite abstract, they cannot be proven, in a strict sense, by empirical facts. When affirming them, we make partly conventional decisions. Acceptance of any scientific proposition, including an experimental one, embraces an element of convention, and is a decision whose rightness cannot be unequivocally proven. An additional source of the conventionality of physics is its link with geometry. Geometry is a kind of framework for physics. We are free to choose any framework. Yet to ensure empirical adequacy of a physical theory, we are compelled to choose certain physical laws in case we use one framework, and different physical laws when using another framework. When empirically testing, we always test a whole framework.

This last idea plays an important role in the philosophy of science of another prominent French scientist—Pierre Duhem. Sometimes he is regarded simply as one more representative of conventionalism. Yet in fact his views differ in some not negligible respects from Poincaré's views on science, and thus deserve a separate examination.

Duhem's Hypothetism

Pierre Duhem (1861–1916) studied mathematics and physics at the École Normale Supérieure, a creation of the French Revolution not less known in the area of education as the École Polytechnique at which Comte and Poincaré studied. After graduation he taught a few years at the Faculty of Sciences of Lille University and later for more than twenty years was professor at the Faculty of Sciences of Bordeaux University. In science he is renowned for his work in physics, particularly on thermodynamics. He was also a distinguished historian of science: his main work in this field is *The World System: A History of Cosmological Doctrines from Plato to Copernicus* (1913–59) in ten volumes, part of which appeared only after his death. He also wrote a three-volume *Studies on Leonardo da Vinci* (1906–13). Without any doubt, he is one of the most celebrated and most erudite historians of science of all times. His main work in philosophy of science is *The Aim and Structure of Physical Theory* (1906). *To Save the Phenomena: An Essay on the Idea of Physical Theory from Plato to Galileo* (1908) also deserves mentioning, although historical analysis in it often prevails over philosophical reflection. *The Aim and Structure of Physical Theory* made a lasting impact on the twentieth-century notion of science. Duhem's status both as a well-known physicist, and especially, as an eminent historian of physics and astronomy made his pronouncements concerning the methods, value, and advance of science very authoritative.

At least in one respect Duhem seems to be closer to Mach then to Poincaré. Namely, he is more interested in physics than mathematics, and does not begrudge efforts to convince us that physics has nothing to do with explanation. His point is as follows. Those who claim that the aim of physical theory is explanation do not want to limit themselves to dealing with sensible appearances given in experience only. They want to strip reality of appearances, covering it like a veil, in order to see reality distinct from these appearances, a *naked* reality. Yet in this case they face the question: what is the nature of the elements composing that reality? Only metaphysics can answer this question. Therefore, those who demand from physics to grant explanations of phenomena and of their (experimental) laws unwittingly subordinate it to metaphysics and deprive it of status as an autonomous science. Moreover, because there are various metaphysical doctrines, the value of a physical theory becomes dependent upon the value of the metaphysical system to which one subscribes. Thus a physical theory based on specific metaphysics becomes unacceptable to proponents of alternative metaphysical systems. On the other hand, even supporters of a given metaphysics cannot be entirely satisfied with any physical theory based on their favorite metaphysics, because in fact it is never possible to draw *all* principles of a physical theory from any supposedly foundational metaphysics.

According to Duhem, physical theory must be grounded on the principles of its own and judged in its own terms. That means that it must not aim at explanation. "A physical theory is not an explanation. It is a system of mathematical propositions, deduced from a small number of principles, which aim to represent as simply, as completely, and as exactly as possible a set of experimental laws" (Duhem 1954 [1906]: 19). There is no need for it to conform to any metaphysical doctrine. "Agreement with experiment is the sole criterion of truth for a physical theory" (ibid., 21). Thus, according to Duhem, physics is completely independent of metaphysics. He was firmly convinced that studies in the history of physics confirm the point that the advance of physics was independent of changes in metaphysics. That does not mean it is dependent only upon sheer empirical facts. Contrary to Mach, Duhem did not think facts are simply *given*. Rather, Duhem contends that experiment in physics is not simply the observation of an occurrence: it is *its interpretation*. Without such a (theoretical) interpretation we could not use our instruments and make necessary measurements. Thus the result of experiment is "an abstract and symbolic judgment."

Duhem accepts, however, Mach's point that the principle of economy of thought plays the role of a governing rule in physical knowledge. Physical theory replaces a great number of laws with a few "fundamental hypotheses" from which mathematical deduction permits us to infer all physical laws. The reduction of a multitude of laws to a few principles affords, according to him, "enormous relief to the human mind," "intellectual economy" in which Mach correctly saw the goal and guiding principle of science. In fact, experimental laws already represent a great degree of intellectual economy, because in place of an imposing mass of specific facts, they elegantly substitute a single proposition.

Mach used his principle of the economy of thought in his theory of perception as well. Mach's epistemological position may be characterized as *presentativism*: objects (elements and their aggregates) are directly presented to us; perception of objects is unmediated by awareness of subjective sensations *different* from the objective properties of things. According to Mach, there is only one kind of entity, and to think so is more economic than to think that there are two different kinds of objects: subjective and objective. Duhem was much less interested in purely epistemological issues including those of the theory of perception. He was never absorbed in the physiology of perception and his philosophical notion of perception is much less advanced than that of Mach. His general position concerning the relation of things and sensations is similar to common sense *representativism*: things are represented by sensations.

Duhem treats a physical theory as an *abstract* system which classifies experimental laws supplied by experimental physics. Physical theory establishes links between these laws and summarizes them. Thus theories play a much more important role in physics than some physicists of the positivist persuasion thought earlier. Mach assumed that they are only of provisional value. Yet he was wrong: theories which contain fundamental hypotheses formulated in mathematical terms are part of end-science. Of course, they do not pretend to disclose an order transcending experience, and by summarizing, represent but do not explain experimental laws.

According to Newton, laws in physics "should be drawn from phenomena and generalized by induction." Duhem emphasizes, however, that in fact, Newton's own theory based on the principle of universal gravitation is not derivable by induction and generalization from Kepler's laws. Moreover, it formally contradicts these laws because gravitational interaction of the planets deflects them from their elliptic Keplerian orbits.

Mach certainly would accept this point of view. But he would reject another conclusion Duhem made from his analysis of methods of physics. According to Duhem, propositions in physics not only cannot be inductively inferred from facts—they cannot be individually falsified either. In Duhem's words, "[a]n experiment in physics can never condemn an isolated hypothesis but only a whole theoretical group" (Duhem 1954 [1906]: 183).

This conclusion is known today as Duhem-Quine's thesis. Some authors are inclined to maintain that it goes beyond the limits of positivist philosophy of science. From our point of view, they are wrong—we will return to this vexed question in the next chapter. Here we would like to emphasize that Duhem draws our attention to the fact that, in the process of inferring predictions and interpreting the results of experiment, we make use of a whole group of theories and hypotheses. Thus, if the predicted phenomenon does not occur, "not only is the proposition questioned [i.e., tested hypothesis] at fault, but so is the whole theoretical scaffolding used by the physicist. The only thing experiment teaches us is that among the propositions used to predict the phenomenon and to establish whether it would be produced, there is at least one error; but where the error lies is just what it does not tell us" (Duhem 1954 [1906]: 185). Thus, according to Duhem, only an "entire system of physical theory" may be submitted to experimental verification; to seek to isolate from this system an individual hypothesis for observational verification is to chase a chimera.

Duhem's approach to an analysis of a physical theory may be called purely holistic. It is this holism which makes Duhem's approach similar to that of Poincaré. Because we can compare with experience only the entire theory, not the individual parts of it, we are free (at least partly) to choose any part of that theory as a subject for modification in case the theory contradicts observational facts. As to the whole of physics, it is, according to Duhem, "a symbolic painting in which continual retouching gives greater comprehensiveness and unity, and the *whole* of which gives a picture resembling more and more the *whole* of experimental facts, whereas each detail of this picture cut off and isolated from the whole loses its meaning and no longer represents anything" (Duhem 1954 [1906]: 204–5).

Duhem's stance as to the role of conventions differs, however, from that of Poincaré and even more from Le Roy. They were convinced that certain fundamental propositions of physics cannot be removed from it in case theory contradicts experience, because these propositions constitute, in fact, definitions. The proposition that freefalling bodies fall with uniform acceleration is, according to conventionalists, a definition of what is meant by "falling freely." Hence, no experience can motivate us to change it. Yet Duhem disagrees: "The history of physics shows us that very often the human mind has been led to overthrow such principles completely, though they have been regarded by common consent for centuries as inviolable axioms, and to rebuild its physical theories on new hypotheses" (Duhem 1954 [1906]: 212).

Thus, according to Duhem, physical theories contain hypotheses and only hypotheses. It is this that motivates the description of his stance in the philosophy of science as hypothetism. And he was very eager to stress that, when formulating hypotheses, we base ourselves on mathematical formulas, not on imagination and visual models. Exactly because atomistic theory was based on such models, he was critical of it not much less than Mach and Poincaré.

Being hypotheses, even the most general propositions of a theory are fallible claims, although sheer experience cannot point out which of them must be replaced if theory contradicts observational facts. His stress on the hypothetical character of physics sets him apart from Mach: theories may be hypothetical but not descriptions of the *given*. The fate of Newtonian physics with its "inviolable axioms" made Duhem's hypothetism hard to ignore. As far as we know, Duhem made no reference to Hume when claiming that physics contains no infallible propositions. Yet he was definitely more akin to Hume's (and logical positivism's) point of view when claiming that all propositions of physics (contrary to those of mathematics) are hypotheses than to Poincaré's conventionalism. And although Duhem did not use formal logical methods of analysis of scientific knowledge, his emphasis on abstraction, mathematical symbolism, axiomatization, and logical deduction, but not on pure descriptions or intuition, also anticipates the typical positivist stance in the twentieth century. It can be argued that exactly he and not Mach or Poincaré came closest to the spirit of logical positivism.

Nearing the end of our examination of the views of Mach, Poincaré, and Duhem, let's revisit this question: is it reasonable to regard them all as positivists? It is difficult to answer this question unequivocally, but we can advance more arguments *pro* than *contra*. Drawing on the definition of positivism presented at the beginning of the book, it is necessary to state the following. Firstly, all three scholars looked at metaphysics critically. Mach presents a direct and open critic of metaphysics. Poincaré and Duhem deed not seem obliged to attack metaphysics directly, but they carefully avoided saying anything positive about it. And Duhem was emphatic in his belief that at least physics is entirely independent from metaphysics.

Secondly, positivism sought to become a scientific philosophy. Here we face some conceptual difficulties, because Mach, Poincaré, and Duhem were first and foremost scientists and not philosophers. In that respect they differ from Hume, Comte, Mill, Schlick, or Carnap. But their understanding of the aims, methods, structure, and foundations of science is very similar to that of many positivists. Their philosophical views were not as elaborated as those of, say, logical positivists. Yet the latter hold them in high esteem as theoreticians of science and were eager to stress that they strongly influenced the philosophy of science of the Vienna Circle. All three are mentioned in the famous manifesto of the Circle, *The Scientific Conception of the World: The Vienna Circle* (cf. Hahn, Neurath, and Carnap 1973 [1929]: 302–4). The quest of positivists to create a scientific philosophy was rather an ambitious aspiration. Following the criteria of scientific knowledge proposed by positivists, it would be difficult to qualify their reasoning concerning science as the science proper. That is quite natural: philosophical reasoning about physics is not physics, and such reasoning cannot be subjected to the same strict requirements as theories of physics may be. Yet even bearing all this in mind, we would like to emphasize that Mach tried at least to create a psychology of science whose scientific status is not more disputable than that of the logic of science logical positivists have been striving to devise.

The third constitutive principle of positivism is the conviction that only scientific knowledge is certain and reliable. From works of Mach, Poincaré, and Duhem we may conclude that they shared that conviction.

Fourth, positivists claim that all certain, i.e., scientific, knowledge about the world relies on experience. No doubt all three thinkers hold the view that experiment and observation make up the foundation of our knowledge. Many positivists treated mathematics as an important instrument of scientific inquiry. Mach, Poincaré, and Duhem are firmly convinced that mathematics plays an important role in scientific research. The fundamental outlook of Mach that science is the mathematical description of facts corresponds to the strictest standards of positivism.

Fifth, positivists believe in the unity of science. Mach upholds this belief. The views of Poincaré and Duhem on the issue are not so clear because in their own philosophy of science they avoided crossing the limits of natural sciences. Yet they never raised doubts about the methodological unity of science of the nature.

Sixth, the dichotomy between facts and values. Mach's position on the issue is absolutely clear: he is one of its most renowned proponents. Poncaré and Duhem upheld it in the area of natural sciences, although, it seems, they did not discuss its role in social sciences.

Seventh, the link between scientific and social progress. Mach emphasizes it, and Poincaré and Duhem also acknowledge that such a link exists.

We must conclude that Mach was no doubt a positivist. In fact, he has always been considered as such. At the time when empiriocriticism flourished, it has always been treated simply as a variety of positivism. Joseph Petzold, a disciple of Mach, regarded usually as his "representative" in Germany, founded there at the beginning of the twentieth century The Society of Positivist Philosophy. Logical positivists named after Mach a society they set up to spread their views. Poincaré and Duhem must be considered as thinkers close to positivism, although they did not even pretend to the title *Sontagsphilosopher* (Sunday philosopher) Mach liked to use when speaking of himself. Yet, as we already mentioned, Duhem's notion of the aims, nature, structure, and foundations of scientific knowledge is most akin to that of logical positivism.

CHAPTER 4

Modern or Logical Positivism

Revolution in Science and Philosophy

Positivism is a philosophy closely related to science. Thus it is quite natural that developments in science influenced the fate of both classical positivism and Mach's philosophy. The classical form of positivism was undermined by the general advance of scientific knowledge. In this process no scientific discipline played a specific, definitive role.

The variety of positivism which dominated at the end of the nineteenth and the beginning of the twentieth century—namely Mach's philosophy—relatively soon also gave way to a new *form* of positivism. Yet the emergence of this new positivism, which superseded Mach's, Poincaré's, and Duhem's views, was closely related to the developments within a *single* discipline. This discipline had made no great progress for more than the two thousand years that had elapsed since the death of Aristotle. Yet in the second half of the nineteenth century and at the beginning of the twentieth century it made a tremendous leap forward. I have logic in mind, of course.

Save for Mill, positivists paid little attention to this discipline. The main reason was, it seems, their treatment of Aristotelian logic as of something scholarly (i.e., Scholastic) and having only tangential significance for the scientific revolution of the seventeenth century and the advance of modern science in general. Aristotle's syllogistics was considered by Comte and other positivists *mainly* as a tool useful in metaphysical deliberations, but of low—if any—value for experiential scientific research.

Mill's attitude towards traditional logic is more sympathetic. In his *System of Logic* he examines the conventional problematic of logic: definition, syllogistic inference, etc. But, it seems, he does this more for systematic reasons and not because he is a great admirer of Aristotelian logic. Mill's main goal in this book is the development of inductive, non-Aristotelian, and nonsyllogistic logic which could be used effectively in empirical research.

Major developments in logic in the second half of the nineteenth and the beginning of the twentieth century took place, however, in the area of deductive, not inductive logic. The great advance in the development of logic was related to logic's overrunning the confines prescribed to it by the Aristotelian tradition. Thus propositions about the *relations* of subjects and not only about their properties became the object of intensive logical inquiry. Yet the most important innovation which broke wide open the compass of logical investigation and brought about interesting and important results was the *mathematization* of logic. The fundamental new idea which gained much support among logicians was that propositions and their logical relations may be described by using formulas in a similar way they are used in describing physical objects and their physical relations. By this procedure two disciplines—mathematics and logic—regarded earlier as quite distinct, were married and the new, mathematical logic was born. This innovation may be compared to the merger of mathematics and physics which opened the way to the scientific revolution of the seventeenth century.

The emergence of mathematical logic gave to researchers and scholars a powerful instrument for the analysis of *language* and first of all of the *language of science*. That opened vast new prospects for philosophy. The logical analysis of language became an extremely important new means for the examination of reason and cognition, which in the last centuries was a major task of philosophy, including positivist philosophy. Hume, Comte, Mill, Mach—all of them were deeply engaged in the examination of the human mind, although their lines of investigation were not exactly the same.

From the new perspective of the logical analysis of language, the mode of inquiry into human cognition characteristic both for Hume and Mach is far from ideal. They regarded cognition as a psychical, mental process which must be investigated in order to identify and examine sensations and to show how our mind produces complex ideas springing from simple ones produced by sensations directly. This process is subjective: we may investigate directly only how our *own* reason works and the main instrument of investigation is introspection. The results of such an investigation are uncertain for at least two reasons. First of all, it is rather difficult to identify elements of thought as mental entities by separating one element from another and keeping them constant in the process of investigation. Secondly, the results of such an investigation are not intersubjectively testable: no one has direct access to anyone else's mind.

Thus inquiry into the process of subjective cognition based on introspection does not meet the requirements of positive, scientific investigation. Within the positivist tradition these arguments were put forward already by Comte. His settlement was to renounce the kind of inquiry in which Hume was interested and examine the historical development of the human mind. Yet Comte neglected many important questions concerning cognition. When Mach started to examine the notion of fact and to purify experience, he faced the same old problems related to the treatment of cognition as a mental process.

As was already mentioned, the subtitle of Mach's *Knowledge and Error* is *Sketches on the Psychology of Inquiry*. He held logic in low regard and doubted its usefulness for the philosophical investigation of cognition. He, however, did not consider the possibility of using it to examine knowledge as it is presented in linguistic, not mental form. As it became clear later, new logic may be used for the clarification of concepts, for making propositions more precise and their relations more transparent. Mach's apprehension of the developments in mathematical logic was, it seems, very limited. Even if he were acquainted with it, he did not realize its significance for philosophy.

Grasping the philosophical importance of new scientific ideas and discoveries requires some time. As the history of non-Euclidean geometry shows, the mathematicians needed a few decades to recognize the significance of the new discovery. The significance of mathematical (formal) logic was widely—although by no means universally—acknowledged by philosophers only after the publication of an epochal work in this area, namely Bertrand Russell and Alfred North Whitehead's *Principia Mathematica*, the three volumes of which appeared in 1910–13. In this work they not only presented the main principles of mathematical logic and systematized its achievements, but also endeavored to show how mathematics, or, at least, arithmetic can be reduced to formal logic.

Russell, who, like Whitehead, was not only a logician but also a philosopher, understood quite well the importance of the new logic both for mathematics and philosophy. Russell put forward the main idea which catalyzed the new form positivist philosophy took in the twentieth century, and which is evidently distinct from the one it had in Mach's works. It is contained in his short but powerful claim that *logic is the essence of philosophy*. In *Our Knowledge of the External World* he expresses this claim in the following words: "[E]very philosophical problem, when it is subjected to the necessary analysis and purification, is found either to be not philosophical at all, or else to be, in the sense in which we are using the word, logical" (Russell 1969 [1914]: 42). He was convinced that the new formal logic may help to achieve *real* progress in philosophy: instead of simply *proposing* a new point of view or new doctrine which philosophers were doing for ages, this progress would lead to overcoming perplexities which tormented generations of philosophers, and eventually to *solving* old problems finally and irrevocably.

Russell cogently made his point by showing how some difficult philosophical problems may be dealt with. Let's consider his most famous example, "[t]he present King of France is bald." This statement is puzzling because France is (as it was in Russell's times) a republic and no longer has a king. Thus the statement attributes a property to a nonexisting entity. Is the statement true or false? If it is a statement, it must be either true or false. Yet may we speak about the properties of nonexisting entities at all?

Today, Russell's solution seems rather simple, as are almost all profound discoveries. Its essence may be best rendered in wording which is usually attributed to Wittgenstein. Both of them were convinced that it is necessary to distinguish between the apparent form of the proposition and its real logical form. Often they coincide, but in the given case they do not. The real form of the proposition is seen most clearly in the symbolic language. We may exhibit the real form of the proposition under examination by rewriting it as follows: "There exists one and only one x such that x is the present King of France and x is bald." This proposition is evidently false. What did Russell do? He transformed the perplexing proposition in which a *description* ("the present King of France") was used into a proposition which does not contain it. Another important feature of the new proposition is that it contains a quantifier of existence. This change of the proposition's form or, as Russell and Wittgenstein would say, bringing to light its real form, makes all perplexities vanish.

Russell presented his theory of descriptions in his famous article "On Denoting" (1905). The transformation of the proposition containing the description "the present King of France" into one which does not contain any description but contains a quantifier of existence is a paradigm case of what is called *logical analysis*: a clarification of the underlying logical structure of a vague expression.

Another of Russell's ideas which must be mentioned in a section called "Revolution in Science and Philosophy" was a bit more technical yet more directly related to the problems encountered by scientists, not philosophers. They faced them in the process of the so-called arithmetization of mathematics, which for many decades was progressing quite independently from the mathematization of logic.

The arithmetization of mathematics was brought about through attempts made by various mathematicians to give a strict mathematical form to the notions of differential and integral calculus, the major achievement of modern mathematics. The original definitions of these notions formulated by Newton and Leibniz were deemed unsatisfactory in the nineteenth century. To overcome this difficulty, the theory of mathematical series and that of limits was developed by Augustin-Louis Cauchy, Karl Weierstrass, and other nineteenth-century mathematicians. Their success showed that arithmetic is the groundwork of differential and integral calculus: the concept of natural numbers, which plays the definitive role in those theories, is a notion central to arithmetic.

After reducing differential and integral calculus to arithmetic, the leading mathematicians did not stop. They decided that for arithmetic itself, a foundation must also be provided, and came to the conclusion that the most fundamental mathematical theory is set theory developed at the end of the nineteenth century by Georg Cantor and others. The notions of a set and of an element (or member) of a set seemed more general than that of a natural number. Hope emerged that all mathematics could be based on set theory. If that could be accomplished, the much-sought *unity of all mathematics* would be achieved at last. The links between different mathematical disciplines would become clear, all mathematical notions would become precisely defined, and mathematical proofs would achieve their long-sought certainty. Without doubt, this would have been a revolutionary breakthrough in the most general and fundamental of all sciences. Yet when this Holy Grail of mathematics seemed to be within arm's reach, unexpected and devastating problems arose: several paradoxes (antinomies) in set theory were discovered.

Probably the most famous among them may be formulated as follows. Let's divide all sets into two classes: those which are not elements of themselves and those which contain themselves among their elements. The sets belonging to the first class are called *normal*: most sets are of this kind. Thus elements of the set of all planets are Mercury, Venus, Earth, etc. but not the set of all planets itself (it is possible to speak about subsets of the set of all planets but this is a different matter). On the other hand, the bibliography of all bibliographies must contain itself as an element: otherwise it will be not a bibliography of *all* bibliographies. The bibliography of all bibliographies is a set which is *not normal*.

Cantor took as a given the view that for every property there is a set of objects which have this property. Of course, in some cases this set is empty, i.e., contains no element because no object has the indicated property, but empty sets do not cause any serious problems. Therefore, it seemed, it is perfectly permissible to speak about "the set of all normal sets." Problems appear when we pose the question: "is the set of all normal sets itself a normal set?" It is not difficult to see that it is impossible to answer this question in a noncontradictory way. If this set is normal, then it is a member of the set of all normal sets, i.e., of itself, and that leads to the conclusion that it is *not* normal. If this set is not normal, it must contain itself among its elements, and because all elements of the set of all normal sets are normal sets, it must be normal. To cut to the chase: if this set is normal, it is notnormal, and if it is notnormal, it is normal. It is a clear-cut paradox. The popular equivalent of the perplexing question concerning the character of the set of all normal sets was formulated by Russell as follows: "Has the village barber who shaves those and only those villagers who do not shave themselves, the right to shave himself?" We cannot answer this question without contradicting ourselves.

The discovery of paradoxes in set theory—and their number snowballed in the first years of the twentieth century—undermined the foundations of the most general, abstract, and exact of the sciences. Some mathematicians, and especially Gottlob Frege, whose contribution to the investigation of the foundations of mathematics was particularly great, were in despair. All the results of their life's work lay scattered in ashes.

Russell's solution of this problem was as follows: it is necessary to make a clear-cut distinction between the different kinds of entities set theory deals with. One type of entity is individuals, another type—sets, the third type—sets of sets, etc. An individual may be an element of a set and a set may be an element of a set of sets but no entity of any type may be an element of an entity of the same type. Thus it is simply *inadmissible* to speak about sets which are (or are not) elements of themselves: neither can we pose any questions concerning such sets, nor make any statements. According to Russell, Cantor and others have been using the fundamental notions of set theory—namely those of an element and of a set—too carelessly. Mathematics (and other sciences also) requires that much more attention be paid to the language of science, to the way different terms may be linked with each other because not everything which, at first glance, seems a proposition is really a meaningful statement.

As we know now, there are different ways to eliminate paradoxes from set theory, not only by dividing all entities into different kinds as Russell proposed. Yet Russell's demonstration of the potency of a logical analysis made a great impression on scholars. The more so that practically at the same time, i.e., in the first years of the twentieth century, revolutionary changes were taking place in another discipline of major importance, namely physics. To be more precise, the crisis in physics broke earlier. It had many facets. Some *empirical* problems that Newtonian mechanics faced had been apparent for some time (there was a small but irresolvable discrepancy between the results of theoretical calculations of planetary motion and the empirical data obtained by observing Mercury's motion). Mach showed that classical physics based on Newton's notions of absolute time, absolute space, and absolute movement faces serious *conceptual* problems as well—it is impossible to define those notions properly. When constructing the special theory of relativity, which revolutionized physics by offering a new understanding of time, space, and motion, Einstein applied conceptual analysis also. He presented an analysis of the notion of simultaneity by showing that the real, empirical content of the notion is different from the content of the usual (Newtonian) idea of simultaneity.

These and other considerations seemed to indicate that conceptual, logical analysis is a powerful tool whose use yields important results. Now it was easy to conclude that if philosophy is not to lag behind science, it must make use of this method to the fullest extent. This tool had to be used for the analysis of scientific language and of language in general. We should recognize that at the beginning of the twentieth century, the awareness of the importance of the examination of language as a means of expression and communication was growing very fast in different areas of investigation.

Analysis as a method of philosophy was widely used in the empiricist-positivist tradition earlier. Yet the analysis practiced by the adherents of this tradition in the twentieth century differed substantially from that which prevailed in the earlier tradition. Firstly, the *objects* of analysis were different: linguistic expressions (concepts, propositions), rather than sensations, ideas, and other mental entities. Secondly, the *method* of analysis was different: logical inquiry, not introspection. Because of this, a hope arose that a new, *truly scientific* philosophy was about to emerge, a philosophy which will not only be oriented towards science, but which will employ a method which itself is an outgrowth of science. The new formal logic on which the method of logical analysis was based was a direct outcome of the advance of science, not of philosophy, as was transcendental or dialectical logic. Thus there were good reasons to expect that the new scientific philosophy based on this method may be of great value to science. The proponents of the method of logical analysis were convinced that in the scientific revolution of the twentieth century, they may play a role no less significant than that which philosophers played in the scientific revolution of the seventeenth century. In fact, both Russell and Einstein acted like philosophers in examining and solving con-

ceptual problems that scientific theories faced. Both were convinced that
their work must be continued. And at least Russell believed that advancing the revolution in philosophy—changing the objects philosophers
examine and the methods they are using—is a precondition of carrying
through the revolution in science.

The character of the new, scientific philosophy had to be radically different from the older types of philosophy. Philosophers of different ages
sought to produce philosophical theories as systems of statements. Yet
this new scientific philosophy does not have to seek to produce specific
philosophical pronouncements. It must serve to clarify problems and analyze existing assertions. In *Our Knowledge of the External World* (1914)
Russell shows full awareness of the significance of logical analysis for
revolutionizing philosophy: the method of clarification represents, according to Russell, an advance similar to that achieved in physics by Galileo.
Kant has compared earlier his exploits in philosophy with those of Copernicus in astronomy, thus Russell had to find a new point of reference.

The Vienna Circle and the Unity of Science Movement

The new understanding of the aims and methods of philosophy suggested
by Russell catalyzed the creation of a new form of positivism which differed greatly from that which we know from the works of Comte and Mill
or those of Mach. Russell's words "logic is the essence of philosophy"
became the credo of the new positivists. In the development of this new
form of positivism, the so-called Vienna Circle played the most important
role. It began its activities after World War I and was associated in a nonformal way with the University of Vienna, being the premises of the Circle's meetings, and many of its prominent members were professors there
too. It is often claimed that the Vienna Circle took shape in 1924 when
Moritz Schlick, who came from Kiel to teach philosophy at the University
of Vienna in 1922, organized a regular discussion circle upon the suggestion of his pupils Herbert Feigl and Friedrich Waismann. If the Vienna
Circle is defined, purely with the aim of identification, as *a circle in Vienna around Schlick*, the founding date of 1924 does not arouse controversy.

Yet there was a circle in Vienna, at which some basically empiricist
and positivist ideas were propounded and the significance of Russell's
new logic emphasized as early as 1907. When recently—because of the
revival of interest in logical positivism—not only the history but also the

prehistory of Schlick's circle became the subject of intense scrutiny, the nonformal group which met in a Vienna café (Café Josephinum) was designated as "the first Vienna Circle." The name "proto-circle" is also in currency. It was not homogenous and included some "catholic philosophers and romantic mystics" (Stadler 1997: 168). Yet its leading members were Philipp Frank, Otto Neurath, and Hans Hahn. All of them later joined the Schlick circle. Thus the designation "the first Vienna Circle" is not a complete misnomer.

Already before the First World War, all of the mentioned members of the "proto-circle" were rather critical towards traditional metaphysics and sought to bridge traditional empiricism and modern logic and mathematics. They were convinced that scientific discoveries are relevant to the fate of philosophical doctrines: the creation of non-Euclidean geometries and theory of relativity made Kant's theory of knowledge, held earlier in high esteem by the representatives of classical natural science and mathematics, rather obsolete. One of the first texts discussed in the proto-circle was Philipp Frank's article, "Law of Causality and Experience" (Frank 1907), in which he tried to synthesize Poincaré's conventionalist interpretation of this law with Hume's notion of causality. The proto-circle was also interested in reinterpreting Mach's views (cf. Stadler 1997: 170–71).

In 1912 Frank—the *spiritus movens* of the proto-circle—left for Prague to replace Einstein as professor of theoretical physics at German Charles-Ferdinand University and remained there until 1938. In 1914 the First World War brought the activities of the proto-circle to an end. Some of its members took direct part in the war. After being seriously wounded, Hahn started to teach at Bonn University and returned to the University of Vienna only in 1921.

Hahn was quite interested in continuing the discussions concerned with the foundations and nature of scientific knowledge which took place in the proto-circle, all the more that plenty of new and important developments appeared in science during the last years. Being a professor of mathematics, he was especially interested in Russell and Whitehead's *Principia Mathematica* and held a university seminar at which the results presented in this great book were discussed. Hahn was also very much interested in filling the chair of philosophy of inductive sciences which was held earlier by Mach and Boltzmann, but left vacant after the death of Friedrich Jodl in 1914. As an influential member of the Faculty of Philosophy, which covered in those years mathematics and physics as well, he was most instrumental in the University of Vienna for sending an official invitation to fill the chair (in spite of the opposition by many conservative professors)

to Moritz Schlick, who taught philosophy at universities in Rostock and
Kiel, was quite knowledgeable about modern science and became well
known after the publication of two books: *General Theory of Knowledge*
(*Allgemeine Erkenntnisslehre*) and *Space and Time in Modern Physics*
(*Raum und Zeit in der gegenwärtigen Physik*). They appeared, accordingly, in 1918 and 1919. It seems that Hahn was especially impressed by
the latter, in which Schlick presents a philosophical analysis—likely the
first—of Einstein's theory of relativity. Hahn's role in bringing Schlick to
Vienna allowed Frank to maintain that, in fact, he was the real founder of
the Vienna Circle.

As was already mentioned, Schlick came to Vienna in 1922 and in au-
tumn of 1924 the discussion group which only later was dubbed the "Vienna Circle" started to meet regularly on Thursdays at the premises of the
Mathematical Institute (Seminar) of the University of Vienna on Boltzmanngasse. Besides its initiators, Schlick and his pupils Feigl and Waismann, the principal participants were Hahn, the economist and sociologist
Otto Neurath, the legal theorist and social scientist with interests in philosophy and mathematics Felix Kaufmann, and the philosopher Viktor
Kraft. The logician Kurt Gödel took part in the meetings of the Circle
also. Even though based in Prague, the physicist Frank was a frequent
visitor and was regarded by all as a member of the Circle. The mathematician Richard von Mises, who moved to Berlin to become professor at
Berlin University, was also a frequent participant.

The circle around Schlick shared many of the interests and precepts of
the proto-circle. We will have the opportunity to examine the views of its
members later. The similarity of the theoretical positions of most of its
members, and especially, the very intensive exchange of the views during
the meetings of the circle, transformed the Circle quite soon into a kind of
philosophical school. Despite Schlick's leadership, though, it would be
wrong to define this school as Schlick's School in the sense the first
Academy was Plato's or the Lyceum Aristotle's. The relations between its
members were not those of a teacher and pupils. Rather, the Circle functioned very much like a contemporary philosophical conference, though
less formal. Presentations were made by different participants and they
were discussed freely. The expert knowledge they possessed in different
fields of science and philosophy was exploited widely, yet nobody had the
last word on any subject. Schlick's contribution was important but not
domineering. Meetings were relaxed and quite cordial. Members of the
circle did not seek to create a stable and rigid philosophical doctrine to
which they swore blood oaths. Rather, they sought a limited and flexible

consensus, and all these circumstances furthered the efficient development of the common body of beliefs which characterized the Vienna Circle.

The Circle's fame spread fast and many like-minded philosophers in Europe and America became interested in it enough to come to Vienna and take part in its meetings. The Circle was open to new participants and new ideas, yet such participation was invitation-only. Thus Popper, who boasted much later that he was *not* a member of the Vienna Circle, it seems, never got an invitation. In 1925, Schlick invited Rudolf Carnap from Jena University to address one of the Circle's meetings. He made a very good impression. The next year Carnap was habilitated at the University of Vienna with the work *Logical Structure of the World* (*Der logishe Aufbau der Welt*) and subsequently started to teach philosophy at the university. He became a member of the Circle and soon acquired the status of its *de facto* leader. He did not challenge Schlick's authority yet became a source of many of the new ideas which fired the imagination of other Circle members. The publication of his ambitious and impressive *Logical Structure of the World* in 1928 was an important event for the entire Circle. Its ideas were intensively discussed at the Circle's meetings, starting with Carnap's first appearance at Boltzmanngasse.

The main part of the work had been written outside the Circle: Carnap came to Vienna with the manuscript and did not change it much before publication. Another important work discussed at length in the Circle's meetings was Ludwig Wittgenstein's *Tractatus Logico-Philosophicus* (*Logisch-philosophische Abhandlung*). Being finished in 1918, it was published for the first time in 1921 in *Annalen der Naturphilosophie*. In 1922 it appeared in book form as a bilingual—German and English—publication. The Vienna Circle started to discuss *Tractatus* thoroughly in 1926, although some members corresponded with Wittgenstein beginning in 1924. Yet he never attended the Circle's meetings.

The relationship between Wittgenstein and the Vienna Circle is interesting and complicated. We cannot deal with it at length. Yet some points (and clarifications) have to be made. Some earlier authors believed that Wittgenstein's early philosophy—that of the *Tractatus* and the philosophy of the Vienna Circle—almost coincide. Even now it is still possible to come across statements that early Wittgenstein was a leading representative of logical positivism. This is not correct even though some logical positivists, especially in the late 1920s, drew heavily on Wittgenstein, and in the manifesto of the Vienna Circle which we shall discuss below, he is clearly identified as an author important for the Circle. Yet he is mentioned (at the end of the text) as only one of the three "leading representatives of the scientific world-conception" together with Einstein and Russell and is not included into two other lists added, that of the "Members of the Vienna Circle" and of "Those sympathetic to the Vienna Circle."

Although Wittgenstein's biography and philosophy are well known since he is one of the leading philosophers of the twentieth century, and we do not feel obliged to treat him systematically, we must recognize that the *Tractatus* was written under the inspiration of Russell and some of its central ideas to a greater or lesser degree parallel those advanced by Russell in his works which preceded the *Tractatus*, especially *The Principles of Mathematics* (1903), "On Denoting" (1905), "Knowledge by Acquaintance and Knowledge by Description" (1911), and *Our Knowledge of the External World as a Field for Scientific Method in Philosophy* (1914). Russell and Whitehead's *Principia Mathematica* also greatly influenced Wittgenstein.

Sometimes the *Tractatus* is regarded as a major work of the *philosophy of logical atomism*, which emerged at the beginning of the twentieth century due to the efforts of Russell and Wittgenstein. We regard this qualification as essentially correct. Anyhow, there is no doubt that the *Tractatus* is more closely linked to Russell's philosophy of the period, especially to "Lectures on the Philosophy of Logical Atomism" which appeared in *Monist* in 1918-19, than to the philosophy of the emerging Vienna Circle. Probably the most important difference between Wittgenstein and his Viennese colleagues is Wittgenstein's preoccupation with ontological questions. For him the analysis of the structure of language is a means to disclose the structure of the world. The members of the Vienna Circle were in general much more interested in the structure of science than in the structure of the world. And it seems they were not interested at all in many of issues which Wittgenstein himself regarded as extremely important, including those on the philosophical subject, the world as a whole, and the nature and status of the propositions concerning the limits of the language (cf. Hodges 1990).

However, Wittgenstein no doubt greatly influenced the Vienna Circle, especially as to understanding the functions of philosophy. We will touch upon these influences later. Yet it is important to note at once that Wittgenstein's impact was not sweeping. He was only one of the many sources upon which members of the Vienna Circle drew. We have already indicated two other sources whose significance is not much less than that of Wittgenstein: the proto-circle, especially Hahn, Frank, and Neurath, and, likely even more important, Schlick's early philosophy. We mean the philosophical views he held before coming to Vienna and establishing the first contacts with former members of the proto-circle and future members of the Vienna Circle. His *General Theory of Knowledge* was published earlier than the *Tractatus* and anticipated some of the ideas presented in Wittgenstein's work. More than once Schlick says in his book, albeit using different terms, things Wittgenstein stated in the *Tractatus*. Since in the late twenties the Vienna Circle mainly adopted Wittgenstein's terms, one can easily get a false impression that the latter is the progenitor of logical positivism. Part of this misunderstanding about Wittgenstein's place in the Vienna Circle may well have to do with Wittgenstein's outsize ego and self-portrayal. He treated his fellow philosophers as "philosophical journalists," sometimes making exaggerated claims of his own innovations, and rarely, if ever, crediting his teachers or anyone else as an influence on his work.

The aims, character, and perspectives of Schlick's and Wittgenstein's philosophy in the 1920s were different. Being a physicist-turned-philosopher, Schlick had a much more profound knowledge and understanding of empirical science than Wittgenstein, who, during his studies of engineering in Berlin-Charlottenburg and Manchester, was much more interested in the application of the theories of natural science than in the epistemic and foundational questions evoked by examining their nature. Wittgenstein's perspective in the *Tractatus* is mainly that of a logician; Schlick's in the *General Theory of Knowledge*—that of a physicist. Many important issues in the philosophy of physics which Schlick examined in his *Space and Time in Modern Science* and which were of major importance for the subsequent development of logical positivism's philosophy of science were practically beyond the ken of Wittgenstein.

After the Circle formed in 1924, correspondence with Wittgenstein was established. At the time, Wittgenstein, after giving away an impressive inheritance, was working as an elementary school teacher in Lower Austria. Direct contacts with him date as late as 1927. At that time the outlines of the philosophy of the Circle were already determined. In Wittgenstein members of the Circle saw a congenial philosopher and not some kind of philosophical prophet whose views were sacrosanct. Thus, in spite of the indisputable fact that the Vienna Circle took over some theses of Wittgenstein and was eager to repeat them in an original form, its views cannot be treated simply as an outgrowth of his philosophy. All the more that a number of thinkers—besides the above-mentioned—contributed to shaping the Circle's views, such as Mach, Poincaré, Duhem, Boltzmann, and Einstein, among others.

Even in cases where Wittgenstein is undoubtedly the direct source of the Circle's ideas, he was never happy with the Viennese interpretation of what he said in the *Tractatus*. This fact is widely known. Yet rather rarely is it seen as evidence of the clear *divergence* of their philosophical positions. Many of Wittgenstein's pronouncements simply did not fit the pre-existing theoretical framework produced by the Vienna Circle *independently* of Wittgenstein.

Some members of the Vienna Circle wanted to rewrite the *Tractatus* in their own language and to make it more amenable to the Circle's thinking. Waismann, who, like Schlick, maintained good personal relations with Wittgenstein, was chosen to perform that project. Waismann's book *Logic, Language, Philosophy* (*Logik, Sprache, Philosophie*) had to be the first in the series of monographs on the scientific world conception, edited by Frank and Schlick and published by Springer from 1928 to 1937. Notwithstanding great efforts made by Waismann, the project failed. According to the plan, Wittgenstein had to approve Waismann's text, but he was unhappy with every interpretation and alteration that Waismann suggested after consultations with Schlick and other members of the Circle. The publishing of the book was delayed many times, until Wittgenstein outright refused even to meet Waismann again to review the latest revisions. Waismann's book was published not in 1928, as was supposed, but only in 1976, after Waismann's death. Wittgenstein never approved the text.

This story tells much about the essential differences between the views of Wittgenstein and the Vienna Circle. They cannot be explained only by variations of their style. And by the 1940s, any kind of cooperation between Vienna Circle and Wittgenstein became impossible because Wittgenstein's views started to evolve towards what is usually called his late philosophy of ordinary language, presented in his posthumously published *Philosophical Investigations* (1953) and other texts issued later. Even still, some members of the Circle were also dissatisfied with Wittgenstein's early philosophy. Neurath once remarked that already in the first proposition of the *Tractatus* Wittgenstein speaks about the world, not science. Thus even though Wittgenstein criticizes metaphysics (albeit rather indirectly) in the *Tractatus*, he does, according to Neurath, what metaphysicians were doing for centuries and what scientific philosophy has no right to do—he seeks to disclose the nature and elements of the world. Some pronouncements Wittgenstein makes in the *Tractatus*, especially at its end, are, according to Neurath, not just metaphysical but, even worse, simply mystical. For a positivist like Neurath this is the strongest possible condemnation.

The period from 1924 until 1929 when Wittgenstein's influence over the Circle reached its zenith, is usually called the constitutional phase of the Vienna Circle. It came to an end with the publication of the text which has few analogies in the history of philosophical movements: a kind of manifesto. In 1929 Schlick received a very attractive invitation to become a professor at Bonn University. The temptation was heightened because far from all philosophers in Vienna were enthusiastic about the activities of the Circle and the ideas of its leader. Yet after some hesitation, Schlick rejected the offer. For members of the Circle it was clear that the main motivation for Schlick's decision was his unwillingness to leave a working and well-integrated collective. The time had arrived to lay dawn the principles upon which the work of the collective was based. While Schlick was a guest professor at Stanford, a text of a pamphlet was prepared by Neurath in collaboration with Hahn and Carnap and published in 1929 under the title *The Scientific Conception of the World: The Vienna Circle (Wissenschaftliche Weltauffassung: der Wiener Kreis)*. It was dedicated to Schlick and presented to him upon his return from the United States (cf. Hahn, Neurath, and Carnap 1973 [1929]: 299). After a short preface, the pamphlet starts with a review of the historical background of the Circle. It begins: "Many assert that metaphysical and theologizing thought is again on the increase today, not only in life but also in science. [...] But likewise the opposite spirit of enlightenment and *anti-metaphysical factual research* is growing stronger today" (ibid., 301).

The authors of the manifesto call this mode of thought which is "grounded in experience and averse to speculation" the "spirit of a scientific conception of the world." Among those who advocate its principles, they mention Russell, Whitehead, and James, and stress Mach and his role in the process of Vienna becoming a center in which the "epistemological and methodological problems of physics" together with "questions about the foundations of mathematics, problems of axiomatic, logistic and the like" were "frequently and assiduously" discussed. Poincaré's and Duhem's ideas as an object of discussion are also explicitly referenced.

Passing onto the question of the philosophical dispositions of the Vienna Circle, the manifesto expressly addresses the fact that a "growing uniformity" of their views was based on "the specifically scientific attitude: 'What can be said at all, can be said clearly' (Wittgenstein)" and "that a position not only free from metaphysics, but opposed to metaphysics was the common goal of all" (Hahn, Neurath, and Carnap 1973 [1929]: 304). The authors of the manifesto claim that the social attitudes of the members of the Circle also show a "noteworthy agreement, although these questions were not in the foreground of themes discussed within the Circle. For these attitudes are more closely related to the scientific world-conception than it might at first glance appear from a purely theoretical point of view. For instance, endeavors toward a new organization of economic and social relations, toward the unification of mankind, toward a reform of school and education, all show an inner link with the scientific world-conception" (ibid., 304–5).

The authors of the manifesto attribute even more significance to the goal for which Circle members strove: *unified science* (*Einheitswissenschaft*). "The endeavor is to link and harmonize the achievements of individual investigators in their various fields of science" (Hahn, Neurath, and Carnap 1973 [1929]: 306). Therefore, the manifesto stresses the importance of collective effort and intersubjectivity:

> [F]rom this springs the search for a neutral system of formulae, for a symbolism freed from the slang of historical languages; and also the search for a total system of concepts. Neatness and clarity are striven for, and dark distances and unfathomable depths rejected. In science there are no "depths"; there is surface everywhere: all experience forms a complex network, which cannot always be surveyed and can often be grasped only in parts, everything is accessible to man; and man is the measure of all things. Here is affinity with the Sophists, not with Platonists; with the Epicureans, not with the Pythagoreans; with all those who stand for earthly being and the here and now. The scientific world-conception knows *no unsolvable riddle*. Clarification of the traditional philosophical problems leads us partly to unmask them as pseudo-problems, and partly to transform them into empirical problems and thereby subject them to the judgment of experimental science. The task of philosophical work lies in this clarification of problems and assertions, not in the propounding of special "philosophical" pronouncements. The method of this clarification is that of *logical analysis*. (ibid.)

Such analysis shows, according to the manifesto, that the metaphysician and the theologian misunderstand themselves by simulating a theoretical content where none exists. Their statements "say nothing but merely express certain mood and spirit" (Hahn, Neurath, and Carnap 1973 [1929]: 307). They make two basic mistakes by producing statements about the world on the basis of the form of *ordinary* language and they wrongly believe that "thinking can either lead to knowledge out of its own resources without using any empirical material, or at least arrive at new contents by an inference from given states of affairs" (ibid., 308). Logical analysis enables philosophy to overcome not only an overt, undisguised metaphysics but also a hidden apriorist metaphysics which admits the possibility of synthetic *a priori* knowledge. "The scientific world-conception knows only empirical statements about things of all kinds, and analytic statements of logic and mathematics" (ibid.).

The publication of the Vienna Circle's manifesto heralded a new phase in the development of positivist philosophy. Although the manifesto itself was a product of the hard intellectual work done by the Vienna Circle during the previous five years, its appearance catalyzed the extremely intense organizational activity of the Circle in propagating its views. Positivism has always been animated by some kind of missionary zeal, something to which even the great skeptic Hume was hardly immune. The Viennese were no less zealous than Comte. And like him, they regarded such organizational activity as extremely important in executing their mission. They were eager to organize as many meetings and conferences as possible.

Yet instead of a Church of Positivism, they founded a more modern organization which was very helpful in popularizing their ideas. It was called the Ernst Mach Society (Verein Ernst Mach) and was founded in November 1928 "to further and disseminate the scientific world-conception" by organizing lectures and publications. It fulfilled its mission brilliantly.

The members of the Circle felt an urgent need to speak not only to a "wider public" but to the academic public as well. The manifesto was published not only because of the Circle's gratitude for Schlick's remaining in Vienna. It seems that another motive was more important: it had to be distributed among the members of the conference of the German Physical Society and the German Association of Mathematicians, which was to hold a conference in Prague in September 1929. The Circle (officially the Ernst Mach Society) decided to hold a conference on the epistemology of the exact sciences at the same time and place and use the opportunity to disseminate their scientific world-conception among the leading physicists and mathematicians, not only with pamphlets and presentations, but also via handshakes and café debates. The already warm relations with some leading German scientists made this "marketing" endeavor rather successful, and the Viennese won over to their side much more of Germany's scientific elite. They continued the practice of propagating their views on scientific conferences held by physicists and mathematicians later on.

On the other hand, they started to stage international conferences and congresses themselves, inviting leading scientists, and not only German ones, to take part. From 1930 until 1941, they managed to organize the Second Conference of the Epistemology of the Exact Sciences (Königsberg, 1930), the Preparatory Conference of International Congresses for Unity of Sciences (Prague, 1934), and six international congresses for

Unity of Science (Prague, 1934; Paris, 1934; Copenhagen, 1935; Paris, 1937; Cambridge, UK, 1938; Cambridge, USA, 1939; Chicago, 1941). All these meetings were convened by the Vienna Circle in close cooperation with other groups of like-minded scholars. The most important of them was the Berlin group, headed by Hans Reichenbach. The analogue of the Ernst Mach Society in Berlin was the Society for Empirical Philosophy founded in 1927 and renamed in 1931 as the Society for Scientific Philosophy. It functioned until 1933. Among its active members who later became famous the most noteworthy is Carl G. Hempel, who as a student in Vienna also took part in the activities of the Vienna Circle. Yet in the late twenties and early thirties among the members of the Berlin group Walter Dubislav and Kurt Grelling were much better known. Active in the Berlin Society were not only philosophers, logicians, and mathematicians but also physicians, psychologists, and psychoanalysts. The most famous among them was the mathematician David Hilbert. He proposed to change the name of the Society to make it more agreeable for logicians and mathematicians (cf. Stadler 1997: 82). Another mathematician, Richard von Mises, formerly a member of the proto-circle in Vienna and a frequent visitor to the Vienna Circle, was an even more active member of the Berlin group. Both the Vienna Circle and Berlin group maintained contacts with the Lvov-Warsaw (or simply Polish) school of logic represented primarily by Stanisław Leśniewski, Jan Łukasiewicz, and Alfred Tarski. Especially the ideas of the latter concerning the definition of truth were of great interest to Carnap and other members of the Vienna Circle. Many scholars from other countries were interested in the intellectual developments taking place in Vienna also, and became collaborators of the Circle. Among such collaborators were Joergen Joergensen from Denmark, Eino Kaila from Finland, A. J. Ayer from the United Kingdom, and W. V. O. Quine and Charles Morris from the United States.

Besides contacts with individual scholars and their groups, staging conferences and congresses, the journal *Erkenntnis* (*Knowledge*) published from 1930 played a crucial role in spreading the ideas of the "scientific world-conception." It was a joint enterprise of the Vienna Circle and the Berlin group. This fact was accentuated by making Carnap and Reichenbach its joint editors. Eight volumes of *Erkenntnis* were published. A significant component of the publishing activity of the Vienna Circle was a series of monographs edited by Frank and Schlick and published by Springer. The title of the series was *Scientific Conception of the World* (*Wissenschaftliche Weltauffassung*); ten books appeared under this title from 1928 to 1937. Another series, *Unified Science* (*Einheitswissenshaft*), was initiated by Neurath in 1933 and continued as the *Library of Unified Science* in 1939–41. Neurath was also editor-in-chief of the *International Encyclopedia of Unified Science*, the publication of which had begun in 1938 and continued until 1968 as *Foundations of the Unity of Science. Toward an International Encyclopedia of Unified Science*. The full list of monographs published in all these series may be found in (Stadler 1997: 656–59).

The reader probably has noted that the phrase "scientific conception of the world" appearing in the titles of the series was replaced after a few years by "unified science" and "unity of science." We came across the notion of *unified science* already in the manifesto of the Vienna Circle. *Unified science* is a key notion of positivism in general, but it played an especially important role in the twentieth century. Unified science was the goal for which members of the Vienna Circle strove, a goal that attracted most collaborators among philosophers, and a goal that drew towards the Circle the most scientists (including such celebrities as Niels Bohr), who took part in the publishing activities, conferences, and congresses we have overviewed.

It is possible to speak in a sense about unified science existing in the eighteenth and the first half of the nineteenth centuries. At that time its theoretical and methodological basis was Newtonian physics. All other sciences sought to imitate it by constructing mechanical models of phenomena under investigation and applying notions and laws of mechanics to explain how they work. Yet at the turn of the twentieth century this mechanical picture of the world was shattered. Logical positivists felt the foundations of a united and consistent world picture crumbling, and saw this as a grave danger to the very goals science was aiming at, and moreover, to the very existence of science as an intellectual endeavor guided by generally agreed principles. They wanted to restore the unity to science which it had lost. This was the great intellectual and practical goal behind all the disparate activities of the unified science movement. Here logical positivists held themselves as the guardians of the centuries-old and extremely important tradition which brought about many achievements of science and of civilization in general. The fragmentation of science into small and closed principalities, the residents of which share no language, was a disaster which had to be averted at any cost.

The creation of unified science was, in fact, one of the two main goals Circle members sought. The second was to exorcise science of all metaphysics. Both were closely related. Firstly, metaphysics, according to the

Circle, was one of the most important stumbling blocks on the road to the unity of science. Various sciences were contaminated with metaphysical ideas of different origin, stemming from different metaphysical systems. Without removing them from science, any effort to establish its unity is doomed to fail. Secondly, the same instrument must be used both to eliminate metaphysics and to unify science. This instrument is logical analysis. With the aid of logical analysis it is possible, as members of the Circle and their collaborators thought, to show that metaphysics has no cognitive meaning, and thus to demonstrate that its ideas are alien to science. On the other hand, by disclosing the real logical structure of scientific notions and propositions, logical analysis must help to reveal their empirical foundations. And because the Viennese were convinced, like Mach or Hume, that human experience is, in fact, integral and that all notions of science might be defined in empirical terms (or, at least, related to them) using the same procedure, establishing the real empirical foundations of scientific knowledge means establishing its *common* basis and through it ensuring the unity of science.

Was not the idea of the unity of science an *ideé fixe* of modern positivism? Does it not resemble a bit the idea of unification and systematization of humankind so cherished by Comte and which Mill viciously derided? It seems that at the beginning of the third millennium we—having survived Nazi and communist totalitarianism—*must* be less eager to unify everything. Yet the authors of the manifesto *Scientific Conception of the World: The Vienna Circle* had no doubt that the unity of science (and that of mankind) is the most noble and important goal. Neurath was the moving force behind this idea, but many others supported it no less vehemently.

The unified science logical positivists spoke about, i.e., science in which various sciences have close links, had to explain or at least systematize and describe all facts in a unified way. At least according to Mach, such description is more economic than any other. At the same time it is more lucid and transparent, thus complying with Russell's and Wittgenstein's ideals of clarity. Moreover, unified science is a science whose methodology is rather simple. In such a science it is not difficult to pursue scientific inquiry in different areas at once, or to jump around from one field of inquiry to another. It is rather simple to apply the results of scientific research within the confines of unified science because of the unification of methodology. This unification gives an opportunity—and this is extremely important—to restructure less developed fields of research, say social sciences, taking as a model the exact natural sciences, thus ensuring the faster growth, precision, and reliability of scientific knowledge. And, last but not least, unified science is a science which overcomes its fragmentation into ever smaller fields with their own scientific "dialects" or "languages" understandable only to the initiated. It is science whose language is unified, thus easily comprehensible. It does not require a high degree of sophistication either on the part those who practice it, or, all the more, on the part of those who want to comprehend it. This is in full accord with the democratic requirement to make science accessible to a broad public. The battle against metaphysics which the positive mind fought for centuries almost always had a tint of the political or a whiff of the missionary about it. The maxim "In science there are no 'depths'; there is surface everywhere" is undoubtedly a democratic, albeit a bit vulgar and ordinary, principle. In the Vienna Circle it matched well with the Marxist predilections of many of its members. No "insight," "perceptivity," or "intuition" is needed in unified science: anyone might become its practitioner. Of course, not everyone involved in the "unity of science" considered all the above-mentioned reasons important or even shared them. Yet meetings and publications under the heading "unity of science" attracted wide attention, which testified that many scientists and philosophers felt that the idea was compelling, important, and worth their dedication.

The movement for "unity of science" with its international congresses and encyclopedia was a specific form of the positivist movement. But many members of the Vienna Circle did not like the term *positivism* and regarded it as rather inappropriate (this was felt less in Berlin where Joseph Petzold in 1912 founded the Society for Positivist Philosophy and won over to his side Hilbert, Felix Klein, Mach, Einstein, Max von Laue, and Freud). The Viennese linked the term "positivism" too closely, it seems, with Comte. They especially wanted to dissociate themselves from all of Comte's social, political, and moral ideas. They had some reservations about his philosophy of science as well. First of all they regarded as totally misguided Comte's antipathy to logic. And they did not want to become one more sect in philosophy, one more "ism." They hoped to overcome all "isms" once and for all. Precisely for this reason the name consistently used in their manifesto is *The Scientific Conception of the World*, which seems neutral and not philosophical at all. The title of this chapter, namely "logical positivism," was used it seems, for the first time in the article published by Albert Blumberg and Feigl in the *Journal of Philosophy* under the title "Logical Positivism: A New Movement in European Philosophy" (Blumberg and Feigl 1930). Logical positivists in Vienna liked this term more than "positivism," yet they used it reluctantly, preferring rather the term "logical empiricism" employed by them quite widely, especially after the end of the Second World War. I will treat the terms *logical positivism* and *logical empiricism* as synonymous, although I fully agree with Thomas Uebel that "there is no systematic unity of usage of the terms 'logical positivism' and 'logical empiricism' even among participants and early critics of the tradition in question" (Uebel 2013: 85). In the last years the term *logical empiricism* is used a bit more often but in this book when talking about the early and later views of the members of the Vienna Circle, the Berlin group, and their collaborators, I will prefer the term *logical positivism*, because my intention is to emphasize that this philosophy is a form of positivism. The term "neopositivism," broadly used in the mid-twentieth century when referring to logical positivism and sometimes also to more or less kindred authors, is rather rarely applied today; it figures, however, in the title of a commendable book written by Rudolf Haller (1993).

As was mentioned already, logical positivists spared no effort in popularizing their ideas both within the confines of the scientific community and among the public at large. Yet until the Second World War, logical positivism was not a widely known and admired philosophy. At least two related factors must be mentioned in this connection. One is linguistic, another political.

At the end of the twenties and early thirties almost all publications by logical positivists were published in German. German was much better known in Europe in those days than now. It was not spoken much, however, outside Central, Northern, and Eastern Europe, and this circumstance limited logical positivism's reception mainly to this region. However, even in this region the spread of logical positivism was thwarted by political forces. Totalitarian regimes: communist in the Soviet Union, national-socialist in Germany, and fascist in Italy were hostile to all alien philosophies. In the late twenties the ideas of the Vienna Circle found some resonance in the Soviet Union, yet the later strangle hold of the Stalinist regime which rejected any unorthodox thought, made their spread—in spite of the sympathy many of the members of the Vienna Circle felt to Marxism—impossible. Hitler's rise to power in 1933 made the continuation of activities of logical positivism in Germany inconceivable. Notwithstanding the fact that Nazism intellectually was a bit more flexible than communism, logical positivism was rejected at once for three main reasons: (1) its opposition to any vague and metaphysical (or mystical) ideas which were abundant in Nazi ideology; (2) its closeness to the Social-Democrat Party and/or "rotten" liberalism; (3) the Jewish origin of many of its representatives. The activities of the Society for Scientific Philosophy in Berlin were abruptly ended in 1933. After 1934 logical positivists faced political difficulties also in Austria, whose authorities terminated the activities of the Ernst Mach Society the same year they banned the Social-Democrat Party. Thus logical positivism had only a few years for its free development. It is really amazing how much it managed to accomplish in those few prewar years in continental Europe. Yet logical positivism acquired its real fame only after the war. And it was achieved mainly not in continental Europe but in the Anglo-American world.

Emigration before and after the Second World War contributed heavily to the shift of activities of logical positivists to the United States and, to a lesser degree, to the United Kingdom. Practically all members of the Vienna Circle (except, it seems, Kraft) left Vienna. The first was Feigl, who left for Harvard on a Rockefeller research fellowship in 1930, and remained in the USA for good. From 1940 he was professor of philosophy at the University of Minnesota, where in 1953 he founded and for many years headed the influential Minnesota Center for the Philosophy of Science. In 1934, Neurath left for the Netherlands and in 1937 founded the Institute for the Unity of Science in The Hague. In 1940 he landed in the UK, were he taught at Oxford from 1941 until his death in 1945. Carnap came to the United States in 1936 to become professor of philosophy at the University of Chicago, and in 1954–61 he was professor at the University of California, Los Angeles. As a visiting professor he taught at Harvard and did research at the Princeton Institute for Advanced Study. Carnap was followed by Frank, who ended up at Harvard, teaching and researching there until 1953. As one of the most dedicated—besides Neurath and Carnap—champions of the unity of science movement, he founded in 1948 the Institute for the Unity of Science, which he headed until 1965. The well-known Boston Colloquium for the Philosophy of Science is the successor to this institute. Waissman came to the UK in 1937, taught two years at Cambridge and later at Oxford. Also in 1937, Hempel came to the United States working, consequentially, at universities and colleges in Chicago, New York, New Haven (Yale), Princeton, and Pittsburgh. From 1933 until 1938, Reichenbach was professor of philosophy at the Istanbul University in Turkey and from 1938 until his death in 1953 at UCLA.

Thus in the postwar years logical positivists found themselves at many leading American and British universities. It is necessary to add to emigrants from Vienna, Berlin, and Prague at least a few of their American and British collaborators: Ayer teaching at Oxford and London, Quine at Harvard, Morris at Chicago, all comprising an even fuller picture of the influence logical positivism exerted on academic circles in the USA and UK. Philosophy of science soon became a standard course taken by many students majoring in various disciplines at American and British universities. Until the late sixties, the philosophy of science taught there was, as a rule, the philosophy of science advanced by logical positivists. Logical positivism was a kindred philosophy to American pragmatism which dominated the American philosophical scene until the war, and was interpreted quite often as a modern scientific form of pragmatism. This explains why logical positivism spread like wildfire in postwar America, a fire which only started to die out in the late sixties.

Now, after inspecting the origins of logical positivism, the circumstances of its emergence, its organizational activities, and reporting its historical destiny, it is time to move on to the next issue: a closer examination of the main doctrines of a philosophy whose representatives quite often denied that they are philosophers and even more often pretended that they do not advance any doctrines. The most usual way to do this is to examine these doctrines as parts of a more or less integral and united philosophy called logical positivism or logical empiricism. This approach has many evident advantages. First of all, most of the members of the Vienna Circle and their collaborators were much more eager to emphasize similarities and not differences of their views. This is quite natural if we take into account the rather hostile—both intellectually and later also politically—environment in which they had to act. Such an environment furthers cooperation between the members of the group rather than strife among them. This situation reminds one a bit of John Stuart Mill's unwillingness to criticize Auguste Comte's views in the 1840s, when positions of classical positivism were still not very strong. Logical empiricists supported the doctrine of unified science; thus, it would be rather strange if they emphasized differences in their own views. Debates on some questions were taking place, but it was expected that eventually the unified position would be achieved. It is important also to recall the fact that from the beginning the Vienna Circle was regarded by its members as a cooperative venture and as the core of the united movement. The tendency to emphasize similarities of their views is evident not only in the Manifest of 1929 (*Scientific Conception of the World: The Vienna Circle*), but also in the later texts written by logical positivists in which they tried to contemplate what logical positivism or scientific philosophy of science is, such as *Kleines Lehrbuch des Positivismus*, published in 1939 by Richard von Mises and translated in 1951 into English as *Positivism* (von Mises 1968 [1939]), Philipp Frank's *Modern Science and Its Philosophy* (1949), Viktor Kraft's *Der Wiener Kreis. Der Ursprung des Neopositivismus*, issued in 1950 and thereafter translated into English as *The Vienna Circle: The Origin of Neo-Positivism* (Kraft 1953 [1950]), and Joergen Joergenson's *The Development of Logical Empiricism* (Joergensen 1951).

The main disadvantage of treating logical positivism in this way is, however, rather significant. I mean the one-dimensionality of such a treatment, and neglecting the tensions which existed among various members of the movement and between notions advanced by them. Therefore, I chose an alternate way of presenting the philosophy of logical positivism. Two figures were most important in it: Schlick and Carnap. The former founded the Vienna Circle and exerted decisive influence on the development of its views in the twenties, and the latter became its *intellectual* leader subsequently, and no one of significance called into question his role as the most representative figure in the movement until his death in 1970. After his death some last Mohicans—Ayer, Hempel—were still expanding the ideas of the movement (they were active in philosophy until the 1990s), yet they lacked, it seems, his vast enthusiasm and exceptional analytic abilities.

Here, it seems, is a suitable place to take a stance on a view recently fashionable among scholars writing about logical empiricism and especially its Vienna Circle phase. This view became influential with the appearance of Thomas Uebel's book *Overcoming Logical Empiricism from Within: The Emergence of Neurath's Naturalism in the Vienna Circle's Protocol Sentence Debate* (Uebel 1992). I agree with Uebel that Neurath's role in *this* debate was really important. But I would completely disagree with the opinion that Neurath was more significant and influential member of the Vienna Circle than Carnap and that Carnap must be regarded as a figure somewhere only in transition *between* Schlick and Neurath. This opinion, it seems, is expressed by Michael Friedman in his book *Reconsidering Logical Positivism* (1999). In the preface of the book he stamps Moritz Schlick as "the founder and guiding spirit of the Vienna Circle," Otto Neurath as "a second leading member of the Vienna Circle," and claims that Rudolf Carnap was "the third leading member" of the Circle (Friedman 1999: xi-xii). From my point of view, Carnap's role in the Circle was much more important than that of "the third leading member." By the 1930s he and not Moritz

Schlick was *the* steering mind of the Circle—Carnap's articles, "The Elimination of Metaphysics through Logical Analysis of Language" (1932), "Psychology in Physical Language" (1933), "Formal and Factual Science" (1935), and his studies *The Logical Syntax of Language* (1934), *Philosophy and Logical Syntax* (1935), and *Testability and Meaning* (1936/37) were of crucial importance for shaping the views of logical positivists and defining the direction of their further development. It seems to me that the role of Neurath in the Vienna Circle, and in logical empiricism in general, having been underestimated in the 1970s and 1980s, recently became evidently overestimated. In fact, Neurath was more interested in economics and sociology (trying to amalgamate in those disciplines positivist and Marxist ideas) than in logical analysis and philosophy in general. Carnap, on the other hand, was one of the greatest philosophers of the twentieth century—in this respect I fully agree with Quine. If I had to write a new book on logical positivism—I wrote two monographs on the subject earlier concentrating mainly on the problem of induction and probability (Nekrašas 1979, 1987)—and not a just a medium-sized chapter of a book as in the present case, I would include much more material on Neurath, but he would *at best* be my number three and not two. And, of course, almost half of the text of this imagined book would have to be devoted to Carnap's contribution to the development of that remarkable philosophy which he did not want to call philosophy at all.

The relation of Carnap to Schlick bears a distant similarity to the relation of Mill to Comte. Both Mill and Carnap acknowledged the authority of the founding father of the school, yet did not feel this authority overweening and sacrosanct. Both developed important new ideas not advocated by, accordingly, Comte and Schlick. Both are now better known, especially in the Anglo-American world, than the founders of social and logical positivism. (Ayer is probably better known than Carnap because of his willingness and aptitude to seek the popularity which Carnap fell short of, yet from a philosophical point of view Carnap was much more innovative.)

The tension between Carnap and Schlick is, however, less intense than the conflict between Mill and Comte. One of the reasons is that Mill developed his positivist ideas at least partly independently from Comte. They never met. And they advanced their ideas in quite different countries. On the other hand, Carnap was invited to come to Vienna by Schlick, they both were Germans, and, more important, they collaborated in the Circle closely for ten years, from 1926 until 1936. And yet their ideas even during their close cooperation were far from identical. An examination of their philosophies gives us an opportunity to form a more vivid picture of logical positivism, to put it in relief and to reveal its two different faces.

When presenting their views I do not seek, however, to describe them in a very systematic and self-sufficient way. My main aim is to present a fair and reliable general picture of logical positivism. Therefore, issues dealt with in the section on Schlick do not entirely overlap those dealt with in the section on Carnap, although both authors had many points of general interest. The issues to be examined in both sections were selected according to the special predilections of both logical positivists.

However, Carnap's name will be mentioned often when examining Schlick's views and vice versa. Other representatives of logical positivism will also not be elided. Thus it is impossible to avoid, say, Neurath, when dealing with the differences between Schlick and Carnap. He played quite an important role in the Vienna Circle as probably the most radical representative of the faction of the Circle, opposing a rather moderate Schlick. Neurath influenced Carnap, who also belonged to that radical wing, although some modern authors such as Nancy Cartwright (cf. Cartwright et al. 1996) overestimate, it seems, his influence on Carnap. The approach chosen in this book allows us to touch upon almost all the main subjects in the philosophy of science discussed at the Circle, and on the other hand, to show that logical positivism was hardly the mutual admiration society like some of its critics portrayed it.

Moritz Schlick

Moritz Schlick was born in 1882 in Berlin into the family of a factory owner of aristocratic descent. He showed interest for philosophy as early as high school, but his main subject of studies at the universities of Heidelberg, Lausanne, and Berlin was physics. In 1904 he received his Ph.D. in physics by defending a thesis on the reflection of light under the supervision of Max Planck. Over the next five years he continued his studies in the natural sciences and psychology in Göttingen, Heidelberg, Berlin, and Zurich. Gradually he became seriously interested in philosophy. The first result of his studies in this field was, however, a work dealing with morals and philosophy of life (*Wisdom of Life*, published in 1908), not with epistemology or philosophy of science. In 1911 Schlick started to teach philosophy at Rostock University in Germany. Over ten years in Rostock, he published a weighty book on epistemology under the title *General Theory of Knowledge* in 1918 (its second edition appeared in 1925) and, seem-ingly, the first monograph in the world dealing with philosophical issues related to Einstein's theory of relativity under the title *Space and Time in Modern Physics* in 1919, based on an article with the same title published two years earlier. He corresponded with Einstein and later they became friends. Schlick maintained very close relations with many other leading scientists of his age.

In 1921 he came to University of Kiel as full professor and a year later accepted an invitation to the University of Vienna where he worked until 1936, when he was shot dead by his former student on a staircase of the university building. The motives of the murder were partly personal (the former student likely also had psychiatric problems), as well as partly ideological and political—Schlick was a well-known liberal. After the *Anschluss*, i.e., Austria's incorporation into Germany, the Nazis paroled the murderer after he served only a small part of the sentence, and he soon became a member of the Austrian Nazi Party (cf. Stadler 1997: 776).

Philosophy as the Pursuit of Meaning

Our aim here is not a full and comprehensive presentation of Schlick's philosophy. Thus his early philosophical views, i.e., those he held before coming to Vienna, will be referred to only occasionally in the context of the examination of his Vienna-period philosophy. Even still, they do over-lap in content with his later views to a rather great degree, although his language changed a bit under the influence of Wittgenstein, through dis-cussions in the Vienna Circle, and with the dynamic developments in the field. Thus the fundamental thesis for logical positivism that all scientific statements are either analytic and *a priori* or synthetic and *a posteriori* was already formulated by Schlick in the *General Theory of Knowledge* (1918), and the equally important issue of the link between verification and meaning was discussed in this book at length also.

Schlick's main contribution to logical positivism does not concern technical issues of the logic of science. He was interested mainly in epis-temology and practically did not use the apparatus of formal logic in his publications. And as Karl Menger, a mathematician and a member of the Vienna Circle, noted, Schlick's interest in logic and mathematics, which was quite strong in the twenties, was fading rather fast by the thirties (cf. Menger 1982: 94–95).

That does not mean that Schlick was unaware that paths which led to, in his own words, a "decisive turning point in philosophy" have their origin in logic. Yet in his programmatic article "The Turning Point in Philosophy" opening the first volume of *Erkenntnis*, he is eager to emphasize that "[h]owever highly the value of the new [logical] methods is to be esteemed, it is plain that nothing so fundamental can be brought about by the mere development of a method. The great turning point is therefore not to be attributed to logic itself but to something quite different which was indeed stimulated and made possible by it, but which proceeds on a much deeper level: the insight into the nature of logic itself" (Schlick 1959 [1930]: 55). This insight requires the introduction of far-reaching modifications into epistemology, and by implication, into all of philosophy.

Of decisive importance is, according to Schlick, the understanding that all our knowledge is such only by virtue of its form, because it represents facts known only through its form. Everything else in the expression is unessential.

> This simple insight, has consequences of the very greatest importance. Above all, it enables us to dispose of the traditional problems of "the theory of knowledge." Investigations concerning the human "capacity of knowledge," in so far as they do not become part of psychology, are replaced by considerations regarding the nature of expression, of representation, i.e. concerning every possible language in the most general sense of the term. Questions regarding the "validity and limits of knowledge" disappear. Everything is knowable which can be expressed, and this is the total subject matter concerning which meaningful questions can be raised. There are consequently no questions which are in principle unanswerable, no problems which are in principle insoluble. What have been considered such up to now are not genuine questions but meaningless sequences of words. (Schlick 1959 [1930]: 55–56).

Juxtaposing the goals of science and philosophy, Schlick emphasizes that philosophy—which he still held to be the Queen of Sciences—is not a system of cognitions or statements, but an activity through which the meaning of statements is revealed. "By means of philosophy statements are explained; by means of science they are verified. The latter is concerned with the truth of statements, the former with what they actually mean. The content, soul and spirit of science is lodged naturally in what in the last analysis its statements actually mean; the philosophical activity of giving meaning is therefore the Alpha and Omega of all scientific knowledge" (Schlick 1959 [1930]: 56).

Holding up the role of philosophy for science, Schlick—unlike Carnap and most members of the Vienna Circle—contended that it is inadmissible to combine them, because philosophy is not a science. He, like Wittgenstein, was convinced that there can be no *science* of language, a science which Carnap was seeking to create in the form of logical syntax and later in the form of logical semantics. In logical syntax and semantics, propositions about language are formulated. Yet by emphasizing the difference between philosophical and scientific attitudes, Schlick negated the very possibility of formulating philosophical propositions at all. In his article "The Future of Philosophy" he wrote: "The view which I am advocating has at the present time been most clearly expressed by Ludwig Wittgenstein; he states his point in these sentences: 'The object of philosophy is the logical clarification of thoughts. Philosophy is not theory but an activity. The result of philosophy is not a number of philosophical propositions, but to make propositions clear'" (Schlick 1979 [1931]: 222).

The difference between Schlick's and Carnap's points of view may be explained by the fact that Schlick drew a much more clear distinction between logic and philosophy than Carnap. According to the latter, there is no sharp line between them. Carnap held that logic in the form of mathematical logic is a science, and believed—with Russell—that "Logic is the essence of philosophy." Thus, Carnap had no means to oppose philosophy and science as sharply as did Schlick. For Carnap, practicing philosophy as logical analysis requires a philosopher to engage into *formulating propositions* and *constructing (scientific) theories*.

Schlick was of different opinion. For him, the object of philosophy is the discovery of meaning, and the object of science is the discovery of truth. "Science should be defined as the '*pursuit of truth*' and philosophy as the '*pursuit of meaning*'" (Schlick 1979 [1931]: 217). He claims that the method of the sciences and the method of philosophy are radically different. The method of logic *is* the method of science. As to the method of philosophy, Schlick refers, very characteristically not to the method of logic, as would Carnap, but to Socrates. According to him, Socrates is an example *par excellence* of a philosopher who clearly understood which is the proper method of philosophy. Socrates asks in *The Republic* "what is justice?" and—Schlick emphasizes—receives "various answers to this question, and in turn he asks what was meant by these answers, why a particular word was used in this way or that way, and it usually turns out that his disciple or opponent is not at all clear about his own opinion. In short, Socrates' philosophy consists of what we may call 'The Pursuit of Meaning.' He tried to clarify thought by analyzing the meaning of our expressions and the real sense of our propositions" (ibid.). Carnap has never referred to Socrates as a philosopher worth imitation. By the way, Carnap sometimes used the expression "scientific philosophy," which Schlick consequently tried to avoid.

In full accord with his concept of philosophy, Schlick has claimed that there can be no science of meaning because it is impossible to formulate true propositions on meaning. Accordingly, he saw only two results to which the "pursuit of meaning" can lead. "The fate of all 'philosophical problems' is this: Some of them will disappear by being shown to be mistakes and misunderstandings of our language and the others will be found to be ordinary scientific questions in disguise" (Schlick 1979 [1931]: 219). Far from all logical positivists would agree that the results of philosophical analysis are purely negative. At least Carnap would deny that his work on logical syntax and semantics (or, later, on the theory of confirmation of scientific hypotheses) is nothing more than "dissolving philosophical problems." He would disagree also with Schlick's claim (echoing in fact Comte's) that such a field of study as ethics may become a part of "the great system of sciences."

We must conclude that Schlick's notion of philosophy as the pursuit of meaning is far from being identical with Carnap's notion of philosophy as logical analysis.

Positivism and Realism

In exploring the nature of positivist philosophy, Schlick paid much attention to explaining its relation not only to science but to realist and idealist philosophy as well. In his seminal article "Positivism and Realism," published, like "The Turning Point in Philosophy," in *Erkenntnis*, Schlick starts with the definition of positivism as every view which denies the possibility of metaphysics. (This definition seems for me a bit too broad.) He acknowledges that in this sense he is a strict positivist. Yet he stresses that this definition may be misinterpreted. Namely, if metaphysics is understood as the theory of "true being," this implies a contradictory and spurious "apparent being." The difference between these two modes of being is usually explained by indicating that "appearances are immediately present, 'given' to us, while metaphysical reality must be inferred from them in some roundabout manner. And thus we seem to arrive at a fundamental concept of the positivists, for they always speak of the 'given,' and usually formulate their fundamental principle in the proposition that the philosopher as well as the scientist must always remain within the given, that to go beyond it, as the metaphysician attempts, is impossible or senseless" (Schlick 1959 [1932/33]: 83).

This point of view looks like a recommendation following from the definition of positivism. Yet according to Schlick, stating the definition in

these terms leads to "dangerous errors" giving false pretense to regard
positivism as a kind of metaphysics, from which the transcendent has been
removed. The most misleading formulation of the positivist view is "Only
the given is real," which might also be rendered as "There is only the
given." It was advocated by some representatives of positivism and—
more often—presented by its critics as a common view shared by all positivists. "We must insist, however," claims Schlick, "that whoever states
this proposition seeks to establish an assertion which is metaphysical in
exactly the same sense and degree as its apparent contradictory: 'There is
a transcendent reality'" (Schlick 1959 [1932/33]: 84–85). Positivism's
stance is different both from that of *metaphysical realism*, which believes
in the reality of the transcendent external, and that of *metaphysical idealism*, which claims that there is no being *outside* of consciousness. Schlick,
who is especially interested in exposing the difference between positivism
and the idealism of a Berkeleyan brand, stresses that the founders of positivism "certainly desired something quite different from a renewal of that
idealism [and] this interpretation is to be rejected as contrary to the antimetaphysical attitude of positivism. Idealism and positivism are incompatible.[...] Ernst Mach especially emphasized that his own positivism
developed in an opposite direction to that of Berkeleyan metaphysics; and
he and Avenarius laid great stress upon not taking the given as a content
of consciousness" (ibid., 85). And, according to Schlick, they were right
because the problem of the reality of the external world is a meaningless
pseudo-problem. Taking for granted that "the peculiar business of philosophy [is] to ascertain and make clear the *meaning* of statements and
questions," Schlick indicates that the meaning of a question is clear to us
"when and only when we are able to state exactly the conditions under
which it is to be answered in the affirmative, or, as the case may be, the
conditions under which it is to be answered in the negative" (ibid., 86). It
is the same with the meaning of every statement: we understand it only if
we may indicate the fact which makes it true. When trying to describe
such a fact or conditions under which a question must be answered in the
affirmative or in the negative, we face the problem of understanding the
meanings of the words which occur in a description of fact or conditions.
Because their definition leads finally to words which cannot be defined,
"they must be indicated directly: the meaning of a word must in the end be
shown, it must be *given*" (ibid., 87). Thus different conditions related to
positive and negative answers to a question "mean difference in the given.
The meaning of every proposition is finally to be determined by the given
and by nothing else" (ibid.). According to Schlick, this insight much better

defines positivism than the claim "there is only the given." Its merit consists in the fact that, contrary to the latter, it cannot be easily misinterpreted as a metaphysical claim.

Realists attribute to positivists the assertion that "there is no external world." Schlick, however, stresses that "if one understands by 'positivism' a view which denies the reality of bodies I must declare positivism to be simply absurd" (Schlick 1959 [1932/33]: 87). It is completely wrong, according to Schlick, to deny reality to physical things in favor of the "content of consciousness." Asserting the reality of sensation does not have different meaning from asserting the reality of a physical body because we do not have different modes of verification of these assertions. Following Mach, Schlick maintains that "[i]t is reality of the *same* sort that one must attribute to the data of consciousness and, say, to physical events. Hardly anything in the history of philosophy has produced greater confusion than the attempts to distinguish one of the two as true 'being.' Wherever the word 'real' is significantly used it means one and the same thing" (ibid., 99).

The external world a *scientist* speaks about when claiming that the nature of extramental things is correctly represented by his equations may be called real only in case "to be real" means to stand in definite relationship to the given, to perceptions or other experiences. Thus positivism or, to be more exact, logical positivism (Schlick uses the words *consistent empiricism* as well)

> does not mean and does not imply that only the given is real. Such an assertion does not make sense. [...] Hence also, consistent empiricism does *not* deny the existence of the external world; it merely points out the empirical meaning of existential proposition. [...] The formulation, used by some positivists, that bodies are only "complexes of sensations" is [...] to be rejected. What is correct is only that propositions concerning bodies are transformable into equivalent propositions concerning the occurrence of sensations in accordance with laws. (Schlick 1959 [1932/33]: 107)

This view was shared by other logical positivists, including Carnap.

The Foundation of Knowledge

After determining the differences between his position and that of idealism and realism (almost all members if the Circle shared his stance on the issue), a modern epistemologist must answer another fundamental question: what is the foundation of knowledge? Schlick is convinced that "[a]ll important attempts at establishing a theory of knowledge grow out of the problem concerning the certainty of human knowledge" (Schlick 1959 [1934]: 209). Schlick points out that all philosophers from antiquity on searched for an unshakable, indubitable foundation, for the bedrock on which the structure of our knowledge may rest. Because of the limited power of human thought, this structure, according to him, cannot be certain, but even uncertain knowledge must have a firm foundation. Characteristically, Schlick does not elaborate *why* a firm base is needed for a shaky structure: he regards the foundational search as something absolutely natural. Although he spoke more than once about the decisive turn in philosophy that logical positivism brought about, he, it seems, did not consider that it might require abandoning the search for a firm foundation of knowledge. Yet some of his colleagues in the Circle were of a different opinion on the question and Schlick felt compelled to confront it.

By the end of the twenties, the standard position in the Circle on the foundation of (empirical, factual) knowledge was as follows: knowledge is based, as empiricists always claimed, on facts which are represented by so-called "protocol statements." As Schlick indicates, "protocol statements" originally meant those statements "which express the *facts* with absolute simplicity, without any molding, alteration or addition, [...] and which precede all knowing, every judgment regarding the world" (Schlick 1959 [1934]: 209–10). Thus it was supposed that if we succeed in expressing facts precisely, protocol statements might be regarded as the firm starting points in constructing an edifice of knowledge. In a sense of firm starting points they had to play a role a bit similar to that of axioms of mathematics: in spite of all the differences between protocol statements as singular statements and axioms as most general statements, both the former and the latter had to serve as foundational statements upon which (using, of course, different logical means) the structure of knowledge is erected.

If, however, protocol statements have the form "N. N. at such and such time observed so and so at such and such place" as some logical positivists supposed and Neurath insisted (cf. Neurath 1959 [1932/33]), then such propositions cannot be regarded as certain. N. N. had many possibilities to commit an error. An observed fact could be described inaccurately, mistakes could creep in when determining the place and time of observation and writing down or printing a sentence. Such a "protocol statement" is no more than a hypothesis, and Neurath, whose side was taken—after some vacillation—by Carnap, openly acknowledged this. Schlick has no doubts that protocol statements of such form may be inaccurate or even outright false and draws the conclusion that "one can use them in the construction of the system of science only so long as they are supported by, or at least not contradicted by, other hypotheses" (ibid., 212–13). This means however, according to him, not only that such protocols cannot play the role of the firm basis of knowledge, but also that the very distinction between protocols and other statements of science becomes meaningless.

This conclusion did not change, however, his firm conviction that knowledge *must* have a solid foundation. In "The Turning Point of Philosophy" he expresses his steadfast conviction (evident also in *The General Theory of Knowledge*) that changes in philosophy do not contravene its relentless quest for certainty. "The idea of claiming only probability for its statements was remote from earlier thinkers. They would have rejected it as incompatible with the dignity of philosophy. In this was expressed a healthy instinct for the fact that philosophy must supply the ultimate support of knowledge. [...] We too believe in the dignity of philosophy and deem incompatible with it the character of being uncertain and only probable" (Schlick 1959 [1930]: 58).

Because, according to Schlick, the momentous aim of science is the *true* description of facts, the problem of the basis of knowledge is nothing other than the question of the criterion of truth. Adoption of Neurath's point of view that protocol statements are freely modifiable hypotheses without a firm foundation leads to the adoption of a coherence theory of truth. If we must forever remain within the confines of a system of statements, the only possible criterion of truth is logical coherence, the consistency of this system. Yet not every coherent system of statements may be treated as true scientific theory: such a system may be just an elaborately consistent fairy tale.

The only way to avoid difficulties is, according to Schlick, to single out some statements which cannot be modified and to add the requirement that all other statements of the system should be compatible with those singled out. The latter cannot be alterable; only the others might be modified in case they are incompatible with the former. Of course, usual protocol statements scientists utter or write down in the process of observation cannot pretend to play the role of unalterable statements upon which the edifice of scientific knowledge rests. We must look for statements which are more basic in an epistemological sense than scientific protocols, statements which *precede* them.

Schlick gives the following answer to the question as to what statements are really *basic*:

> In the ordering of statements according to their origin which I undertake for the purpose of judging their certainty, I start by assigning a special place to those that I make

myself. And here a secondary position is occupied by those that lie in the past, for we believe that their certainty can be impaired by "errors of memory"—and indeed the more so the farther back in time they lie. On the other hand, the statements which stand on the top, free from all doubt, are those that express facts of one's own "perception," or whatever you like to call it. (Schlick 1959 [1934]: 218)

Schlick's contention that one's *own* statements play a decisive role in the system of knowledge is worth special attention. It is by no means new in philosophy and Schlick is perfectly aware of this. Yet after mature consideration he chooses precisely this view by claiming that he can accept only such a system of knowledge into which his own observation statements fit unmodified. This traditional (for an empiricist) point of view differs from the one towards which Neurath and Carnap began to move already in the early thirties. They were interested not in personal but impersonal knowledge. Statements printed in a book represent such knowledge.

In the discussion on protocol statements which played an important role in the development process of logical positivism, issues discussed were the form of protocol statements and their degree of certainty. Yet the real crux of discussion was the question about whose knowledge must be given a preeminent role. Schlick's position was a traditional one, to my *own* knowledge. Starting from the early 1930s, Neurath and Carnap held a rather different view: they were concerned only with *intersubjective* knowledge. The latter exists only in the form of statements. Thus, for them only statements as linguistic constructs are legitimate objects of analysis. From Schlick's point of view, in our analysis we must go beyond statements and to search for grounds to adopt them. Precisely because of this, Schlick did not use much logic in his analysis of knowledge: it is not very useful at the extralinguistic level.

My own simplest statements expressing my own perceptions, free from any doubt and thus suitable to play the role of the bedrock on which the structure of knowledge rests, are always, according to Schlick, "of the form: 'Here now so and so,' for example 'Here two black points coincide,' or 'Here yellow borders on blue,' or also 'Here now pain,' etc." (Schlick 1959 [1934]: 225). Because it is rather difficult to find such statements in scientific protocols, instead of the term "protocol statements" Schlick uses the term "basic statements." He calls such statements also "observation statements" or "confirmations" (*Konstatierungen*). Common to all such statements is the fact that they include demonstrative terms. Thus "to understand the meaning of such an observation statement one must simultaneously execute the gesture, one must somehow point to reality" (ibid.). When grasping the meaning of observation statements we simultaneously verify them. In this respect they resemble analytic assertions and like the latter are absolutely certain. According to Schlick, they are the only synthetic assertions that are not hypotheses.

They are certain, however, only at the moment when they are pronounced. If written down or memorized, they lose both definitiveness and certainty. In fact, they are not full-fledged statements at all: when we write down the demonstrative words "here" and "now," they lose their meaning. And if we look for the firm foundation of knowledge, we cannot replace these important demonstrative words by space and time coordinates, because in such cases we substitute for the observation statement a protocol statement which is no more than a hypothesis.

What role do observation statements with demonstrative words play in the process of cognition? Confirmations stand temporally at the beginning of the cognition, "stimulating it and setting it going." Yet Schlick does not regard this function of confirmations to be very important because protocol statements stand "in a less close connection with the observed" as might be thought. The second function of confirmations is more important: they are the absolute end of the process of verification of hypotheses. When making an anticipated confirmation, we obtain thereby a feeling of joy, satisfaction, and fulfillment of our true mission. Confirmations, according to Schlick, ensure

the unshakable point of contact between knowledge and reality. [...] They do not, in any way lie at the base of science; but like a flame, cognition, as it were, licks out to them, reaching each but for a moment and then consuming it. And newly fed and strengthened, it flames onward to the next. These moments of fulfillment and combustion are what is essential. All the light of knowledge comes from them. And it is for the source of this light the philosopher is really inquiring when he seeks the ultimate basis of all knowledge. (Schlick 1959 [1934]: 227)

These striking metaphors and yearning desires are, to put it mildly, a bit uncommon coming from the pen of a logical positivist. It is not surprising that Schlick did not find many supporters among the members of the Vienna Circle (except the faithful, as always, Waismann) who shared his position in the dispute on protocol statements. In the mid-thirties most of the logical positivists were already convinced, likely more by Carnap then by Neurath, that in applying the method of logical analysis we must take care not to cross the boundaries of language. By the way, in "The Foundation of Knowledge" Schlick notes that "in a quite definite sense we cannot 'go beyond language' as Wittgenstein puts it" (Schlick 1959 [1934]: 214). Yet there is no doubt that the most ardent—beside Wais-

mann—adherent of Wittgenstein in the Vienna Circle goes in his epistemological analysis of the foundations of knowledge precisely beyond
language. Moreover, he refers not only to sensations or perceptions, to
which older positivists referred quite often, but to feelings of different
kind: joy, satisfaction, a sense of fulfillment. These feelings were sometimes mentioned by the empiricists and positivists when inquiring into
morals, but never when inquiring into truth. Schlick finds them appropriate to refer to when explaining the foundations of knowledge. This is no
accident. Joy, satisfaction, and sense of fulfillment play an important role
in the ethics and the philosophy of life of a philosopher who used to describe himself as a "consistent empiricist."

Philosophy of Life and Ethics

Schlick was a man of two minds. In Waismann's words: "Those who see
in Schlick merely an acute thinker, are acquainted only with the surface;
they do not discern the other side of his character. [...] In his whole attitude Schlick was in truth a mind attuned to poetry and metaphysics"
(Waismann 1979 [1938]: xv–xvi). That does not mean that the founder of
the Vienna Circle was a crypto-champion of transcendent metaphysics.
Yet as Morris indicated a year after Schlick's death, referring to the words
of Schlick himself, his "conception of philosophy is nearest to the old
conception of philosophy as 'wisdom of life'—for the sage is the person
who sees most clearly the wider meaning of his words and activities"
(Morris 1937: 14). The framework of logical analysis was always a bit too
narrow for Schlick, and he did not shy away from making pronouncements that little resemble the statements of an analytical philosopher.

We encounter such pronouncements in his first philosophical work, a
small book, *The Wisdom of Life* (*Lebensweisheit*), published in 1908. In it,
Schlick maintains that a person realizes himself most fully not in purposive activity but in joyful play. Thus the supreme wisdom of how to live
one's life is to love and do whatever love inspires you to do: an aesthetic,
even Dionysian, vision of the ideal life.

Schlick's early inquiry into the philosophy of life may seem of no concern to us, because we are interested in Schlick's role in the Vienna Circle
and his later philosophy. Yet we find very similar ideas in his article "On
the Meaning of Life," published in 1927. In it, Schlick, the logical positivist, speaks with a voice not to be expected of the founder of the most important antimetaphysical movement of the twentieth century. He treats
here philosophy more as an empirical axiology than logical analysis.

In that publication Schlick returns to his earlier idea that purposive activity is not the most important thing in life. Reaching a goal produces new longings and "existence seems doomed to be a restless swinging to and fro between pain and boredom, which ends in the nothingness of death" (Schlick 1979 [1927]: 113). With Nietzsche, Schlick claims that we cannot hope to find an ultimate meaning of life if we view it as goal-directed. And if by work we mean any activity commenced solely to reach some goal, it cannot give meaning to life. If we wish to find such meaning "we must seek for *activities* which carry their own purpose and value within them, independently of any extraneous goals; activities, therefore, which are not work, in the philosophical sense of the word" (ibid., 114). Play as free, purposeless action is precisely such an activity. Its effects may be useful but it must be taken for its own sake, not because of expected consequences.

According to Schlick, play and joy are inherently linked: play is joy in sheer creation. Thus the creation of the artist is a characteristic example of play. Yet cognition is pure play also. The work of a craftsman and even that of a farmer tilling his fields may be play as well. Joy and devotion to any activity transforms work into play. "The last liberation of man would be reached," writes Schlick, "if in all his doings he could give himself up entirely to the act itself, inspired to his activity always by love. The end, then, would never justify the means, he might then exalt into his highest rule of action the principle: "What is not worth doing for its own sake, don't do for anything else's sake! All life would then be truly meaningful, down to its ultimate ramifications; to live would mean: to celebrate the festival of existence" (Schlick 1979 [1927]: 118).

We can learn from the child how to celebrate it, because a child is capable of the purest joy. Joy, enthusiasm, youthfulness are the most precious things in life. The meaning of life is precisely youth, not just as time of growing, as a particular span of years but as a time of play, as a way of acting, as a devotion to deed which has nothing at all to do with years and their accumulation.

In his "On the Meaning of Life" Schlick sounds, at least sometimes, like a moralist. Of course, nothing prevents a philosopher from being at times a moralist. Yet in his main work devoted to theory of morality, namely the *Problems of Ethics* (1930), Schlick tries to separate these two roles of a man dealing with morals more carefully. He emphasizes that for the philosopher "there is no greater danger than to change from a philosopher into a moralist, from an investigator into a preacher. [...] Of course, the prophet and the investigator can be one and the same person; but one cannot at the same moment serve both interests, for whoever mixes the two problems will solve neither" (Schlick 1962 [1930]: 1–2). The task of ethics which is a part of philosophy is to understand what is morally valuable, what serves as the standard or the norm of human conduct, and not to produce or establish morality. Ethical questions concern the "good," but its final task, according to Schlick, is not to define what is morally good but to describe those conditions under which the word "good" is actually used. Ethics, like any other science, concerns questions of fact and finds the material it works with in experience.

It may seem that such a notion of ethics is inconceivable: how is it possible to deal with norms grounding one's arguments exclusively on facts? Yet for Schlick this is no great problem: he interprets norms as standards characteristic for society and thus as (social) facts. To seek to ascertain norms, we must discover the conditions under which an act or intention is actually called "good."

Hence although ethics deals with norms, it can do nothing more than to disclose social facts and unveil the hierarchy of norms of behavior. It can do nothing more than explain: it justifies value judgments "only to the extent that it shows that the judgment corresponds to a certain norm; that this norm itself is 'right' or justified, it can neither show nor, by itself, determine. [...] The question regarding the validity of valuation amounts to asking for a higher acknowledged norm under which the value falls, and this is a question of *fact*. The question of the justification of the highest norms or ultimate values is senseless, because there is nothing higher to which these could be referred" (Schlick 1962 [1930]: 17–18). Ethics may discover the highest norms but cannot determine if they are right.

Valuations are psychic occurrences. The science which examines psychical processes in which man makes valuation is psychology. Thus explanatory ethics which seeks to determine the laws of moral behavior is, according to Schlick, a part of psychology.

According to Schlick, moral demands are expressions of the desires of society, and these demands are established only because they appear to be useful for it. Yet he rejects the standard utilitarian formula that the good is what brings "the greatest happiness of the greatest number of human beings" and replaces it by a more cautious formula: "In human society, that is *called* good which is *believed* to bring the greatest happiness" (Schlick 1962 [1930]: 87). By this, Schlick avoids difficulties involved in the calculus of happiness.

Schlick's ethics differs substantially from the emotivist one which is most often associated with logical positivism. From the emotivist point of view, most cogently expressed by Alfred J. Ayer and Charles L. Stevenson when uttering statements in which we meet the adjective "good" used in the ethical sense (or "right," "wrong," etc.) and which are, in fact, equivalent to rules of moral behavior, we express no more than our emotional approval of certain behavior or disapproval of a behavior of opposite kind. Because "is" cannot lead to "ought," it is impossible to find a criterion for determining the validity of such rules (or ethical judgments) and that means that they cannot be regarded as true or false. Thus it is impossible to argue rationally over ethical judgments. Carnap also shared that view. According to emotivists, what a theoretician can do in ethics is no more than analysis of ethical terms and of ethical judgments. Yet, as we tried to demonstrate, Schlick regarded ethics as an empirical science and not an applied conceptual analysis; moreover, he was convinced that ethical statements have cognitive meaning and are not just expressions of our emotions.

In general the founder of the Schlick Circle was not the most typical representative of it as regards both style of philosophizing and content of some specific conceptions. By many other members he, at least in the nineteen-thirties, was regarded as a bit too old-fashioned, sympathizing with ancient epistemology and even metaphysics, not enthusiastic enough for the new formal logic, not a revolutionary in philosophy, and also conservative-leaning in social matters. He was of aristocratic origin and behaved—notwithstanding all his democratic manners—somewhat aristocratically. The middle-class "plebeians" in the Vienna Circle were not thoroughly fond of him. Schlick's close relation with Wittgenstein they hardly regarded as commendable. Schlick was venerated; Carnap more so. Carnap embodied the spirit of the new philosophy: studious, prolific, innovative, inquisitive, precise, scientific. As was hinted already, Carnap's relation to Schlick somewhat parallels that of Mill to Comte. Like Comte, Schlick was something of a guru and at times, albeit almost imperceptibly, sentimental. He was not averse to using metaphors, as we have seen. Carnap, however, without the slightest pretension to be treated as a sage, evokes Mill. He, akin to Mill, was more sober, dry, sustained, and self-controlled. His philosophical impact was greater than that of Schlick and his views are much more representative for the Vienna Circle and logical positivism in general.

Rudolf Carnap

Rudolf Carnap was born in 1891 in Ronsdorf (now part of Wuppertal), Germany. From 1910 until 1914 he studied mathematics, physics, and philosophy at Jena and Freiburg Universities. One of his professors was Gottlob Frege. Carnap received his Ph.D. in 1921; the title of his doctoral thesis was "Space" (Der Raum). Thereafter he continued his philosophical studies at Jena. In 1925 he made his first presentation to the Vienna Circle and a year later started to teach at the University of Vienna. From 1931 until 1936 he was professor of natural philosophy at the German Charles-Ferdinand University in Prague. After emigration to the USA, he taught at the University of Chicago from 1936 till 1952 and at the University of California at Los Angeles from 1954 till 1961. He spent two years as a fellow at the Institute for Advanced Studies in Princeton and lectured at Harvard as a visiting professor. Carnap died in 1970.

Carnap was a rather prolific writer. Among his most important books it is worth mentioning *Logical Structure of the World* (*Logische Aufbau der Welt*, 1928), *The Logical Syntax of Language* (*Logische Syntax der Sprache*, 1934), *Meaning and Necessity* (1947), and *Logical Foundations of Probability* (1950). The majority of his innovative ideas concerning the philosophy of science—and we intend to pay most attention exactly to them—are scattered in his numerous articles, and not concentrated in his books. We will center on Carnap's philosophy of science and not on his other areas of pursuit for two main reasons. Firstly, he interests us mainly as a philosopher, not a logician. Secondly, positivist philosophy (and especially logical positivism) was primarily a philosophy of science, and Carnap is vital for our study precisely as a positivist.

As we mentioned earlier, in the Vienna Circle Carnap soon emerged as its intellectual master. Even when lecturing in Prague, he rarely missed the Circle's meetings. Carnap was an original thinker and his ideas influenced greatly the direction of the development of logical positivism. After Schlick's death in 1936, Carnap's position as *the* leader of logical positivism became uncontestable. It was he who shaped the philosophy of logical positivism to the greatest extent. Also, he was a pioneer in developing new logical techniques of analysis, and nobody equals him in applying them to solve the problems in which logical positivists were interested. Many other prominent members of the movement spoke enthusiastically about logical analysis, but applied it themselves mostly in an informal or ad hoc manner. It is rather difficult to find elaborate formal logical models and constructions in the works of Schlick, Neurath, or Feigl. Carnap used them abundantly and in a rather precise, sometimes even pedantic manner. Because of that, many of his works are not very easy to read. He also wrote many lucid nontechnical articles but, it seems, was not very interested in gaining popularity by writing books in which he could address a wider circle of philosophers or even the general public not knowledgeable in formal logic. It seems that except for the early *Philosophy and Logical Syntax* (1935), practically the only relatively popular book he wrote is *Philosophical Foundations of Physics* (1966), edited by Martin Gardner from transcripts of Carnap's classroom lectures and discussions.

It would not have been difficult to edit a volume of Carnap's more popular articles, yet he showed little enthusiasm for the project, and after his death some unexpected problems concerning copyrights emerged. For many years, the most easily accessible book in which a few of Carnap's articles were published was *Logical Positivism*, edited by Ayer in 1959. Out of seventeen articles published in it only three, however, were from Carnap's pen. Therefore, for quite a long time the leader of a major, well-known philosophical movement was, paradoxically, not well-known to a wider public interested in philosophy. In recent years the situation has somewhat changed. *Philosophical Foundations of Physics* was republished using for the title the former subtitle as *Introduction to the Philosophy of Science* (Carnap 1995 [1966]), a monograph, *The Unity of Science* (Carnap 2011 [1934]), was reissued, and a valuable collection of texts under the name *Logical Empiricism at its Peak: Schlick, Carnap, Neurath* (Sarkar 1996a) was edited. Books written by Carnap are reappearing, plenty of articles and some monographs on him have been written, and a prestigious *Cambridge Companion to Carnap* (Creath and Friedman 2008) was published. Carnap is being treated as one of the most important thinkers of the twentieth century. At the end of that century Quine defined his rank in a very direct manner: the two greatest philosophers of the century are Wittgenstein and Carnap. To tell the truth, Qunie held the latter in higher esteem.

Philosophy as Logical Analysis

When explaining the functions of philosophy understood as logical analysis, Carnap emphasizes the major results the application of logic in philosophy can yield:

> The researches of applied logic or the theory of knowledge, which aim at clarifying the cognitive content of scientific statements and thereby the meanings of the terms that

occur in the statements, by means of logical analysis, lead to a positive and a negative result. The positive result is worked out in the domain of empirical science; the various concepts of the various branches of science are clarified; their formal-logical and epistemological connections are made explicit. In the domain of *metaphysics*, including all philosophy of value and normative theory, logical analysis yields the negative result *that the alleged statements in this domain are entirely meaningless*. Therewith a radical elimination of metaphysics is attained, which was not yet possible from the earlier antimetaphysical standpoints. (Carnap 1959 [1932]: 60–61)

The positive results: clarification of meaningful concepts and propositions, and laying foundations for the sciences are, according to Carnap, more fruitful although the negative application of logical analysis is also important, especially in situations when metaphysics thrives. As to the logical character of statements obtained through the result of logical analysis, they are, he claims, partly analytic—when the results have a general form—and partly empirical—when, say, the meaninglessness of a specific sentence in a specific text is established.

Carnap's works exemplify in the clearest way what was understood in logical positivism by logical analysis. His notion of logical analysis was not, however, inflexible. The most important change occurred in the mid-thirties.

Carnap quite often emphasized that logical analysis is a *translation*. It is a translation of expressions of the logically imperfect language used in everyday life, philosophy, and science into expressions of language having a clearly defined logical structure and vocabulary which are preferable from a logical and philosophical point of view: more simple, exact, unequivocal, and empirical than the language in which an analyzable expression was formulated. In the first phase of the development of Carnap's notion of logical analysis he—like other logical positivists—presupposed that there is only one perfect language into which all expressions might be translated. The vocabulary of this language had to be purely empirical, based on terms in which immediate experience is described. The underlying structure of this language had to yield the logical system of the *Principia Mathematica*. The uniqueness of this ideal language was supposed to be a consequence of the uniqueness of logic.

When Carnap fully grasped the importance of the fact that there are various logical systems, this conception of analysis had to be given up. In *The Logical Syntax of Language* Carnap sets down his famous principle of (logical) tolerance. According to it, "[i]n logic, there are no morals. Everyone is at liberty to build up his own logic, i.e. his own form of language, as he wishes. All that is required of him is that, if he wishes to discuss it, he must state his methods clearly, and give syntactical rules instead of philosophical arguments" (Carnap 1937 [1934]: 52). Syntactical rules govern how words are arranged into phrases and sentences. After Carnap adopted the principle of tolerance, it became clear that logical analysis means rather the *reconstruction* of an analyzed expression which may be conducted by building a freely chosen logical model of it, and there is more than one way to do it.

Logical analysis had to be used primarily in replacing the vague concepts of special sciences by logically precise ones. Yet it must be applied when examining philosophical expressions as well. One of the important procedures to be used in the process of analysis of philosophical expressions is, according to Carnap, their translation from one mode to another, to be more exact, from the material mode to the formal. Let us come to the crux of the matter.

In the twenties and early thirties, Carnap believed that logical analysis may help to eliminate metaphysics altogether, not merely to weaken it. Metaphysics had to be eradicated with the aid of an empirical criterion of meaning. This criterion deserves special consideration, and we shall examine it later. Yet, according to Carnap, "even after the elimination of metaphysics" some philosophical sentences continue to cause perplexities. They concern—at least seemingly—"the structure of space and time, the relation between cause and effect, the relation between things and their qualities, the difference and the real relations between the physical and the psychical, the character of numbers and numerical functions, the necessity, contingency, possibility or impossibility of conditions and the like" (Carnap 1935: 59–60). He came to the conclusion that such sentences are deceptive: they seem to concern the objects mentioned but in fact they concern linguistic forms.

Sentences which concern the form of linguistic expressions (say, "such and such sentence is analytic") are called by Carnap *syntactic sentences*. "With these," writes Carnap, "are to be contrasted those sentences which concern not linguistic expressions but extra-linguistic objects; they may be called *real object-sentences*. There is also a third, an intermediate kind of sentence. Sentences of this kind are, so to speak, amphibious, being like object sentences as to their form, but like syntactical sentences as to their contents. They may be called pseudo-object-sentences" (Carnap 1935: 60).

An example of a real object-sentence is "The rose is red." The sentence "The word 'rose' is a thing-word" is a syntactical sentence. And "The rose is a thing" is a pseudo-object-sentence. Carnap stresses that "[i]n the mode of speech applied in pseudo-object sentences there are used words which designate objects or matter, while the words used in syntactical sentences obviously concern form" (Carnap 1935: 64). Therefore, he calls pseudo-object sentences also sentences of the *material mode of speech* and the syntactical sentences of the *formal mode of speech*. Most of the sentences of philosophy deceive us because they are pseudo-object sentences. They seem to concern objects mentioned in them but they really concern linguistic forms. The use of a material mode of speech leads in philosophy to confusion and controversy. These "idle" controversies may be removed in a simple way: theses defended fiercely by various philosophers must be translated into the formal mode. Then it may become visible that there is no ground whatever for caustic debate.

One example of such an idle controversy which owes its origin to the use of a material mode of speech is, according to Carnap, the dispute between phenomenalism and realism. The theses "A thing is a complex of sense data" and "A thing is complex of atoms" seem incompatible assertions leading to interminable dispute about what a thing actually is. Yet let's translate these sentences to the formal mode of speech. The first sentence may be rendered as "Every sentence in which a thing-designation occurs is equipollent [i.e., equivalent] to a class of sentences in which no thing-designations but sense-data designations occur." The second may be translated to formal mode of speech in the following way: "Every sentence in which a thing-designation occurs is equipollent to a sentence in which space-time coordinates and certain descriptive functors (of physics) [i.e., physical functions] occur" (cf. Carnap 1935: 301). Because it is possible to reconcile these two claims, the whole controversy between phenomenalism and realism is unavailing. Another example of an idle controversy which can be settled by translating seemingly opposing theses into the formal mode is the dispute concerning the reality of numbers.

Even the first proposition of Wittgenstein's *Tractatus*, "The world is the totality of facts, not of things" may be expressed, according to Carnap, in the more precise formal mode of speech as "Science is a system of sentences, not of names." Therefore, he regards translatability into the formal mode of speech as "the touchstone for all philosophical sentences" and claims that they may be discussed fruitfully only in the formal mode of speech. If they cannot be translated into this mode of speech, they must be left aside. "Sentences which do not give even slight indication to determine their translation are outside the realm of language of science and therefore incapable of discussion, no matter what depths or heights of feeling they may stir" (Carnap 1935: 313).

At the time when he wrote *The Logical Syntax of Language*, Carnap was convinced that the logic of science is the logical syntax of language. Although he disagreed with Wittgenstein, who held the opinion that it is impossible to speak in a meaningful way *about* language, Carnap believed at that time that logical analysis must be limited to examining the relations between linguistic expressions. Yet in discussions with Tarski and Gödel, he realized that the syntactical mode is not the exclusive mode in which it is possible to speak about language. In *The Logical Syntax of Language*, Carnap claimed that the problems of philosophy are merely syntactical problems, but it soon became clear to him that this formulation is too narrow and that these problems are actually metatheoretical problems. "The narrower formulation," asserted Carnap in his "Intellectual Biography," "is historically explained by the fact that the syntactical aspect of language had been the first to be investigated by exact means by Frege, Hilbert, the Polish logicians, and in my book [*Logical Syntax of Language*]. Later we saw that the metatheory must also include semantics and pragmatics; therefore the realm of philosophy must likewise be conceived as comprising these fields" (Carnap 1963a: 56). Supplementing syntax by semantics in which the problems of designation and truth may be examined was a very important step in the development of logical analysis. It widened its possibilities greatly. It became possible to build semantical reconstructions (models) of those notions, *probability*, for example, which were not amenable to syntactical analysis.

Translation from the material to the formal mode of speech is, of course, far from being the only form of logical analysis. Yet as was already mentioned, logical analysis is always a translation, even if the mode of speech is not being changed. According to the principle of tolerance, first of all we need to choose (or to build) a logical system on which translation is based. Subsequently we face another choice because we must select not only logic but vocabulary as well. Specifically, like other logical positivists, Carnap faced an alternative: to choose either phenomenalist or physicalist language as the basic language into which expressions under analysis must be translated. By phenomenalist language he meant "one which begins with sentences about sense data, such as 'There is now a red triangle in my visual field'. The sentences of the physicalist language or thing-language speak [on the contrary] of material things and ascribe observable properties to them, e.g. 'this thing is black and heavy'" (Carnap 1963a: 50).

Carnap always emphasized that we are free to choose either language. However, in the twenties he obviously preferred the phenomenalist language. Most clearly this preference is reflected in *The Logical Structure of the World*, which was written mainly before Carnap joined the Vienna Circle (and thus cannot be regarded, speaking strictly, as a work of a logical positivist), but which deserves some attention as the first serious attempt to carry out the phenomenalist program of translation of all kinds of expressions into phenomenalist language. From later attempts only Nelson Goodman's *The Structure of Appearance* (1951) matches it.

Carnap's main goal in *The Logical Structure of the World* is to construct on a firm basis the system of concepts to be used in various sciences when describing the world. Carnap did not intend to describe the actual process of formation of concepts. He wanted to furnish a rational reconstruction of it, to present a "constructional system," i.e., an epistemic-logical system of objects or concepts—it can be taken in either sense—which enables a derivation or "construction" of all concepts from a few fundamental ones.

By reducing all concepts of science to a simple and reliable basis, this reconstruction had to demonstrate the essential unity of science and to ensure the soundness of the whole conceptual structure of our scientific knowledge. "Under the influence of some philosophers, especially Mach and Russell," writes Carnap in his "Intellectual Autobiography," "I regarded in the *Logische Aufbau* [*The Logical Structure of the World*] a phenomenalist language as the best for a philosophical analysis of knowledge. I believed that the task of philosophy consists in reducing all knowledge to a basis of certainty. Since the most certain knowledge is that of the immediately given, whereas knowledge of material things is derivative and less certain, it seemed that the philosopher must employ a language which uses sense-data as a basis" (Carnap 1963a: 50).

Yet he never doubted that it is possible to carry out the construction of the world on the basis of physicalist language also. When writing *The Logical Structure of the World*, he believed that the choice of the phenomenalist language can ensure the certainty of the foundations of our knowledge. Some years later, however, he concluded that this certainty is rather an illusion, and propositions of no kind may form a firm basis for the edifice of science. (We shall discuss the grounds for this conclusion later.) Thus, as a means of ensuring certainty, phenomenalist language is not much better than physicalist. On the other hand, a great advantage of the latter over phenomenalist language is its intersubjectivity. That is why Carnap decided that a physicalist basis is more suitable for constructing a system of all scientific concepts than a phenomenalist one and lost interest in the specific problems considered in *The Logical Structure of the World*.

From the early thirties he was convinced in the truthfulness of the thesis that physical language is a universal language. In Carnap's words, "[t]o every sentence of the system [of science] language there corresponds some sentence of the physical language such that the two sentences are inter-translatable" (Carnap 1959 [1932]: 166). This thesis is known as the thesis of physicalism.

Carnap understood very well that physicalism has far-reaching implications. "If the physical language, on the grounds of its universality [and intersubjectivity], were adopted as the system language of science, all science would become physics. Metaphysics would be discarded as meaningless. The various domains of science would become parts of unified science. In the material mode of speech: there would, basically, be only one kind of object—physical occurrences, in whose realm law would be all-encompassing" (Carnap 1959 [1932]: 166). At the same time, he continued to claim that when conducting logical analysis we are free to choose one or another language and this choice does not imply any ontological commitments. To emphasize that option for one or another language has nothing to do with ontology, he used to speak about *methodic* solipsism (which gives preference to the phenomenalist language) and *methodic* materialism (which prefers physicalist language). We may opt for one or another *façon de parler* by using phenomenalist or physicalist language, but this choice is a methodic or pragmatic one, not ontological. The position taken by Carnap on the issue reminds us of Mach's stance on the possibility of different interpretations of neutral elements and Poincaré's position on the choice of one or other geometry.

Far from all philosophers agreed with Carnap that the decision to use specific language has nothing to do with ontology. He however was firm: there are two kinds of questions concerning the existence of entities, be they concrete material things or classes and numbers.

If someone wishes to speak in his language about a new kind of entities, he has to introduce a system of new ways of speaking, subject to new rules; we shall call this procedure the construction of a linguistic *framework* for the new entities in question. And now we must distinguish two kinds of questions of existence: first, questions of existence of certain entities of the new kind within the framework; we call them internal questions; and second, questions concerning the existence of reality of the system of entities as a whole, called external questions. (Carnap 1956 [1950]: 206)

The answers to internal questions may be found by logical or empirical methods, and these theoretical questions do not pose serious philosophical problems. External questions, on the other hand, are noncognitive or practical. We are free to choose any linguistic framework. Thus "if someone decides to accept the thing language, there is no objection against saying that he has accepted the world of things. But this must not be interpreted as if it meant his acceptance of a *belief* in the reality of the thing world; there is no such belief or assertion or assumption, because it is not a theoretical question. To accept the thing world means nothing more than to accept a certain form of language, in other words, to accept rules for forming statements and for testing, accepting or rejecting them" (Carnap 1956 [1950]: 207–8). The same may be said about external questions concerning the reality of numbers or, say, the reality of physical space and time. External questions are a matter of decision. If they are interpreted as questions of the type "Do things really exist?" or "Do numbers really exist?" which must be answered by "yes" or "no," they become pseudo-questions. Of course, we are free to ask: "Are our experiences such that the use of the linguistic forms in question will be expedient and fruitful?" This is a theoretical question of a factual, empirical nature. But it concerns matter of degree; therefore a formulation in the form "real or not?" would be inadequate (cf. Carnap 1956 [1950]).

Formal and Empirical Sciences

Logical analysis is primarily a means of clarifying the cognitive content of scientific statements. Yet according to Carnap—and this was a general point of view of all logical positivists—not all scientific propositions have cognitive content. Some even lack it completely, such as propositions of logic and mathematics. They do not have cognitive content because they do not describe the world. Both propositions of logic and mathematics are tautologies: they say nothing about the objects we intend to discuss but concern only our way of speaking about them. Tautologies rule out no possibilities, they are true under all circumstances, they are compatible with any state of affairs, and as such, they say nothing about the world.

Later Carnap abandons all talk of tautologies and starts to speak instead about analytic statements. A statement is called analytic if it is unconditionally valid, i.e., valid independently of the truth-value of other statements of language. A statement is inconsistent (self-contradictory) if it is unconditionally invalid. A statement is called synthetic if it is neither analytic nor inconsistent (cf. Carnap 1953 [1935]: 124).

For Carnap, as well as other logical positivists, the distinction between analytic and synthetic statements is of the foremost significance. This distinction is the basis of the differentiation between the formal and factual sciences. The first contain only analytic, the second include synthetic statements. The nature of these sciences is radically opposed. The factual sciences—physics, biology, sociology, history—have their objects and they describe them in synthetic statements which have factual content. These sciences use analytic statements as well, but it is in principle possible to eliminate them altogether by replacing them with some rules.

The formal sciences—logic and mathematics—play a subsidiary role in relation to the factual sciences: they contain auxiliary statements. "These auxiliary statements have indeed no factual content or, to speak in the material idiom, they do not express any matters of fact, actual or non-actual. Rather they are, as it were, mere calculational devices, but they are so constructed that they can be subjected to the same rules as genuine (synthetic) statements. In this way they are an easily applicable device for operations with synthetic statements" (Carnap 1953 [1935]: 126). Thus, formal sciences themselves have no independent significance. They serve as components of science introduced for technical reasons in order to facilitate linguistic transformations of the synthetic statements contained in the factual sciences.

As an empiricist Carnap has not the slightest doubt that, in Kant's terms, all synthetic statements are *a posteriori*. In this respect he is in full accord with Schlick, and agrees with him that in case someone needs a concise definition of logical positivism (or logical empiricism), it may sound as follows: *logical positivism is the denial of the existence of synthetic* a priori *statements.*

Because of the general consensus of the members of the movement that formal sciences play only an auxiliary role, the philosophy of science of logical positivism is primarily a philosophy of factual sciences. What Carnap was interested in regarding the formal sciences was mainly the relation of mathematics to logic. We do not intend to go deeply into the details of this rather complicated and partly technical question, and will be content with the brief outline of Carnap's position. It is similar to that of Frege and Russell: mathematics is a part of logic. In his "Intellectual Autobiography," Carnap writes: "I had learned from Frege that all mathematical concepts can be defined on the basis of the concepts of logic and that the theorems of mathematics can be deduced from the principles of logic" (Carnap 1963a: 46). Precisely this led Carnap—like Frege and Russell before him—to the concept of mathematical truths as truths based on logic alone. He supposed that all valid mathematical statements are analytic in the sense that they hold true in all possible (from the logical point of view) cases or, in Leibniz's terms, in all possible worlds and thus have no factual content.

Strangely ignoring Hume, whose notion of mathematics was, as we saw, quite similar to that of Carnap, the latter claims that he and other members of the Vienna Circle combined for the first time the main thesis of empiricism with a passable explanation of the nature of logic and mathematics. By mathematics Carnap understands theory of numbers and their functions, abstract algebra, abstract group theory, and the like, but *excludes* geometry. To be more specific, pure or mathematical geometry is a part of mathematics, yet physical geometry is a part of physics. Questions concerning space become empirical questions once physical definitions, say that of a line in terms of a ray of light, and physical rules, say that for the determination of congruence, are laid down. Carnap's notion of geometry coincides with that of Schlick and Reichenbach.

The view that mathematics is a part of logic, i.e., mathematical concepts can be defined on the basis of concepts of logic, and theorems of mathematics can be deduced from principles of logic, is known as *logicism*. It faces serious difficulties. According to Carnap, the definition of mathematical concepts on the basis of logical concepts poses no great problems. The crux of the matter is the question whether set theory—in terms of which the most important mathematical concepts including that of the number can be defined—is a part of logic. Carnap agrees that the purely logical character of such axioms of set theory as the axiom of infinity and the axiom of choice is dubious. And if they cannot be deduced from principles of logic, they, it seems, cannot be regarded as analytic. That would mean that the claim that all statements of formal sciences are analytic must be jettisoned. Therefore, Carnap made a great effort to show that they may be interpreted as analytic statements. However, most mathematicians were not very happy with Carnap's interpretation of axioms of set theory. Today logicism has few supporters. Two other major conceptions concerning the foundations of mathematics, namely formalism and intuitivism (and a version of the latter, constructivism) are much more popular. Carnap's idea that the conflict between logicism and formalism is reconcilable is also not shared by many specialists.

The very notion of analyticity which is central to Carnap's interpretation of formal sciences was much discussed in the postwar years. We postpone, however, the examination of the results of this discussion in which Carnap and Quine were the main protagonists to "Positivism and the Analytic Tradition" in Chapter 5, because it was a major point of critique of logical positivism inside analytic philosophy.

According to Carnap, theories of formal sciences are built (in the ideal case) as deductive axiomatic systems with clearly defined rules of formation and transformation which enable the strict proof of theorems. From a methodological point of view such theories are uncomplicated. Their ultimate basis is logic. And because statements of these sciences do not have factual content and do not concern facts, our experience, whatever it may be, cannot amend the results achieved in the formal sciences. The world may vanish but this fact will not change the validity of the statements of logic and mathematics.

From a methodological point of view, the factual sciences are much more complicated. Therefore, the philosophy of science of logical positivism deals almost exclusively with factual science. We shall address their methodological problems soon. But before that we must address another question of fundamental importance for Carnap and logical positivism in general.

This question concerns the very notion of a factual statement, i.e., a statement with factual content. In purely negative terms a statement is synthetic if it is neither analytic nor inconsistent. But can any series of words which is neither analytic nor inconsistent pretend to be regarded as a full-fledged synthetic statement? As we already mentioned, logical positivism denies the possibility of *a priori* synthetic statements. Its representatives demonstrated (or were convinced that they demonstrated) that those statements which Kant regarded as synthetic *a priori* are either analytic—as are all statements of mathematics—or, else, synthetic *a posteriori* as are Newton's laws. Thus all synthetic statements must have factual content. Only facts may reveal if these statements are true or false. But Carnap as well as other logical positivists were convinced that not every series of words which (from a purely grammatical point of view) seems to be an assertion and is neither analytic nor inconsistent has factual content indeed. Some series of words resembling assertions are not real statements. Thus before inquiring into the methodological problems of the factual sciences it is necessary, according to Carnap, to determine which series of words have factual content and can be considered in the factual sciences, and which must be rejected at once because they only simulate factual content yet possess none.

The Criterion of Empirical Significance

Thus, a logical positivist faces the following task: to discover a criterion which would permit him to separate statements based on facts and expressing them directly or indirectly from those which do not have factual content but do not belong to the area of competence of formal sciences, and, therefore, must be excluded from the domain of science altogether. The field of science must be clearly demarcated from the area of non-science. Statements with factual content are also formulated quite often in everyday life. In fact, such statements form the basis of science. Admittedly, it is often necessary to make them more precise if we want to use them in science. But this is not a very complicated task. The major task, according to Carnap, is to demarcate science from metaphysics, and not from everyday experience.

Here it is worth briefly returning to Hume and other predecessors of logical positivism. Hume was firmly convinced that drawing a clear-cut line between "reasoning" appertaining to scientific inquiry and "sophistry and illusion" is of major importance for the well-being of inquiry. Comte draws a line between the positive mind on the one hand and the metaphysical and theological mind on the other hand in quite different terms. Yet no less than Hume, he believes that this line is extremely significant. When purifying experience of error and illusion, Mach pursued a similar goal. Singling out pure facts from experience, he sought to forge a solid framework for scientific inquiry by demarcating the area of science from that of nonscientific (especially metaphysical) pursuits.

Logical positivism continues this tradition. What makes its efforts different from those of Comte and Mach, and to a lesser extent from those of Hume, is, firstly, the ambition to demarcate positive science from metaphysics in a more rigorous way than earlier positivists managed to do, and secondly, to do this with the aid of a major instrument of logical analysis, namely the criterion of meaning, which would leave no doubt that everything which is beyond the pale of demarcation is not only irrelevant to science, but lacks any cognitive meaning whatsoever. Logical positivism, at least early logical positivism, harbored the hope that it is possible through this effort to do away with metaphysics, to wipe off this persistent rival of science from the surface of the earth finally and permanently.

The criterion we are talking about had to play two major roles: firstly, to set down the main requirements with which statements of (factual) sciences must comply, and secondly, to compromise statements which do not conform to it. The criterion had to be applied to statements or series of words which look like statements. This criterion was known at different times and in the works of different authors under different names: the criterion of verification, the criterion of (cognitive) meaning, the criterion of empirical significance, etc. These different names reflect changes which occurred both in the formulation of the criterion and in the elucidation of its status. Yet, originally Carnap, following in this respect Hume, was more interested, it seems, not in statements or sentences either conforming or not conforming to this criterion, but in concepts which either have or do not have cognitive significance and meaning.

At least his famous article "Elimination of Metaphysics through Logical Analysis of Language" (1932) starts from the examination of words, not sentences, although it clearly shows the link between the meaning of words and the sentences in which these words are included. To be more exact, he shows that there are two kinds of pseudo-statements: those which contain meaningless words, and those that contain meaningful words but which violate the rules of logical grammar. Carnap finds plenty of words playing an important role in *metaphysics* but having no meaning: "principle of the world," "God," "the Idea," "the Absolute," "the Unconditioned," "the Infinite," "the being of being," "non-being," "thing-in-itself," "being-in-itself," "being-in-and-for itself," "objective spirit," "essence," "emanation," "manifestation," "articulation," and so on (cf. Carnap 1959 [1932]: 67). Let "a" be any word and "S(a)" the elementary sentence in which it occurs having the form "x is a." Then the sufficient and necessary condition for a being meaningful may be given by each of the following formulations, which ultimately say the same thing:

1. The *empirical criteria* for "a" are known.
2. It has been stipulated from what protocol sentences "S(a)" is deducible.
3. The truth conditions for "S(a)" are fixed.
4. The method of verification of "S(a)" is known." (ibid., 64–65)

Thus if we replace "a" by "stone," all these conditions are fulfilled. Yet it is not the case if we replace "a" by any of the above metaphysical words. That means that neither these words nor sentences of any form containing these words are meaningful. Such sentences are pseudo-sentences: they look like normal sentences, yet convey no meaning and have no significance. In this respect they are no better than those sentences which clearly violate the rules of combination of words like sentences "Caesar is and" or "Caesar is a prime number."

As a characteristic example of a text whose sentences are almost all pseudo-statements, Carnap quotes an excerpt from Heidegger's "What Is Metaphysics?" boldly culminating with the words "The Nothing itself nothings," and shows that most of them violate logical syntax in spite of the fact that they accord with historical-grammatical syntax. Even seemingly innocuous sentences like "I am" violate the rule established by modern logic according to which the sign for existence cannot be predicated of signs for objects, but only to predicates (in the sentence "I am," "am" is not a copula prefixed to a predicate but designates existence). The sentence "I am a European" is meaningful but the sentence "I am" is not. The meaninglessness of the sentence "I am" means that the famous dictum of Descartes "*Cogito, ergo sum,*" and a simple statement "God is" are equally meaningless also. Another quite common source of the meaninglessness of sentences is, according to Carnap, the "type confusion" of concepts. The names of persons and names of numbers belong to different logical types, and precisely because of this the sentence "Caesar is a prime number" violates the rules of logical syntax and has no meaning. Although confusions of type occur more often in conversational language, Carnap detects a profusion of pseudo-statements of this kind in the writings of Hegel and Heidegger.

Carnap claims, moreover, that since metaphysics seeks to discover a kind of knowledge which is not accessible to empirical or factual science, it is impossible to find meaningful metaphysical statements. "Therefore a sentence can be used only to assert an empirical proposition, if indeed it is used to assert anything at all. If something were to lie, in principle, beyond possible experience, it could be neither said nor thought nor asked" (Carnap 1959 [1932]: 76).

The meaninglessness of all metaphysics does not mean, however, that its statements lack any content whatsoever. Yet the content they have is neither theoretical nor cognitive. Metaphysics is one of the forms of expression that people have towards life. As such it is legitimate. Yet by simulating theoretical content where there is none is a misleading way to express feelings. A much more appropriate way to express this attitude towards life is music.

> The harmonious feeling or attitude, which the metaphysician tries to express in a monistic system, is more clearly expressed in the music of Mozart. And when a metaphysician gives verbal expression to his dualistic-heroic attitude towards life in a dualistic system, is it not perhaps he lacks the ability of Beethoven to express this attitude in an adequate medium? Metaphysicians are musicians without musical ability. Instead they have a strong inclination to work within the medium of the theoretical, to connect concepts and thoughts. (Carnap 1959 [1932]: 80)

By confusing knowledge and art, metaphysicians achieve nothing in the name of knowledge and little more in the expression of attitude.

Seeking a criterion clarifying the meaning of factual statements and demarcating those statements from the pseudo-statements of metaphysics lacking cognitive content, Carnap and other logical positivists proceeded from Wittgenstein's idea according to which to understand a statement means to identify the case when it is true (cf. Wittgenstein 1961 [1921]: 4.024). By slightly modifying Wittgenstein's notion of meaning, logical positivists decided that a statement is meaningful (has cognitive significance) if it is verifiable, i.e., if it is possible to determine its truth-value. Waismann, likely the first in the Vienna Circle, gave to it the following form: *the meaning of a statement is the method of its verification.* This effective formula became famous and we found its variations in some publications of logical positivists. Yet many colleagues of Waismann from the Vienna Circle and the Berlin group were much more interested in the question of which statements, sentences, or propositions are meaningful and not in the question of what meaning is, which sounded to them either a bit abstract or tinged with metaphysics, or both. At least Carnap and Reichenbach did not regard the second question as essential.

The criterion according to which only those statements are meaningful which can be verified, i.e., the truth-value of which can be determined, is known as the *verifiability criterion of meaning*, or simply the *principle of verification*. This criterion itself needs clarification because the phrase "can be verified" is not entirely clear. The possibility of verification may be understood variously: as a technical, physical, or logical possibility.

Limiting the possibility of verifying scientific statements with the capabilities of contemporary scientific devices and instruments was not seriously considered by logical positivists. Such a view was clearly too narrow. Thus all logical positivists agreed that the statement "There are mountains on the dark side of the moon" is perfectly meaningful in spite of the fact that in the 1930s and 1940s it was technically impossible to verify it.

Two alternatives remain: the physical or logical possibility of verification. According to the first alternative, the process of verification must be compatible with the laws of nature. Thus, for example, the instruments of measurement used in the process of verification cannot move faster than light. And, of course, an observer who is involved in the process of verification must be treated as a real observer who can make only a finite number of observations, although a statement under verification may cover an infinite number of cases. The choice of the alternative "physical possibility" requires the reformulation of the verifiability criterion of meaning in the following way: "A statement is meaningful if and only if, after a finite

number of observations or experiments, it is possible to determine unequivocally its truth value." Schlick and Waismann shared—at least initially—precisely this strict interpretation of the principle of verifiability.
According to this version of the principle, it is necessary to admit that
universal laws of nature concerning infinite classes of objects are meaningless statements. The way out of this predicament—following Wittgenstein's suggestion—was to treat them not as statements but as rules of
inference serving for the transition from one empirical statement to another. Schlick was interested in this interpretation. Yet in this case the
laws of nature become empirically untestable because facts can neither
confirm nor disconfirm the rules. Hence in the second half of the thirties
he came to the conclusion that verifiability as a necessary and sufficient
condition of meaningfulness must be understood as a logical possibility
(cf. Schlick 1979 [1936]). That means, however, that only those sentences
are meaningless which violate the logical rules of language, say, confuse
logical types of the words. For Carnap such a solution of the problem was
unacceptable.

He was convinced that to demand that all meaningful statements be
fully and directly verifiable is to demand too much. In his extensive and
important study *Testability and Meaning* (1936/37), Carnap examined the
difficulties earlier versions of the principle of verification faced, and
modified the concept by introducing the notions of complete and incomplete, direct and indirect verification thus easing the strictures to which
statements must conform.

In it Carnap claims that Schlick's notion of the procedure of verification is inappropriate. According to Carnap, by answering the question of
what Schlick means by "possibility" and explaining that he means "logical
possibility," he does not answer in a satisfactory way the question of what
"verification" means. Rather, Schlick examines the fact which is described by a statement and not the procedure of verification of a statement. According to Carnap, a statement is verifiable not because it describes a *logically* possible fact but because verification of a statement is
physically feasible.

The examination of the process of verification must concentrate, according to him, on the investigation of the possibilities of a real investigator. He cannot determine precisely and without any doubt the truth-value
of universal sentences.

If verification is understood as a complete and definitive establishment of truth then a
universal sentence, e.g. a so-called law of physics or biology, can never be verified, a

fact which has often been remarked. Even if each single instance of law were supposed to be verifiable, the number of instances to which the law refers—e.g. the space-time points—is infinite and therefore can never be exhausted by our observations which are always finite in number. We cannot verify the law, but we can test it by testing its single instances, i.e. particular sentences, established previously. If in the continued series of such testing experiments no negative instance is found, but the number of positive instances increases, then our confidence in the law will grow step by step. Thus, instead of verification, we may speak here of gradually increasing confirmation of the law. (Carnap 1950 [1936/37]: 425)

Knowing whether a sentence is verifiable (in a weaker or stronger sense) and thus meaningful is a different matter from knowing its truth-value, however. Although in the Vienna Circle much effort was spent on perfecting the criterion of meaning based on Wittgenstein's idea which was considered insufficiently precise, another of his ideas according to which we may know the meaning of a statement without knowing its actual truth-value was adopted without discussion. Thus, according to Carnap, the very possibility of the confirmation of a universal sentence makes it meaningful even if we may not know its truth-value. However, in the development process of the philosophy of logical positivism, the emphasis has been shifting gradually from the questions related to the problem of meaning to those of methodology, especially those related to the evaluation of scientific hypotheses.

One of the reasons for this shift was meeting the practical needs of the factual sciences. Another was the difficulties logical positivists faced when confronted with the question of the very status of the criterion of meaning. Opponents of logical positivism were quick to ask "Is the principle of verification itself verifiable?" It was clear that it is neither an analytic, nor contradictory, nor a factual statement. Yet, critics claimed, logical positivists admit of no other category of statements. There were some possibilities to counter this claim, say by referring to the differences between object language and metalanguage. Yet even in this case the problem remained: determining the grounds for adopting this principle. The principle was directed against metaphysics. But it was rather difficult to convince metaphysicians to adopt it by proving this principle in one or another way. And if it is only a kind of code suggested by positivists, then metaphysicians may simply reject it.

The fledgling logical positivism of the twenties and early thirties was a very militant philosophy having ambitions to eradicate metaphysics altogether. Yet soon it became clear that metaphysics is much more resilient than logical positivists surmised. Logical analysis proved a double-edged sword: the supposedly deadly weapons were all too deadly for science, as was the case with the most radical version of the principle of verifiability which was incompatible with the interpretation of laws of nature as assertions. Yet turned against metaphysics they were unable to strike a death blow. Thus logical positivists had to acknowledge that they did not manage to win the battle against metaphysics by *proving* its meaninglessness. At least Carnap in *Testability and Meaning* made exactly this conclusion.

First of all, he emphasizes that "the question about the criterion of meaning has to be constructed and formulated in a way different from that in which it is usually done. [...] A clear formulation of the question involves reference to a certain language; the usual formulations do not contain such reference and hence are incomplete and cannot be answered" (Carnap 1950 [1936/37]: 3). He is convinced that it is necessary to distinguish a historically given language-system and a language-system which is yet to be constructed. Questions concerning the first kind of languages are theoretical, those concerning the second kind are practical.

Furthermore, when discussing questions concerning historically given language-systems Carnap suggests that we do not ask about the meaningfulness of a sentence of a given language at all. "It would be advisable," writes Carnap, "to avoid the terms 'meaningful' and 'meaningless' in this and similar discussions—because these expressions involve so many rather vague philosophical associations" (Carnap 1950 [1936/37]: 3). Instead of speaking about meaningful and meaningless sentences, Carnap proposes to differentiate sentences of a given language according to their methodological status by distinguishing completely or incompletely confirmable, or completely or incompletely testable sentences.

Later, however, when the distinction between the observation language, which uses terms designating observable properties, and the theoretical language, which contains terms referring to unobservable events and entities such as electrons or electromagnetic fields, became standard in logical positivism, it became clear that notions of confirmability or testability easily applicable to observation language cannot be transferred—without alterations—to theoretical language. Hempel, who proposed the *Translatability Criterion of Cognitive Meaning*, according to which a sentence has cognitive meaning if and only if it is translatable into an empiricist language (cf. Hempel 1959 [1950]: 116–17), has not been able to show how to apply it to all sentences which contain theoretical terms because some terms of advanced scientific theories such as "electric field" cannot be defined on the basis of observational predicates. The criterion of (empirical) significance for theoretical languages was formulated by Carnap in his article "The Methodological Character of Theoretical Concepts" (1956). At first glance, it is very simple: an expression of a theoretical language is a significant sentence if it satisfies the rules of the formation of this language, and every descriptive constant in the expression is a significant term (cf. Carnap 1956: 60). Yet the definition of a significant term is rather complicated because the concept of significance must be relativized to the theoretical language, the observation language, the set of postulates of the theoretical language, and the correspondence rules linking the terms of observation and theoretical languages.

In spite of all modifications in the empirical criterion of significance introduced by Carnap and other logical positivists, it continued to face grave criticism. Critics managed to find examples of sentences regarded by all competent researchers in one or another field of scientific inquiry as perfectly meaningful but not conforming to one or another version of the criterion. Thus it has been attacked as too narrow. Other critics attacked it as too wide. As we mentioned, not only the form but the very legitimacy of the criterion of significance was also put under question. And as time passed, more and more scholars doubted that the question of meaningfulness may be raised for single sentences. They thought that this question could be raised only for the whole system consisting of a theory and correspondence rules. And they claimed that in such a case it would be much more sensible to speak about the explanatory or predictory power of a theoretical system and not about its meaningfulness. Thus the problem of meaning which dominated logical positivism in the prewar years was gradually left aside by members of the movement, who took up other, more methodologically important issues.

The Structure of Scientific Knowledge

One of those issues was the problem of the structure of scientific knowledge. It concerns mainly two different but related questions. The first one is the mode of defining the terms used in scientific theories. Another question is the relation between scientific propositions which includes theoretical terms of a relatively high degree of abstractness, and those which describe the properties of phenomena under observation in empirical terms.

The general tendency of empiricism displayed through the centuries was a flattening out of scientific knowledge. Francis Bacon was convinced that science must remain as close to empirical facts as possible: the human mind needs lead weights to keep it down to earth, else it would lose itself in wild, airy speculations. Hume spoke about complex ideas being formed of simple ones, i.e., in fact of perceptions. (In a more modern way of expression this would mean the requirement to define all terms of a scientific theory on the basis of terms in which immediately observable things are described.) The tradition of empiricism was a tradition of strict reductionism. Mach, who boiled down the function of science to the sediment of simple description, was antagonistic to all abstract theories. Clearly, there are a few positivists who fall out of this tradition. It would be difficult to blame Comte for the tendency to reduce all terms of scientific theories to empirical ones. Duhem's conception of the structure of scientific theories is even more remote from naive reductionist empiricism.

The strict verificationism most logical positivists adhered to in the twenties and early thirties, however, reproduced reductionist tendencies in the elucidation of the structure of scientific theories. The results of the examination of the structure of mathematical knowledge, which was a reference point for logical positivists in dealing with issues of the methodology of factual sciences, suggested that in theories of factual sciences, all the terms must be defined—like in mathematics—on the basis of primary ones. And because primary terms—in the epistemic sense of this word—are for positivists the terms in which the results of observation are described, all other terms of factual sciences had to be reducible to the terms of observation. In his early years, Carnap held precisely this view, and *The Logical Structure of the World* represents such a position. The main reason why early logical positivism adhered to reductionism was the following. For propositions being verifiable—and only in that case they were regarded as having cognitive meaning—they had to be reduced to empirical propositions (or protocol sentences) directly describing given facts. And that meant that the terms in which meaningful propositions were formulated had to be definable in terms in which directly given facts are describable.

The impossibility to recast scientific and first of all physical theories according to reductionist prescriptions and the gradual liberalization of the criterion of empirical significance changed the picture considerably. The replacement of the criterion of verifiability by the criterion of confirmability presented an opportunity to give up not only the requirement to reduce all meaningful propositions to empirical propositions, but also the requirement to define all scientific terms on the basis of empirical terms.

In *Foundations of Logic and Mathematics* (1939), Carnap expands his principle of tolerance formulated in *The Logical Syntax of Language* and claims that there are two possibilities to proceed in the construction of a system of terms which do not have uniform degrees of abstractness: some are more elementary than others and may be directly applied in concrete cases of observation, while others are more abstract. "The *first method* consists in taking elementary terms as primitive and then introducing on their basis further terms step by step, up to those of highest abstraction" (Carnap 1939: 205). Typical elementary terms, according to Carnap, are such properties of things as bright, red, warm, sweet, hard. For the introduction of terms of a higher level of abstractness on the basis of elementary terms, explicit definitions are not always necessary; so-called reduction sentences may be also used.

Carnap notes that in its first phases of development, science limited itself with the formulation of laws cast in terms of a low level of abstractness and proceeded by defining other terms on the basis of elementary ones. "But it turns out," says Carnap,

> that it is not possible to arrive in this way at a powerful and efficacious system of laws. [...] The higher the physicists went in the scale of term the better did they succeed in formulating laws applying to a wide range of phenomena. Hence we understand that they are inclined to choose the *second method*. This method begins at the top of the system, so to speak, and then goes down to lower and lower levels. It consists in taking a few abstract terms as primitive signs and a few fundamental laws as axioms. Then further terms, less and less abstract, and finally elementary ones, are to be introduced by definitions; and here, so it seems at present, explicit definitions will do. (Carnap 1939: 206)

Examples of abstract terms are *electric potential, electric resistance, wave function*. Unfortunately, not all the definitions of less abstract terms on the basis of abstract terms are as yet known. Thus the second method requires you to take also some less abstract terms as primitive. Accordingly, not all particular laws can be proved on the basis of laws formulated in abstract terms only.

The chains of definition of elementary terms on the basis of abstract terms give partial (empirical) interpretation of the latter. Yet not all abstract terms are linked with elementary ones through such chains of definition. Some abstract terms get their meaning only through their link with other (empirically interpreted) abstract terms, i.e., through laws of a high degree of generality which are taken as axioms implicitly defining terms contained in them which do not have more direct empirical interpretation. Thus, the meaning of the abstract terms always remains incomplete.

In the late thirties Carnap still spoke about elementary and abstract terms of a theory as terms of one language system. The ideal of a unified science—and *Foundations of Logic and Mathematics* appeared as part-three of volume one of the *International Encyclopedia of Unified Science*—was still guiding his thought. Yet in the fifties he came to the conclusion that it is more advisable—for gaining more clarity—to make a distinction between observation language and theoretical language as sublanguages of the general language of science. "The first contains sentences such as 'This thing is hard, white and cold'; the latter such as 'At the space time point with the coordinates x, y, z, t the components of the electric field have such and such values" (Carnap 1975 [1958]: 75).

The observation language involves only elementary logic. Theoretical language contains very extensive logic which includes all of classical mathematics (let's recall that Carnap did not change his earlier views on the relation of logic to mathematics). According to Carnap, the language of mathematics is the basis of the theoretical language. The latter is obtained from the first by adding the descriptive (i.e., nonlogical or non-mathematical) terms. Such terms are *indispensable* for the formulation of laws of physics and other factual sciences, and Carnap notes that Galileo's old assertion that the book of nature is written in the language of mathematics is obviously wrong.

Carnap calls theoretical terms "T-terms" and observational terms "O-terms." The language of theoretical science is constructed by laying down two kinds of postulates. "First the theoretical postulates, or T-postulates, which contain no O-terms but only T-terms, and secondly, the correspondence postulates, or C-postulates, which contain both O-terms and T-terms. The T-postulates might be fundamental laws of the scientific theory concerned. The C-postulates link the T-terms with the O-terms and thereby give certain empirical, 'operational' rules for the use of the T-terms" (Carnap 1975 [1958]: 79).

Thus in the fifties Carnap clearly opts for the second method of constructing the language of science, a method he mentioned years before in the *Foundations of Logic and Mathematics*. Mathematical calculus forms the backbone of a scientific theory. It is first constructed "floating in the air, so to speak; the construction begins at the top and then adds lower and lower levels" (Carnap 1939: 207). By adding descriptive T- and O-terms to the logico-mathematical signs of the calculus and introducing theoretical and correspondence postulates, we change calculus "floating in the air" into a theory of factual science which is anchored at the ground of observable facts.

Yet even after that, not all sentences of a theory of factual science acquire "fully established" meaning. One reason is the following: not every

T-term is linked by correspondence postulates with O-terms and these postulates to not have the character of explicit definitions. Thus correspondence postulates of the type "when one body is warmer than another the temperature of the first is higher than that of second" do not establish fully the meaning of the term *temperature* (it is worth noting that later Carnap regards even such terms as *length, temperature, mass,* and *pressure* as theoretical). Another reason why sentences of a theory of factual science do not have completely established meanings is the fact that the meaning of a sentence may be remolded by adding further C- or T-postulates. Moreover, in factual sciences there is no certainty that basic laws taken as theoretical postulates will be in accord with all future observations. Thus a physicist or biologist must always be prepared to modify his system and partially change meanings of its theoretical terms.

After the observational and theoretical languages of science were finally separated and the fact that in advanced sciences some empirical terms are defined on the basis of theoretical ones was acknowledged, the *movement for the unity of science* suffered a quite serious blow. Theoretical terms of various disciplines are very different and their list can never be completed. They cannot play the role of relatively small, clearly identifiable, and homogeneous set of empirical terms on the basis of which it is possible to define all *other* scientific terms. The idea of conceptual bedrock of science on which all its conceptual structure rests and which is able to ensure its conceptual unity had to be dropped. From the end of the fifties the unity of science movement has been guided only by the ideal of unity of the *method* of science and not by that of the unity of its *language*.

Constructing a scientific theory and establishing the meaning of its terms is one of the two most important and related issues the philosophy of science has to deal with when examining scientific knowledge. Another one is testing a theory. On the whole, Carnap accepts the general scheme of the hypothetico-deductive method. All the propositions of a theory of factual science are hypotheses. From the basic laws of a theory, the laws of a lower degree of abstractness and generality are derived, and from them propositions of an even smaller degree of generality, until propositions formulated in observational terms are finally reached. In the process of deduction additional presuppositions of a different kind are, of course, used. Elementary propositions are compared with observed facts. If they match them, the whole system of propositions making up a theory is confirmed. If not, the theory is disconfirmed.

Carnap does not juxtapose the confirmation and falsification of a theory as sharply as does Popper. Neither confirmation nor disconfirmation is

conclusive. New facts may change our opinion as to the empirical ad-
equateness of a theory, i.e., its matching facts. The propositions in which a
researcher formulates the results of his observations or experiments are
hypothetical as well: Carnap adopted this viewpoint already in *Testability
and Meaning*. Because of the hypothetical character of observational
statements and due to the complicated process of deduction of propositions of the lowest level of generality, the fact of contradiction of the actual results of observation and the results expected on the basis of the theory cannot be interpreted as falsification of the theory in a Popperian
sense. Empirical facts neither verify nor falsify (in the strict sense) a theory, they just confirm or disconfirm it. And, of course, it is really difficult
to single out a specific proposition used in the process of deduction which
is confirmed or disconfirmed by the results of observations. Carnap was
aware of the problem Duhem faced: the propositions of a scientific theory
cannot be tested independently from other propositions. He did not try to
make specific methodological recommendations on how to deal with the
problem of the incompatibility of the results of observation and of an abstract scientific theory. Carnap believed that the logic of science is not up
to such a task. At the level of pragmatics, he suggested that the hypothetical propositions involved in the process of deducing troublesome (i.e.,
incompatible with results of observation) conclusions should be retained,
modified, or rejected on the basis of the considerations of convenience.

The spectrum of the choices involves, according to Carnap, the possibility of replacing some hypotheses but also replacing the *whole language
system* or linguistic framework, thus introducing a new way of speaking
subject to new rules, and a wholly or partially new system of theoretical
concepts. When or how this is to be done, Carnap does not try to specify.
He does not think, it seems, that such a universally applicable rule may be
proposed at all. The decision to change a linguistic framework depends on
the evaluation of the results of observation and, maybe, the exhaustion of
possibilities to modify an old theory (conceptual system) as Kuhn and
Lakatos suggested later.

However, according to Carnap, the logic of science may be used for
examining the results of testing *some* hypotheses. These hypotheses must
be of a rather low level of generality and abstractness and have to be formulated in the same terms that empirical evidence for (or against) them is
formulated. Normally they are not part of some wider theory or, at least,
testing them does not bind you to some specific theoretical presuppositions. Such hypotheses play an important role in the sciences, whose degree of theorization is substantially lower than that of physics, especially

in biology, medicine, and sociology. According to Carnap, it is possible to create a logical theory which may be used for determining the degree of confirmation of such hypotheses by empirical evidence. According to the version of the criterion of empirical significance proposed by Carnap in *Testability and Meaning*, all scientific hypotheses must be confirmable by empirical evidence. It is not so easy to determine a numerical value of the degree of its confirmation by available empirical evidence. Yet at least in some cases the latter problem may be solved and Carnap devoted a great part of the time he spent in America to create a logical theory which could be used for the appraisal of scientific hypotheses of a relative low degree of abstractness.

The Probabilistic Appraisal of Hypotheses [1]

It seems that Carnap always was inclined to treat the laws of the factual sciences as hypotheses. Duhem's influence on Carnap was stronger than on Schlick, and Carnap never entertained the possibility of considering the universal laws of nature which do not meet the requirements of the criterion of full verifiability as rules of inference. Yet *The Logical Structure of the World* reflects the belief he held in the 1920s that even universal laws which have the character of hypotheses must be grounded on a firm *foundation* of empirical facts. Later, however, discussions in the Circle concerning the character and nature of protocol sentences and, especially Neurath's position on the issue, forced Carnap to acknowledge that at least *within* science it is impossible to find its solid foundation. And he was critical as to Schlick's attempts to look for such a firm foundation of scientific knowledge outside science, in personal experience and in confirmations as Schlick understood them. At least from the early thirties Carnap, unlike Schlick, took the thesis that philosophy is logical analysis at face value: there are no meaningful philosophical problems outside the applicability—at least in principle—of the methods of logic. The fact that Carnap quite often called epistemology an applied logic is illustrative of his attitude. And the use of logical methods presupposes dealing with concepts and statements, not with sensations or perceptions.

---

[1] The subject of probabilistic appraisal of hypotheses in (Reichenbach's and Carnap's) logical positivism is dealt at length in my book *Probable Knowledge: The Rise and Development of the Program for Probabilistic Appraisal of Scientific Knowledge in Logical Empiricism*, published in Russian and Polish (Nekrašas 1987), a few ideas of which are presented in this subsection.

The researchers express the results of their observations in statements. They do not try to express their sensations—they formulate results of observation by describing properties of objects under observation. When making observations, researchers do not scribble anything into their notebooks resembling Schlick's confirmations. And real observation (or protocol) statements are fallible and hypothetical. Even the best observers sometime make mistakes. In *Testability and Meaning* after establishing that complete verification of the laws of science is impossible, Carnap writes:

> Now a little reflection will lead us to the result that there is no fundamental difference between a universal sentence and a particular sentence with regard to verifiability but only a difference in degree. Take for instance the following sentence: "There is a white sheet of paper on this table." In order to ascertain whether this thing is paper, we may make a set of simple observations and then, if there still remains some doubt, we may make some physical and chemical experiments. Here as well as in the case of law, we try to examine sentences which we infer from the sentence in question. These inferred sentences are predictions about future observations. The number of such predictions we can derive from the sentence given is infinite; and therefore the sentence can never be completely verified. (Carnap 1950 [1936/37]: 425)

The truth-value of a sentence which can never be completely verified is never known with absolute certainty. Thus it must be treated as a hypothesis. If subsequent test-observations confirm it, it is possible to speak about its "increasing confirmation," if not, its negation is confirmed. What to do, however, if we do not have one but more hypotheses competing each with other? It would be desirable to know the degree of confirmation of each hypothesis to be able to compare them and select the best on the basis of one or several criteria—taking into account maybe not only the degree of confirmation but, say, simplicity as well. Yet how to measure the degree of confirmation?

In *Testability and Meaning* Carnap provides no answer to this question. At that time, Reichenbach was developing his probabilistic logic in which probability conceived statistically as relative frequency (more exactly, as the limit of relative frequency) was interpreted as a truth-value of a hypothetical statement. Carnap held the view that a two-valued logic is sufficient for the analysis of scientific statements. We do not know the exact truth-value of a hypothesis, but every hypothesis is either true or false. The *degree of confirmation* may have, however, more than two numerical values. Already in the mid-forties, Carnap came to the conclusion that the degree of confirmation can be interpreted as a probability taking any value between zero and one. He conceived this probability as a logical, not a statistical concept (cf. Carnap 1945).

In the first half of the twentieth century, probability became one of the most popular and important concepts of science. Many scholars became aware that probability belongs to the same tier of principal scientific concepts as space and time and is of major interest to philosophers. A rapid dissemination of probabilistic conceptions and the wide use of statistical methods in various branches of science and practice (physics, biology, insurance) influenced the comprehension of the nature of the laws of science. The account of a natural or social process in terms of probability started to be regarded as a perfectly legitimate form for the results of scientific inquiry. The probabilistic picture of the world became entrenched in the natural sciences. At the same time it became clear that the classical conception of probability as the ratio of favorable to all possible chances is unsatisfactory for many reasons. It is easy to count equal chances in a game of cards (in the sixteenth and seventeenth centuries the fledging mathematical theory of probability dealt almost exclusively with the problems of games of chance), but how to count them when determining the probabilities of natural occurrences? Already for this simple reason classical probability cannot be widely used in science. Thus the problem of the nature of probability or, in other words, that of its interpretation, became an urgent question both for scientists and philosophers. As was already mentioned, the problem of the strict definition of the concept of differential which came into wide use already in the eighteenth century became a serious challenge to scientists of the nineteenth century. In the first half of the twentieth century, scientists and philosophers faced a similar challenge: the concept of probability was used increasingly widely in various branches of natural and social science, but its adequate definition was clearly lacking.

When discussing the nature of probability, most of the members of the Vienna Circle tended to the frequency or statistical interpretation of probability which characterizes probability as the limit of relative frequency. According to this interpretation, the probability that a newborn baby is a boy is equal to the limit of frequency (or, simplifying a bit, frequency in the long run) of newly born boys among all newly born babies (this frequency is about 0.51). This interpretation, which goes back to the nineteenth century and often is ascribed to John Venn, was advanced in the twentieth century by Richard von Mises, who took part in the activities of both the Vienna Circle and the Berlin group. Logical positivists contended that the frequency interpretation of probability is adequate wherever the apparatus of probability theory and mathematical statistics is used for the analysis of natural and social events.

In the *Tractatus*, Wittgenstein gave a different definition of probability which was based on the concept of the logical range of proposition. In sharp contrast to frequency theorists, he conceived probability statements as analytical, not factual. His conception—unelaborated as it was—was similar to that developed at the same time by John Maynard Keynes (cf. Keynes 2010 [1921]) which in turn recalled a conception presented a few years earlier by Jan Łukasiewicz (cf. Łukasiewicz 1913). And when Waismann, inspired, it seems, mainly by Wittgenstein, published in 1930 his article on the logical analysis of the concept of probability, he started from the declaration that, like Leibniz and Bolzano, he regards the theory of probability to be a branch of logic (cf. Waismann 1930/31: 228).

Initially Carnap looked critically at developments in the logical theory of probability. Yet the conviction that no factual statements can be completely verified led him after some hesitation to conclude that probability can be ascribed not only to (random) events but also to propositions, and that in the latter case it must be interpreted as a logical notion.

The origins of the idea of a probabilistic evaluation of knowledge are mainly related to the tradition of skepticism, most likely starting with Karneades. Within the positivist tradition Hume, of course, is the main source. Yet the idea that our factual knowledge is only hypothetical had to compete hard with the much stronger notion of the absolute certainty and perfect reliability of scientific knowledge. The latter ideal, originating from the ancient opposition of *episteme* and *doxa*, played an important role in the formation of modern science. It was closely related to the firm belief in the unlimited capacities of science. Thus it was quite natural that it was rather well entrenched in positivism. Specifically, Comte, Mill, and Mach shared this ideal. And as we saw there were some proponents of this ideal even inside the Vienna Circle although the scientific revolution of the twentieth century decisively shattered, it seemed, the last bastion of absolutely certain knowledge in the natural sciences, i.e., Newton's mechanics.

With hypothetism finally prevailing, the preconditions for carrying out the idea of probabilistic appraisal of scientific knowledge within logical positivism were created. Of course, hypothetism does not imply probabilism. Yet already the classical conception of probability links probability with knowledge or rather the lack thereof. Thus the interpretation of probability as an *epistemic* notion was not something unexpected. What was new for Carnap's approach to the problem of probability was a proposal to use two different interpretations of probability. He was not very interested in the *pure* mathematical theory of probability which, as he knew, may be built on the basis of axioms proposed by Andrey Kolmogorov in 1933, but begs the question of how to measure the value of probability.

Those two interpretations of probability are quite disparate. Probability$_2$ is a notion used in factual sciences which defines an event's probability as the limit of its relative frequency in large number of trials. On the other hand, probability$_1$ is not a characteristic of random events, but a degree of confirmation of one proposition by another, i.e., the degree of support one proposition lends to another.

Probability$_1$ as a degree of confirmation of one proposition (hypothesis) by another (evidence) is a logical relation between two propositions which has some similarity to logical implication and may be treated so to say as partial logical implication. This probability is determined by the meanings of these propositions and thus represents a semantic notion of confirmation. Carnap's notion of logical probability is directly relevant to the analysis of inductive reasoning. He begins his main treatise on probability and induction, *Logical Foundations of Probability*, with the following words:

> This book presents a new approach to the old problem of induction and probability. The theory here developed is characterized by the following *basic conceptions*: (1) all inductive reasoning, in the wide sense of nondeductive or nondemonstrative reasoning, is reasoning in terms of probability; (2) hence inductive logic, the theory of the principles of inductive reasoning, is the same as probability logic; (3) the concept of probability on which inductive logic is based is a logical relation between two statements or propositions; it is the degree of confirmation of a hypothesis (or conclusion) on the basis of some given evidence (or premises); (4) the so-called frequency concept of probability, as used in statistical investigations, is an important scientific concept in its own right, but it is not suitable as the basic concept of inductive logic; (5) all principles and theorems of inductive logic are analytic; (6) hence the validity of inductive reasoning is not dependent upon any synthetic presuppositions like the much debated principle of the uniformity of the world. (Carnap 1951 [1950]: v)

All these six conceptions correlate well with the fundamental principles of the philosophy of logical positivism. Crucially, the distinction he draws between logical and statistical probability enables you to juxtapose two kinds of statements in terms of probability as follows: statements in terms of logical probability are analytic; statements in terms of statistical probability are synthetic. And it is quite natural for Carnap to claim that principles and theorems of the theory of logical probability, i.e., of inductive logic, are analytic. Logic cannot be based on nonanalytic principles. The theorems of inductive logic are *deductive* consequences of these principles, thus they are analytic statements as well. However, the cogency of all six basic conceptions of Carnap's theory of probability and induction were questioned by various researchers, and as time passed, Carnap gave up some of them himself.

Carnap's analysis of the concept of probability is a classic example of a logical analysis in the strictest sense of the term. The best form of the logical analysis of a concept is, according to Carnap, the logical construction of an adequate *explicatum* of a concept. The task of explication consists, in Carnap's words, "in transforming a given more or less inexact concept into an exact one or, rather, in replacing the first by the second" (Carnap 1951 [1950]: 2). The voluminous *Logical Foundations of Probability* supplemented two years later by the *Continuum of Inductive Methods* is devoted to presenting an *explicatum* of a concept of logical probability.

Its significance is wider than that of an explication of one of the important notions used by contemporary science. Carnap and other logical positivists were convinced that his inductive logic substantially widens the area of problems that logic may solve. Deductive logic analyses the notion of deductive inference which leads from more general statements to those more specific and closer to experience. Thus it serves the deductive systematization of knowledge. Inductive logic, in a Carnapian sense, serves the inductive systematization of knowledge, i.e., its systematization in the opposite direction: from experience towards general statements because inductive (logical) probability "characterizes the status of any scientific hypothesis, e.g. prediction or law, with respect to given evidence" (Carnap 1951 [1950]: viii). By clarifying the concept of logical probability or degree of confirmation, Carnap seeks to clarify the logical nature of induction. The problem of induction is, according to him, the same "as the problem of the logical relation between a hypothesis and confirming evidence for it. [...] While deductive logic may be regarded as the theory based upon the concept of logical deductibility, inductive logic is the theory based upon what might be called the degree of inducibility, that is, the degree of confirmation" (ibid., 2).

Carnap's explication of the concept of logical probability is based on the assumption, opposite to that of Keynes, that this concept is not unanalyzable. The basic notion used by Carnap while explicating this concept is that of a state-description. This notion is widely used by Carnap in his semantics. It is a formal analogue of Leibniz's "possible world" and a notion used by Wittgenstein. State-description in a given formal language is a conjunction indicating, for each individual, which qualities expressible in a given language it has. Thus it is the most complete description of the state of the universe in terms of a given language. Based on the notion of the state-description, it is not difficult to introduce a semantic notion of the range of a proposition as a set of state-descriptions for which that proposition is true. Consequently, an additive measure for such sets is introduced and logical probability is defined as the ratio of the measure of the conjunction of hypothesis and evidence $m(h\&e)$ and the measure of evidence $m(e)$. Thus degree of confirmation is defined in a similar way as in the mathematical theory of probability in which a conditional probability is defined on the basis of an unconditional probability. The main difference is the following: Carnap's inductive logic enables you to calculate numerical values of m for all propositions expressible in the given formal language.

It is not possible to present, even in outline form, the ingenious logical construction Carnap developed in the process of erecting the architecture of his inductive logic. *Logical Foundations of Probability* is the biggest book ever written by Carnap. Those readers who would like to become acquainted with its main logical ideas may refer, for example, to Henry E. Kyburg's *Probability and Inductive Logic* (1970).

The main goal Carnap aimed at in the *Logical Foundations of Probability* was to define the *correct* measure m and thus find correct (or the best) inductive method, i.e., first of all the function c [$c(h/e) = m(h\&e)/m(e)$]. Yet he soon dropped the idea and instead of the second volume of the *Logical Foundations of Probability*, he published the *Continuum of Inductive Methods*, in which he presents a set of different functions c out of which every researcher interested in an inductive appraisal of his hypothesis may choose one and use it consequently. (This move may be regarded as a deepening of his logical principle of tolerance.) Yet later he came to the conclusion that even his continuum of inductive methods is not wide enough, and the possibilities of its application for the appraisal of hypotheses are too limited, and in "A Basic System of Inductive Logic" published posthumously in 1971, he presented an essentially new system which became known as his "new inductive logic." This new point of view originated by Carnap in the sixties placed the problems of inductive logic within the context of analyzing rational decisions. Under pressure of criticism, he decided that it is rather wrong to identify logical probability with the degree of confirmation (the latter may be defined, however, in terms of probability) and started to call logical probability the degree of firmness, not of confirmation. The most important element of the new approach was a quasi-psychological construct of the rational agent whose behavior is based on purely intellectual dispositions which do not depend on acquired experience although the decisions an agent makes depend upon experience acquired in a specific empirical situation.

In the sixties Carnap's views concerning both the goals and foundations of inductive logic suffered important changes. From the thirties to the fifties, Carnap's methodology (or philosophy of science) was mainly apragmatic, i.e., it concentrated on the structure of such cognitive items as concepts, propositions, and their relations. It described the structure of science in purely logical terms and did not take into account the pragmatically motivated behavior of the researcher. Apragmatic theory does not involve any references to the *subject* of knowledge and is not interested in his *actions*. In this respect, both Carnap (from the thirties to the fifties) and Popper were engaged in a similar enterprise: they were seeking an epistemology without a subject of knowledge.

The program of constructing a probabilistic inductive logic was undertaken in logical positivism in the forties as an extremely important part of the comprehensive project of establishing an apragmatic theory of scientific knowledge. The need for a probabilistic inductive logic was conditioned by the patent insufficiency of deductive means for a complete description of relations among propositions of different epistemic status, mainly those between observational statements and hypotheses. Being convinced that philosophical investigation consists in rational reconstruction, Carnap saw in the forties and fifties the main goal of inductive logic in the explication of the semantic concept of confirmation, specified as a partial logical implication relating evidence and hypothesis.

Although at that time Carnap believed that inductive logic might become a basis of such a theory of science, eventually he concluded that it is impossible to determine which inductive method is the best for all researchers and research situations, and impossible to indicate how to precisely calculate the numerical values of the degree of confirmation and firmness. Moreover, he decided that even epistemic decisions (the choice of a hypothesis) involve preferences which are of a personal nature. Accordingly, in the sixties cardinal changes took place in Carnap's attitude towards the goal of rational reconstruction in general, and of inductive reconstruction in particular. The object of reconstruction was taken to be the behavior of a rational individual (not necessarily of a scholar or researcher). In this and other respects "A Basic System of Inductive Logic" and Carnap's late works on the methodology of inductive logic substantially differ from *The Logical Foundations of Probability* and *The Continuum of Inductive Methods*. The far-reaching revision of the goals, and in consequence, of the foundations of inductive logic, was caused by a joint action of both internal and external factors.

The main internal catalyst was the awareness that the results obtained in the process of developing inductive logic had not lived up to the very goals in the pursuit of which the project of probabilistic inductive logic had been undertaken. The hope of constructing an inductive logic not merely for the "simple language forms" for which it was actually developed, but also for the far more complex ones characteristic of modern science, all turned out to be futile, despite some increase in the descriptive power of the language of inductive logic. The gradual extension of the class of admissible inductive methods in the process of the development of inductive logic and the consequent handing over to the specialists in specific sciences, not the logicians, the task of choosing a particular inductive method from a broad class of such methods, testified to the importance of the role subjective attitudes and personal valuations play in the application of inductive logic.

Having admitted that a philosopher of science cannot abstract from the fundamental role that the individual plays in making decisions on the basis of his own pragmatic considerations, Carnap had made revisions not only to the goals of inductive logic but also to its basic concepts, primarily that of logical probability, by reinterpreting it in personalist terms (interpreting logical probability as a specific kind of personal probability) and claiming that the real meaning of the concept is to be understood only within the context of the analysis of decision-making procedures. Carnap always made a distinction between formal inductive logic and discussion of problems related to its application. Yet in his late years, he decided that if we want to know why inductive logic is being constructed on the basis of particular axioms, we must turn to the notion of a rational *subject* and to examine his *decisions*. Thus an earlier apragmatic theory must be given up and replaced by more adequate pragmatic one.

A reinterpreted inductive logic could easily be used as a logical basis for the theory of rational action. The new theories of rational choice evolving independently from inductive logic and in need of a foundation, as well as the shift of interest in analytic philosophy towards various aspects of human activity, were important external factors conducive to such a reinterpretation of inductive logic.

A methodological reflection on the capacities of his own variant of inductive logic had forced Carnap to conclude that the main domain of its applicability was the domain of practical, as opposed to theoretical, activity, for it is the former, in distinction to the latter, that involves singular, not universal hypotheses. In spite of the promising results obtained by Jaakko Hintikka and other scholars in exploring the possibilities of assigning positive probability to the universal hypotheses, Carnap firmly held the view that their probability must be zero. The null probability of universal hypotheses may be regarded as a serious obstacle for their inclusion in the body of accepted knowledge. Yet it is worth mentioning that Carnap, the later Carnap in particular, was opposed to attempts at constructing any rules whatever for the acceptance of hypotheses, though he did not think acceptance or rejection of a hypothesis is an illegitimate act. One of the reasons was the fact that in the practical domain, as opposed to the theoretical one, there is no need to accept hypotheses. In Carnap's view, accepting a hypothesis in the practical domain is counterproductive, since putting us under obligation to *always* act in such a way as if we were certain of the hypothesis's truth may lead to catastrophic results. In practical action it is always advisable to take into account even those hypotheses (e.g., a total failure of some technical complex such as a space ship or an A-plant) whose probability is very low.

Carnap, the logical positivist, long held the view that philosophical reflection might lead to a modification of the results of the nonreflective intuition. Yet in the sixties, in "Inductive Logic and Inductive Intuition," he concluded that the results of intuition are irrefutable. Carnap changed his mind in the process of reexamining his earlier solution of the problem of the justification of induction. Both for empiricism and positivism this problem caused endless complexities. The solution of the problem proposed by Carnap was, as we saw, rather simple: the principles of inductive logic are analytic statements. The link between justifying induction and establishing the principles of inductive logic seemed obvious: if we know the numerical values of the inductive probability of a hypothesis on the basis of given evidence, we know the degree of (inductive) confirmation of a hypothesis by the given (empirical) data. The essential point in inductive reasoning is, according to Carnap, the determination of probability values. Hence the problem arises on what ground we may calculate them.

They are determined on the basis of axioms of an inductive logic. Why do we accept some axioms and reject their alternatives? In the fifties Carnap, who was a pioneer in building a probabilistic inductive logic able to calculate precisely the degree of confirmation of any proposition by any other proposition (with the stipulation that both propositions belong to the same formalized language), was not much interested in an alternative probabilistic logic of such a kind, because there were simply no other such logics available. Yet in the late sixties, after creating more of such logics, the problem of choosing their axioms emerged; it recalled the problem of choosing axioms of geometry which surfaced in the nineteenth century. In "Inductive Logic and Inductive Intuition," Carnap concluded that the real basis of inductive reasoning is inductive intuition.

Carnap writes the following:

> I think that it is not only legitimate to appeal to inductive reasoning in defending inductive reasoning, but that it is indispensable. This is a rather audacious statement to make, because it looks like defending a vicious circle. But I believe all procedures of self-clarification, of making clear to ourselves what it is we have in mind, are in a way circular. We clarify B through A, and then we turn around and explain A with the help of B. I think we cannot do without it. If a person were unable to distinguish valid from invalid steps in inductive reasoning, even in the simplest cases, in other words, if he were inductively blind, then it would be hopeless to try to convince him of anything in inductive logic. In order to learn inductive reasoning, we must have what I call the ability of *inductive intuition.* (Carnap 1968: 265)

According to Carnap, intuition is not a source of infallible knowledge. Yet we do not have a better one. Thus we must base ourselves on intuition when choosing axioms of inductive logic.

This conclusion is of major importance because it is incompatible with the basic principle of logical positivism. The members of the movement, including Carnap himself, long held the opinion that only logic and experience are sources of genuine knowledge. Stunningly, in the late sixties, the leader of the movement contested its basic principle. Yet the significance of the notion of inductive intuition he introduced is very doubtful: it does not explain why different logicians—in spite of relying on an allegedly universal inductive intuition—came forth with widely differing systems of inductive logic.

Scientific Humanism and Socialism

Contrary to Schlick, Carnap was not interested in the problems of ethics. He was much more fascinated by politics. In his "Intellectual Biography" Carnap testifies: "I have not been active in party politics, but I was always interested in political principles and I have never shied away from professing my point of view. All of us in the Vienna Circle took a strong interest in the political events in our country, in Europe, and in the world" (Carnap 1963a: 82). Matters related to political principles were discussed in the Circle privately because all of its members—with the exception perhaps of Neurath—held the view that the Circle must remain a place in which only theoretical questions must be discussed. Carnap also regarded political issues as practical questions and not theoretical ones. Yet he emphasized that in the Vienna Circle at least three political—in the broad sense of the word—views hardly needed any discussion because they were shared by all members of the Circle.

> The first is the view that man has no supernatural protectors or enemies and that therefore whatever can be done to improve life is the task of man himself. Second, we had the conviction that mankind is able to change the conditions of life in such a way that many of the sufferings of today may be avoided and that the external and the internal situation of life for the individual, the community, and finally for humanity will be essentially improved. The third is the view that all deliberate action presupposes knowledge of the world, that the scientific method is the best method of acquiring knowledge and that therefore science must be regarded as one of the most valuable instruments for the improvement of life. In Vienna we had no names for these views; if we look for a brief designation in American terminology for the combination of these three convictions, the best would seem to be "scientific humanism." (ibid., 83)

Besides these views shared with others members of the Circle, Carnap indicates in his "Intellectual Biography" a few others which he did not share with Schlick but which are close to the views of Neurath. Some other logical positivists, and first of all, Reichenbach shared these views in many essential features. Carnap claimed that he held these views at least since his Vienna years. Let him speak at length for himself because his "Intellectual Biography" is the only publication in which he presented them to the wider public.

> It was and still is my conviction that the great problems of the organization of economy and the organization of the world at the present time, in the era of industrialization, cannot possibly be solved by "the free interplay of forces," but require rational planning. For the organization of economy this means socialism in some form; for the organization of the world it means a gradual development toward a world government. However, neither socialism nor world government are regarded as absolute ends; they are only the organizational means which, according to our present knowledge, seem to give the best promise of leading to a realization of the ultimate aim. This aim is a form of life in which the well-being and the development of the individual is valued most highly, not the power of the state. Removing the obstacles, the main causes of suffering, such as war, poverty, disease, is merely the negative side of the task. The positive side is to improve and enrich the life of the individuals and their relations in family, friendship, professional work, and community. (Carnap 1963a: 83)

It is not certain when Carnap wrote these words. The publication of Schilpp's volume on Carnap was delayed several times: it should have appeared in 1954 but was published only in 1963. Carnap's phraseology looks characteristic of the fifties. Yet, as he claims himself, he had not changed his political views from the thirties, thus he hardly changed them in the sixties. Hence we can assert that politically Carnap was a socialist. Admittedly, he understood well that there are grave dangers in the increasing the power of the state. Both rational planning at a societal level and a world-government level implies a strong state. Such a state imperils civil liberties and democratic institutions. Carnap was convinced, however, that it is possible to find a way of organizing society "which will reconcile the personal and cultural freedom of the individual with the development of an efficient organization of state and economy" (Carnap 1963a: 84). Yet, regrettable as it may be, humanity has not yet found such a way of organizing society, and moreover, probably lost forever the hope that "rational planning" on the state level is a means of "efficient organization of state."

PART TWO

Impact

CHAPTER 5

Positivism,
Its Critics and Rivals

Positivism and Two of its Adversaries: Nietzsche and Heidegger

Positivism's place in philosophy—its relation to other trends—is far from being finally determined. It is very likely that many philosophical movements are indebted to it more than they may admit. There is a clear need to define more precisely positivism's place in, and influence on, the philosophy of the nineteenth, twentieth, and twenty-first centuries.

Analyzing the relation between an influential philosophical movement and other movements is not an easy task. Yet recently the need for such analysis has become especially acute because of the sharp postmodernist critique of positivism. According to many postmodernists, positivism is the most dangerous embodiment of the diabolical spirit of modernism. The comparison of positivism with present-day postmodernism will be postponed until the end of this chapter. In this section I would like to concentrate on the relation with positivism of two philosophers active in the last third of the nineteenth century and in the first two-thirds of the twentieth century, and thus to begin the project of reflection and analysis which a movement as influential as positivism has long deserved.

Positivism's influence upon the nineteenth century is attested by the name attributed to it by some authors: *the positivist century*. During the century bearing its namesake, positivism had many friends. Yet its foes were by no means few. The positivist movement was very critical of many traditional and newer philosophical schools; therefore, it is quite understandable that it brought upon itself an open and sometimes passionate critique.

Positivist philosophy was a preeminent trend of post-Hegelian philosophy (the year Hegel died—1831—is often considered to be a relative borderline between classical and modern philosophy). Yet, at the same time, it was only one of a group of rather kindred trends of thought. The general trait or constitutive feature uniting this family of trends was the sympathetic attitude of all trends towards science, its methodology, and promises of objectivity and certainty, an attitude characterized by their aspiration to be or to become scientific philosophies, even though different trends treated the latter differently. This group includes Marxism, pragmatism, critical rationalism, and some other less important schools. Analytic philosophy, a very important member of this group, is a special case because logical positivism itself was one of its forms.

Sharing some important attitudes and principles, members of this group, nevertheless, frequently are rivals who are (or were) competing for dividends to be earned from close association with modern science which, as many would claim, has been shaping and continues to shape Western civilization. The relations between positivism and kindred philosophies included elements of both sympathy and rivalry, even when, in some cases, this sympathy was carefully concealed. Younger philosophies close to positivism had an interest in emphasizing their differences from their philosophical forebear and did their best to prove their superiority. At the same time, they were either unwittingly or unadmittedly dependent upon it to a rather significant degree.

The relations of philosophical trends outside this group with positivism were substantially different. The trends of philosophy which did not regard science as their beacon, were, in general, more adverse to positivism—after all, positivism treated them as forms of metaphysics and hence strived to destroy them, or at the very least, to belittle their significance, as they had nothing to do with knowledge and merely simulated theoretical content while it was completely absent. Positivism had quite a few philosophical adversaries. However, most would agree that the most prominent figure among those in the nineteenth century was Friedrich Nietzsche, and in the twentieth century—Martin Heidegger.

Heidegger once called Nietzsche the last metaphysician. Yet Nietzsche, as is well known, was rather critical of traditional, especially Platonic, metaphysics. In the first pages of *Beyond Good and Evil*, Nietzsche asks rather ironically:

> *How could* anything originate out of its opposite? For example, truth out of error? Or the Will to Truth out of the will to deception? Or the generous deed out of selfishness? Or the pure sun-bright vision of the wise man out of covetousness? Such genesis is impossible; whoever dreams of it is a fool, nay, worse than a fool; things of the highest value must have a different origin, and origin of *their* own—in this transitory, seductive, illusory, paltry world, in this turmoil of delusion and cupidity, they cannot have

their source. But rather in the lap of Being, in the intransitory, in the concealed God, in the "Thing-in-itself—*There* must be their source, and nowhere else!"—This mode of reasoning discloses the typical prejudice by which metaphysicians of all times can be recognized, this mode of valuation is at the back of all their logical procedure; through this "belief" of theirs, they exert themselves for their "knowledge," for something that is in the end solemnly christened "the Truth." (Nietzsche 1917 [1886]: 2)

At least as a critic of traditional metaphysics, Nietzsche may be regarded not as an opponent but as an ally of positivism. Moreover, his sarcasm and irony is, likely, a more effective weapon than the criterion of empirical significance. However, Nietzsche's attitude to positivism has been inconsistent and has varied with time. Nietzsche has been closest to positivism in his so-called Middle Period—between 1876 and 1882. "It is indisputable," notes Heidegger, "that prior to the time of his work on the planned magnum opus, *The Will to Power*, [...] Nietzsche went through a period of extreme positivism. [...] Such positivism, though of course transformed, also became a part of his later fundamental position" (Heidegger 1991 [1961]: 154). Thus, in applying the methodology of his genealogy, he aspires to be a researcher of the kind scientists are, interested in examining without sentiment or bias the development of morals and explaining the consequences of adopting different systems of morality.

This, obviously, does not mean that Nietzsche was an objective investigator of the evolution of morals. The partisan moralist (or antimoralist) in him quite often takes the upper hand over the impartial researcher. But the same may be said about nineteenth century positivists, especially Comte, even though their moral views have been directly opposite to Nietzsche's. It seems that one of the main differences between Nietzsche and the nineteenth century positivists is his denial of any link between the development of science and social progress. Opposing positivism, Nietzsche claims that the liberation from oppression of social ideals is a condition for the freedom and autonomy of science, and hence for its successful development. In any case, Nietzsche was by no means a supporter of positivist social ideals, which can be seen, for example, from these words from *The Gay Science*: "We hold it absolutely undesirable that a realm of justice and concord should be established on earth" (Nietzsche 2008 [1887]: 241).

Nietzsche was very critical of the alleged positivist tendency to reduce almost all philosophy to the history of scientific method. However, despite regarding Comte as the main representative of this tendency, Nietzsche inscribed Comte's name (next to Aristotle, Bacon, and Descartes) into his short list of the greatest methodologists of all time. According to him, even though positivism has groundlessly ignored the multifacetedness and multidimensionality of human life, it has had a favorable role in overcoming Platonism and the very idea of the *real world*, which Nietzsche took to be the main error of philosophy. In the *Twilight of the Idols* Nietzsche presents a concise history of overcoming Platonism. His account is peculiarly similar to Comte's scheme of the progressive development of the human mind, but contains six, rather than three, stages. Positivism, due to its insistence that only phenomena are knowable, is ranked rather highly there. It immediately precedes Nietzsche's own philosophy which he (like Hegel earlier) obviously regards as the final word in the development of philosophical thought.

Nonetheless, Nietzsche was rather critical of positivism's epistemology and moral theory. Of decisive importance in Nietzsche's argument *against* positivism is his critique of the positivist notion of facts. "Against positivism," says Nietzsche, "which halts at phenomena—'There are only facts'—I would say: No, facts is precisely what there is not, only interpretations. We cannot establish any fact 'in itself': perhaps it is folly to want to do such a thing. 'Everything is subjective,' you say; but even this is interpretation. The 'subject' is not something given, it is something added and invented and projected behind what there is" (Nietzsche 1968 [1886/87]: 267). From Nietzsche's point of view, facts in the positivist sense (i.e., facts as the solid foundation of knowledge) are expression of the longing for support, the firm basis needed for any believer. They are needed for people who crave for something steadfast, unshakable, immovable. Such people strive to be strong and firm. However, this striving is, in fact, an expression of their weakness, not firmness.

> Christianity, it seems to me is still needed by most people in old Europe even today; hence it still finds believers. [...] Metaphysics is still needed by some, but so is that impetuous *demand for certainty* that today discharges itself in scientific-positivistic form among great masses—the demand that one *wants* by all means something to be firm (while owing to the fervour of this demand one treats the demonstration for this certainty more lightly and negligently): this is still the demand for foothold, support— in short, the *instinct of weakness* that, to be sure, does not create sundry religions, forms of metaphysics, and convictions but does—preserve them. Indeed, around all these positivistic systems hover the fumes of a certain pessimistic gloom, something of a weariness, fatalism, disappointment, fear of new disappointment [...]. (Nietzsche 2008 [1887]: 205)

With Nietzsche's claim that every positivist has strived to find a firm foundation for knowledge I cannot agree and I have already discussed the erroneousness of such judgment. At this point, however, it is important to emphasize that according to Nietzsche the instinct of weakness manifests itself not only in positivist epistemology, but also in positivist morals, which conforms to the needs of the weak-minded and weak-willed. He claims that from a moral point of view, positivism does not differ from socialism which, in turn, is a continuation of the Christian tradition. In *The Dawn of the Day* (*Morgenröthe*) Nietzsche claims that Comte "outchristianised" Christianity with his famous formula *vivre pour autrui*—live for the other (cf. Nietzsche 1911 [1881]: 156). Mill's doctrine on sympathetic affection, compassion, and utility, according to Nietzsche, springs out of the same source as Comte's formula. Positivism gives priority to the community over the individual. Thus it inevitably leads to the weakening and overcoming of the individual. He is treated by positivists only as an element and instrument of the Whole, be it State, Nation, or something else. Positivist collectivism (even in its liberal version defended by Mill) is, of course, repulsive to Nietzsche. His moral views, requiring one to contrast the slave morality (based on the Christian tradition) to that of masters, are very different from the altruism of classical positivism.

However, it's necessary to notice one respect in which positivism differs from Christianity. God's children are equal, at least before God. Meanwhile positivistic collectivism does not mean egalitarianism. Equality was not a positivistic credo. Comte has clearly indicated that while analyzing the role of the scientific and political elite in a perfect positivist society. The characters of Comte's *High Priest* and Nietzsche's *Übermensch*, opposites at first sight, have some intriguing common traits.

In spite of certain shared points, the basic assumptions, principles, and especially the style and spirit of Nietzsche's philosophy differ greatly from those of positivism. Martin Heidegger, who openly acknowledged his indebtedness to Nietzsche, was even more remote from positivism. There are almost no points of contact between him and positivist philosophy. Nevertheless, positivists felt open enmity towards him. For positivists, he was the incarnation of metaphysical thinking in the twentieth century. Exactly because of this he was chosen by positivists as the principle target of the critique against metaphysics. Carnap, in his well-known article "The Elimination of Metaphysics through the Logical Analysis of Language," devotes much effort to show that Heidegger's pronouncements break the fundamental rules of logical syntax and lack any cognitive significance. With Nietzsche, positivists were much more forbearing. In his works they discerned few metaphysical meanderings in the genre of Heidegger's "The Nothing itself nothings," and praised him for avoiding the major error of confusing science with art. "We find there," wrote Carnap, "for instance, historical analyses of specific artistic phenomena, or an historical-psychological analysis of morals. In the work, however, in which he expresses most strongly that which others express through metaphysics or ethics, in *Thus Spoke Zarathustra*, he does not choose the misleading theoretical form, but openly the form of art, of poetry" (Carnap 1959 [1932]: 80).

Heidegger's early works were written in a rather intricate theoretical language, not characteristic of Nietzsche. Yet, in his later years, Heidegger leaned toward a more poetic language and style. Thus were Carnap to evaluate the later works of Heidegger, like *Hölderlin and the Essence of Poetry*, his judgment might have been softer. However, in the early thirties, Carnap's verdict was merciless: after analyzing in detail an excerpt from Heidegger's "What Is Metaphysics?," Carnap decided that "a metaphysician himself here states that his questions and answers are irreconcilable with logic and the scientific way of thinking" (Carnap 1959 [1932]: 72).

Heidegger, however, was not struck by Carnap's critique. His calm reaction is easy to explain. In the text analyzed by Carnap he states himself that the very idea of logic is not as fundamental as is often thought—it dissolves in the whirl of a more basic questioning. In another text which contains an answer to Carnap, although his name is never mentioned, namely in *An Introduction to Metaphysics*, Heidegger repeatedly claims that this "more basic questioning" begins with the question "Why are there beings at all instead of nothing?" (Heidegger 2000 [1953]: 1).[1] Not being chronologically first, this question is first in rank for philosophy, because it is the most far-reaching, the deepest, and most fundamental. Basic philosophical questioning is inquiry into the *extra*-ordinary, it is itself "out of order," a "mystery of freedom."

Obviously, positivists never even asked the question why there are beings at all instead of nothing, for, in their opinion, such question is clearly metaphysical. According to Heidegger, in this way positivism is betraying the real aims of philosophy. He claims that asking the question about Being is a part of humanity's history, although the connection of this question to decisive historical questions may seem very remote and indirect. Yet while the human spirit was strong it saw this connection.

Unfortunately the spirit and the world—which is always the world of spirit—is darkening. "The essential happenings in this darkening are: the

---

[1] The text was written in 1935 but published (with slight corrections) for the first time only in 1953.

flight of the gods, the destruction of the earth, the reduction of human beings to a mass, the preeminence of the mediocre. [...] The darkening of the world contains within itself *a disempowering of the spirit*, its dissolution, diminution, suppression, and misinterpretation" (Heidegger 2000 [1953]: 47).

According to Heidegger, this disintegrative process had begun long ago. Yet Heidegger emphasizes that the first half of the nineteenth century was of special importance in this enfeeblement of spirit. Sometimes this (real or supposed) degradation of the spirit is linked with the collapse of German idealism. But Heidegger's point of view is different:

> For it was not German idealism that collapsed, but it was the age that was no longer strong enough to stand up to the greatness, breadth, and originality of that spiritual world—that is, truly to realize it, which always means something other than merely applying propositions and insights. *Dasein* began to slide into a world that lacked that depth from which the essential always comes and returns to human beings, thereby forcing them to superiority and allowing them to act on the basis of rank. All things sank to the same level, to a surface resembling a blind mirror that no longer mirrors, that casts nothing back. The prevailing dimension became that of extensions and number. *To be able*—this no longer means to spend and to lavish, thanks to lofty over-abundance and the mastery of energies; instead it means only practicing a routine in which anyone can be trained, always combined with a certain amount of sweat and display. (Heidegger 2000 [1953]: 48)

Heidegger, like Nietzsche previously, links the degradation of the human spirit with the advance of industrial mass society fostering equality, and subordinating science to the needs of technology. The spirit, according to him, becomes reinterpreted as intelligence, or mere cleverness in examining and calculating things given and the possibility of changing them and complementing them to make new things. Cleverness, practice, division of labor, and organization mark this falsification of the spirit into intelligence, which can be taught, learned, and used as a tool by others.

It's not difficult to notice a connection between positivistic views and this "deterioration of the spirit." But what was positivism's particular role in this process? Was it one of the causes which brought about the deterioration of the spirit, or was it merely a consequence of its emasculation? Heidegger does not give a clear-cut answer to this interesting question. He is more interested to emphasize positivism's narrow-mindedness, stemming from its being only concerned with ordering and explaining everything that is present and already posited in time. Thus science is deprived of its metaphysical roots, and spiritual energies become subject to deliberate cultivation and planning. The spirit misinterpreted as utilitarian intelligence degenerates and loses its ability to stand in openness to the essent. Genuine, nonpragmatic knowledge is unattainable for such spirit.

By dispraising positivism as one of the forms of the emasculation of the spirit (Marxism is another one mentioned), Heidegger dismisses Carnap's criticism contained in "The Elimination of Metaphysics." According to Heidegger, our concern for logic is exaggerated because of the failure to understand the question about the essent and its significance. And this only means that the forgetting of Being is ever increasing.

In a clear reference to Carnap's contrast between the logical form of the statements like "Rain is outside" and pseudo-statements about the Nothing, Heidegger writes:

> One cannot, in fact, talk about and deal with Nothing as if it were a thing, such as the rain out there, or a mountain, or any object at all; Nothing remains in principle inaccessible to all science. Whoever truly wants to talk of Nothing must necessarily become unscientific. But this is a great misfortune only if one believes that scientific thinking alone is the authentic, rigorous thinking, that it alone can and must be made the measure even of philosophical thinking. But the reverse is the case. All scientific thinking is just a derivative and rigidified form of philosophical thinking. Philosophy never arises from or through science. Philosophy can never belong to the same order as the sciences. It belongs to a higher order, and not just "logically," as it were, or in a table of the system of sciences. Philosophy stands in a completely different domain and rank of spiritual Dasein. Only poetry is of the same order as philosophical thinking, although thinking and poetry are not identical. (Heidegger 2000 [1953]: 27–28)

Thus Heidegger's stance towards positivism and its practice of logical analysis is rather clear. Metaphysics and philosophy is not science at all, and the logical requirements that are applicable to the sciences are completely out of place in the context of basic questioning.

Heidegger's attitude towards positivism has slightly changed with time, and in some later texts he seems to take a more conciliatory stance. He emphasizes positivism's favorable role in the development of Western civilization, rather than in its weakening as he previously claimed. On the other hand, he blames even more vigorously all classical metaphysics for the forgetfulness of Being.

According to the later Heidegger, philosophy ends in the present epoch, finally taking the shape of scientific theory and science-based social activity. The end of philosophy means the triumph of science and technology, and, at the same time, the triumph of the social organization adequate to it. The end of philosophy means the beginning of a world civilization based on Western, namely European, thinking. Even though this triumph, for Heidegger, is more of a loss, positivists could only applaud such claims.

However, they would be less happy to hear from Heidegger that science, even science denying its philosophical roots, can never get rid of its philosophical origin and always bears its mark. And because philosophy is, as he maintains, metaphysics, clearly then science cannot rid itself of metaphysics. Moreover, Heidegger repeatedly claims, what is for him self-evident but with what positivists would obviously disagree, that positivism itself is a kind of metaphysics. To be more precise, he admits that positivism is an antipode of metaphysics, but maintains at the same time that the antipode of metaphysics is also a kind of metaphysics. Carnap and other positivists would regard this claim as additional evidence confirming that Heidegger's pronouncements transgress the boundaries of logic. Heidegger, however, makes this claim almost in passing: for him it is self-evident that the denial of metaphysics is itself metaphysics. It is obvious that indignant positivists could reply: "We do not *negate* metaphysics in the sense the term is used in logic: our statements are not negations of the statements made by metaphysicians." To this Heidegger would say that he is using the word *negation* with a sense different from *their* logic. To Carnap's proposal for Heidegger to explicate this sense or meaning and to lay down the rules of his *own* logic (or language), Heidegger would be likely to retort that he is not interested in formal logic. This would exhaust the possibilities of a fruitful discussion.

Claiming that positivism is pure metaphysics, Heidegger maintains at the same time that it is poor metaphysics. In his *Nietzsche*,[2] Heidegger asserts that positivism is *inferior to* Hegel's metaphysics, while being at the same time dependent upon it. In post-Hegelian philosophy Heidegger holds Marx, and of course, Nietzsche, in higher regard. However in later texts, especially those in which he deals with the problem of the *end of philosophy*, he maintains that there is no reason to speak about one philosophy being superior or more perfect than another.

At the same time he claims that the end of philosophy does not mean that no work remains for intellectual endeavor. On the contrary, there is one extremely important task which was accomplished neither by philosophy in the form of metaphysics nor by the sciences originating from it. A global civilization is emerging. Its present basis is the scientific world-outlook. Yet it is quite possible that in the future the global civilization will overcome the scientific conception of a person's place in the

---

[2] The book published in German in 1961 contains lecture courses and notes, dated from 1936 to 1946.

world. Heidegger is convinced that the main task of thinking is to clear the ground for such an overcoming.

It is precisely the task of overcoming of traditional metaphysics, accused by Heidegger of the forgetting of the Being, which unites Nietzsche, Heidegger, and positivism despite all differences in their views. Positivism took up this task first. All later efforts to overcome metaphysics are of an epiphenomenal nature. Heidegger may believe that in overcoming metaphysics he is more radical than positivists who oppose metaphysics while remaining at the same time within its confines. Positivists, however, are of different opinion. They hold that his repudiation of classical metaphysics and resounding return to pre-Socratic philosophy do not make him better than post-Socratic metaphysicians. By deliberately breaking the rules of logic—no matter *why* he does that—Heidegger places himself beyond the boundaries of meaningful discourse and deprives himself of the possibility to say anything of cognitive value.

This positivist judgment about Heidegger may be too severe. It seems, however, that Heidegger's hope that the character of the global civilization which has been shaped by the positive mind may radically change, has so far been proven futile.

Let us raise a simple and straightforward question: did positivism's adversaries win a decisive victory over it? The answer is twofold. On the one hand, the adversaries succeeded in revealing positivism's imperfections and weakening its position. But did its adversaries obliterate it? Did they drive out positivist thoughts, beliefs, and views from philosophy, science, politics, and everyday life? These seminal questions we will attempt to answer in the final chapter.

Positivism, Marxism, and Critical Theory

We should begin the analysis of positivism's relations with kindred currents of thought by attempting to examine positivism's relationship to Marxism. I perfectly understand that the reader may still remain skeptical about the validity of the claim that positivism and Marxism are congenial philosophical trends. He may point out that positivism has been an object of severe Marxist critique and recall Lenin's ruthless attack on positivism in *Materialism and Empirio-criticism* (1909), which spared no invectives and maledictions. And he could add that most historians of philosophy treat Marxism and positivism as currents that are alien to each other rather than akin.

Nevertheless, I claim that the two movements are closely related, although neither positivists nor Marxists were eager to emphasize the similarities of their positions concerning significant issues, such as the importance of science for social progress and the role of the proletariat in establishing the future social order. It seems that Marxism is closer to positivism than it is usually assumed. In this section I intend to present arguments supporting this point of view.

It must be acknowledged right away that both Karl Marx and Friedrich Engels mention positivism and its representatives rather rarely. Both of them think about positivists, and especially Comte, rather critically. When Marx makes more or less favorable reference to J. S. Mill, he has in mind his economical, not philosophical views.

Although Marx had some knowledge of Comte's main ideas already in the 1840s, it looks as if it was rather perfunctory for at least two decades. On July 7, 1866, Marx wrote to his friend and collaborator Engels that he is studying Comte and gave the main reason for his interest: the enormous popularity of Comte among the English and French. His opinion that Comte is far inferior to Hegel, whom he revered much, was expressed clearly in this letter (cf. Marx 1987 [1866]). One of the few remarks on Comte to be found in his books is contained in the postface to the second edition of the first volume of *Capital* (1873). Engels's attitude towards positivism was, it seems, a bit more favorable. Yet he mentions positivists as rarely as does Marx.

Thus if the presence or absence of specific citations and references is regarded a sufficient evidence of intellectual influence on an author or lack of it, it is quite natural to conclude that positivism did not exert any noticeable influence on Marx and Engels. In his well-known book on the development of Marxism, *Main Currents of Marxism: The Founders, the Golden Age, the Breakdown* (1976), Leszek Kołakowski reaches exactly this conclusion which, I believe, is unsound. At least one of the reasons both positivists and Marxists ignored each other was manifestly a political one. Both philosophies were competing for dominance among the workers' movements of the second half of nineteenth century, and, it seems, proponents of both found that an extended critical examination of the main doctrines of an opposing movement is not furthering their political interests.

The hypothesis that Marx and Engels strove to disparage the significance of positivism for political reasons can be indirectly confirmed by the fact that nothing but political motives prompted Lenin to devote considerable effort and time to the critical analysis of a philosophy which had, it seems, a very remote relation to the social and political problems

Lenin and the Bolsheviks were facing in the first decade of the twentieth century. Yet, Lenin's vehement attacks on Ernst Mach, Richard Avenarius, and their followers in his *Materialism and Empirio-criticism* (1909) were driven precisely by his political aims.

At the beginning of the twentieth century, Mach's ideas were very popular in many countries. They were also popular among leading Russian Bolsheviks, most of whom were quite well educated. They, and especially Alexander Bogdanov, Anatoly Lunacharsky, and Vladimir Bazarov, regarded Mach and Avenarius as the foremost representatives of progressive, scientific philosophy. They took seriously Engels' claim that Marxist philosophy must change its form following every important scientific discovery. Being convinced that more than enough such discoveries have been made in recent years, and realizing that Marxist social and political philosophy lacks a solid epistemological foundation, they sought to merge empirio-criticism and Marxism.

Lenin regarded this tendency to be politically dangerous. Georgi Plekhanov, a leading theoretician of the rival Menshevik fraction of the Russian Social Democratic Party, used the fact that empirio-criticist ideas were popular among Bolsheviks to blame them for moving away from Marx, as well as to maintain that only Mensheviks are consistent Marxists. This accusation looked quite convincing for the Russian Social Democrats because it was known that the Bolsheviks were in the process of revising some principles of Marx's philosophy of history, including the idea that a socialist revolution was impossible in an economically backward country. Because of Marx's incontestable authority within the Russian Social Democratic Party, any well-founded accusations of deviance from Marxism threatened to diminish the Bolsheviks' political influence within the party and beyond its limits.

Lenin decided to eliminate this danger. His criticism of empirio-criticism is subordinated to this task. He aims not so much at deep theoretical argument against empirio-criticism but at the discreditation of Mach's philosophy in the eyes of Marx's devotees. With that aim in view, Lenin concentrates on demonstrating that there is little or no difference between Mach's philosophy and that of Hume's or Berkeley's. Proceeding further and using abundant quotations from Marx and Engels, he shows that the latter were resolute opponents of Hume's agnosticism and Berkeley's immanent (subjective) idealism. Thus, concludes Lenin, Marxism is irreconcilable with Machism.

So much as to the general scheme carried out in *Materialism and Empirio-criticism*. The text is, of course, more complex. Thus, in implementing this scheme, Lenin faces some problems related to the exact determination of Mach's philosophical position. He discerns the two main opposing parties in philosophy: materialism and idealism (Lenin intentionally uses the word "party"—thus philosophical distinctions acquire political content or at least political gloss). Yet Mach explicitly claimed that his philosophy is neither materialist nor idealist. Accordingly, Lenin maintains that both Mach and his followers are a contemptible middle party in philosophy, who confuse the materialist and idealist ideas on every question. The most important of them, Lenin believes, is the fundamental question of epistemology: are our sensations images of bodies, or are bodies complexes of sensations? Although Mach claims being above materialism and idealism, in Lenin's opinion, he's constantly sliding towards idealism; fighting against materialism is a favor to idealism. Moreover, by defending idealism, which Lenin takes to be ideological ground for political domination of the ruling class, Mach—being aware or not—supports the bourgeoisie in their fight against the proletariat.

Most of Lenin's arguments against Mach, as a rule, look convincing only to the eyes of those who already are materialists. Some of them sound quite witty and ironical as, for example, Lenin's rhetorical question: is it more economical to think of the atom as being divisible or indivisible? Yet it is easy to see that this question is beside the point and poses no threat to Mach's principle of the economy of thought.

Although Lenin's theoretical arguments were rather weak, he did achieve his main objective. Almost all Russian Marxists, including the leading Bolsheviks, accepted his claim that Mach's conceptions are irreconcilable with those of Marx. And because the affinity of some of Mach's ideas with those of Hume, Comte, and Mill was evident, positivism became a favorite object of Marxist critique. Marxists often quoted Comte's words that positivism is radically opposite to materialism, both regarding its philosophical nature and political purpose. They blamed positivism for renouncing dialectics, for trivial practicality, and its antirevolutionary character. They would assail it for a deficiency of value orientation, lack of faith in the human cognitive abilities, and even for alleged efforts to undermine science by eliminating from it causal *explanations* of phenomena.

Notwithstanding all the invectives Marxists used against positivism, their positions on many important theoretical issues are similar, and in some cases, simply identical. Lenin's criticism of Mach concentrates on epistemology. Yet, firstly, Lenin often misinterprets Mach's position by simply identifying it with that of Berkeley or Hume. Secondly, Mach's

stance connecting epistemological phenomenalism with ontological phenomenalism is not characteristic of all of positivism. Positivism was
showing little interest in strictly epistemological issues, i.e., the issues
which require an analysis transcending the limits of philosophy of science.
Thirdly, neither Marx nor Engels were preoccupied with epistemological
problems. Thus, in his attempts to elaborate on them, Lenin leans on the
eighteenth century materialists whose philosophical positions were cen-
sured by Marx in his famous *Theses on Feuerbach* (cf. Marx 1998
[1888]). Thus, it seems, Lenin could be accused of misunderstanding
Marx. Marx's epistemology is difficult to reconstruct from his very fragmented remarks, yet it undoubtedly has a few elements in common with
pragmatism.

Some similarities between positivism and Marxism may be explained
by the fact that both were heavily influenced by Saint-Simon. I do not
intend to claim that Comte's *direct* influence on Marx was vast. For our
purposes it is sufficient to establish the fact that on many important issues
positivists and Marxists have been in agreement.

Both movements were opposed to metaphysics and both affirmed that
their notions and theories were scientific in character. Both Comte and
Marx were adherents of historical determinism, admired scientific and
technological progress, and believed that the future social order which
each called socialism will differ substantially from that of the present.
Of course, this does not mean that they agreed on everything. Contrary
to Marx, Comte opposed the abolition of private property. There were
also important differences between them regarding the means to establish the socialist order. Comte preferred the consensus of minds
achieved by the advance of science and morals, and not class struggle.
He belittled the role of political struggle in general and renounced violent action completely. Yet both Comte and Marx believed that some
organized and systemic action is necessary for establishing the new order, and both ignored the fundamental tension between this activist attitude and the claim that the coming of socialism is inevitable. Both emphasized the role of the proletariat for achieving this order, assuming
that in a socialist society, the individual must be subordinated to the
whole and undergo a process of thorough socialization. They were adherents of a stern organization of society, scientific planning, and of
societal control of property and production. Though both praised the
proletariat, both believed that only those who are able to grasp, develop,
and apply the scientific social theory have the right to really master and
control society. At least in one respect Marx had been, so it seems, more

positivist than Comte himself. Namely, Marx after 1848 no longer referred to himself as a philosopher, preferring to be called a social scientist or an economist. Comte, on the contrary, never renounced the title philosopher, in spite of the fact that philosophy was not even included in his system of positive sciences.

The connections between Marxism and positivism in the twentieth century are even more compelling. It happened that some positivists of that century claimed that they were Marxists or at least felt sympathy towards Marxism. Such a combination would have been inconceivable in the nineteenth century, because Comtean and Marxist social philosophies, though having some common traits, were hardly compatible. Yet, as I already mentioned, some Russian Social Democrats tried to combine Mach's epistemology with Marx's social philosophy. The prevalence in modern positivism of methodological problems and a rather strict juxtaposition of fact and value made the combination of logical positivism and Marxist social and political philosophy (or, at least, of corresponding political attitudes) feasible. The initial affinities between the social philosophies of positivism and Marxism, including progressivism, collectivism, and a scientific orientation, made it possible for some representatives of logical positivism to adopt Marxism as their political credo. They were not very happy with Comte in general and with his political philosophy in particular, and shared the opinion prevailing among politically conscious academicians in the twentieth century that this political philosophy was outdated. Comte's inclination to specify every last element of the future social order and his lachrymose sentimentalism produced a comic effect in the more skeptical and cynical twentieth century. Mill's liberalism was a clear alternative to Comte's political doctrine, and Schlick opted for the former. Yet for those members of the Vienna Circle and the Berlin group who, being buffeted by the economic turbulence of the interwar period, invested their hope in scientific planning, Marx's socialism was more attractive than Comte's socialism.

Logical positivists admired Marxist appreciation of observable facts and scientific methods of inquiry. They endorsed Marx's claim that human history is a proper part of natural history, and that the future natural science will include the science of man in the same way as the science of man will include natural history. Marx's idea of a unified science was closer to their own corresponding concept than Comte's doctrine of the hierarchy of sciences.

Marx regarded his theory as a powerful instrument both of social critique and social change. It may seem that the requirement of logical positivism for confirmability and intersubjectivity, as well as the principle of fact-value dichotomy precludes the use of social theory as a tool of social critique. Yet as Robert S. Cohen correctly states, logical positivists sought "to help furnish a stocked toolhouse for those, socialist or others, who were trying to construct a world that would have the material and spiritual requisites for social justice and human freedom. Not all positivists were equally specific in their social and economic views, but it seems clear that all recognized that theirs was a radical movement, which ultimately would replace the emotional and intellectual foundations of the existing political and social order" (Cohen 1963: 157).

Among logical positivists, most radical in his social and economic views was, without doubt, Neurath. He was clearly the most devoted adherent of Marxism. He not only based his sociological writings on Marxist notions, but took an active part in the practical implementation of the Marxist ideal of a centrally planned economy while he was serving in 1919 as the head of the central planning office of the Bavarian Soviet Republic (seeking, in his own words, "full socialization of Bavaria"). After the fall of its communist government, he was convicted of high treason and spent some time in prison. Thus it was rather natural that he laid stress not only on the radicalism of Marx's and Engel's social views corresponding to the revolutionary role of logical positivism, but also on the closeness of positivism's and Marx's views on the nature of social theory. In *Empirical Sociology*, bearing the subtitle *The Scientific Content of History and Political Economy*, Neurath emphasizes that the founders of Marxism never argue with reference to injustice of the present order, or to an ideal social order, but investigate, on the basis of empirical data, how, under certain conditions, people would behave. Neurath acknowledges that sometimes they formulate value judgments, yet he claims that these do not play an essential role in their argument, being mere accompanying pronouncements, and as such may be omitted without changing the real content of Marxist theory. Moreover, he asserts that, "of all the attempts at creating a strictly scientific unmetaphysical physicalist sociology, Marxism is the most complete. [...] Compared with Marx, men like Buckley, Comte, Spencer and others are less consequential, important and influential" (Neurath 1973 [1931]: 349–50).

The question whether Marx's social and political theory contains no irremovable elements of values or metaphysics is rather complex. Some of my university colleagues, say Alvydas Jokubaitis, are convinced that in Marxism the positive, scientific components are interwoven with romantic, metaphysical ones. However, as we shall see later, Jürgen Habermas answers this question almost like Neurath and stresses the similarities between Marxism and positivism.

Respecting Marxism as he did, Neurath was very interested in establishing collaboration between the Vienna Circle and the Frankfurt School—the most important school of neo-Marxism which was based in the Institute for Social Research at Frankfurt University. This initiative was supported by Carnap and Reichenbach; each felt sympathy towards Marxism. It is worth mentioning that, contrary to Popper, logical positivists did not regard Marx's claims to be unscientific or void of empirical significance. Marx's name is mentioned immediately after Comte's in the manifesto of logical positivism, *The Scientific Conception of the World*, where both are being praised for laying the foundations of social sciences based on empiricist, antimetaphysical tenets. At the same time, the manifesto places the competition between metaphysics (allied with theology) and the scientific world-conception in the context of "the fierce social and economic struggles of the present" and links the tendency of the masses to lean towards a "down-to-earth empiricist view" with "their socialist attitudes" (cf. Carnap, Hahn, and Neurath 1929: 317).

Members of the Frankfurt group welcomed suggestions for collaboration with logical positivists, although they understood that methodologically the Frankfurt School, with its characteristic emphasis placed on the role of social critique, moves away from classical Marxism with its stress on an objective and positive social theory so praised by Neurath. According to the representatives of the Frankfurt School, human beings are not mere objects of social theory, for they are also subjects of social appraisal and social action. Treating them as mere objects incapacitates philosophy in its role of social critique. The importance of this is clearly reflected in the name *critical theory* adopted by the representatives of the Frankfurt School to designate their own approach. However, since the intellectual climate in Germany and Austria has been hostile to logical positivism and the Frankfurt School (idealism dominated philosophy departments in both countries, and the wider public was increasingly seduced by fascist ideas), the benefits of cooperation was obvious to both.

Not of minor significance in establishing contacts between both groups were their parallel political views. Decidedly leftist attitudes clearly dominated both the Frankfurt and Vienna schools. It is noteworthy that members of the Vienna Circle were even more active in the leftist movement than members of the Frankfurt School. Neurath was the most conspicuous figure in this respect. Yet other members of the Vienna Circle, notably Edgar Zilsel, Herbert Feigl, and Friedrich Waismann, engaged in leftist political and social actions as well. One of the members of the Ernst Mach Society, used by logical positivists for the dissemination of their ideas, was the leading representative of Austro-Marxism, Otto Bauer. This society had been regarded by the then rightist Austrian authorities as social democratic, and was banned in 1934.

Max Horkheimer, the director of the Institute for Social Research from 1931 and the leader of the Frankfurt School, began inviting logical positivists to lecture in Frankfurt (the first guest of the Institute in 1928 was Hans Reichenbach). Those lectures were followed by lively discussions.

Contacts between the two groups were not broken even after Max Horkheimer, Theodor Adorno, Herbert Marcuse—the most prominent members of the Frankfurt School—had moved to the United States, where the institute that had been closed in Frankfurt for political reasons was reopened under the name of the New School for Social Research in 1934. Horkheimer invited Neurath to visit him in New York in 1936. Their discussions in 1936 prompted Horkheimer to write his critical article "The Latest Attack on Metaphysics," published in 1937 in *Zeitschrift für Sozialforschung* which was the tribune of the Frankfurt School. It is an important text that clarifies the relation between twentieth century Marxism and twentieth century positivism, although aiming to highlight the methodological views of the Frankfurt School, Horkheimer mostly concentrates upon the issues that reveal differences rather than agreement between the views of logical positivists and members of the School.

Some of Horkheimer's arguments are weak or obviously incorrect. For example, his claim that physicalism is a form of solipsism. Horkheimer's assertion that dialectics improves upon formal logic is purely declarative and lacks proof. Yet some of his critical observations are interesting and significant. One of them is Horkheimer's claim that the notion of unified science is related to the naive harmonious view of the world.

This claim is sometimes regarded as unfounded because the unified language of science serves to eliminate meaningless problems and solve meaningful ones, but not to cultivate harmony (cf. Dahms 1994: 103–4). Yet I doubt that Dahms is right: one of the pillars on which the notion of unified science rests is the belief that even persons or groups with conflicting interests must hold the same theoretical views, including the view on the social world and divisions in it (or lack of them). This belief is linked to the belief in the value-neutrality of social science and may be questioned. Besides, not only positivism, but traditional, orthodox Marxism that has been challenged by the critical theory also postulates the existence of a one and only true social theory (which, as Marx would add, adequately defines and explains conflicting social interests). Horkheimer's position on this issue, later developed by Jürgen Habermas (cf. Habermas 1968), is different: our social interests inevitably influence all our views, the theoretical included.

The second serious point Horkheimer makes regarding positivism concerns its social role. Representatives of logical positivism were interested in this role not less than the representatives of classical positivism and regarded the question about the character of this role as significant and to be decided by empirical investigation. Although no such investigation has ever been carried out, and, to tell the truth, it is not precisely clear how it could be pursued, they believed that this role is a positive one. They simply assumed that positivism furthers social progress. Yet Horkheimer's opinion on this question was different: positivism rather serves the *status quo*.

Horkheimer must have had in mind totalitarian Germany. Thus he was actually accusing positivists (whom he was calling liberals) of intentional or unintentional support for totalitarianism. This accusation seems almost absurd in the face of the fact that almost all logical positivists left Germany or Austria whose annexation (*Anschluss*) in 1938 was a consequence not only of Hitler's policy of expansion but of the pro-Nazi political development of Austria itself. Most of them voiced their protest against the Nazi's New Order, which they rejected, not to speak about its active support.

Nevertheless, Horkheimer touches upon two serious issues. The first one concerns the internal tension arising within a philosophy which regards itself as progressive, but finds no means to philosophically justify the policy or morals it deems progressive and to reject those that it takes to be regressive. The second issue concerns the positive character of the political impact exerted by positivism in general, which can also be questioned.

Neurath, as the representative of logical positivism in their contacts with the Frankfurt School, attempted to correct what he considered to be the misrepresentation or, at least, misleading interpretation of the views of logical positivism, and to develop the position of logical positivism regarding some issues raised in Horkheimer's article. However, Horkheimer refused Neurath's suggestion to publish his response to criticism in *Zeitschrift für Sozialforschung*. This had a negative impact upon the relations between the two groups, even though some contacts occurred later.

The most important text in which a leading representative of the first generation members of the Frankfurt School takes a stance towards classical positivism is Herbert Marcuse's *Reason and Revolution*, which appeared in 1941. In it a special emphasis is laid on a direct link between the positivist notion of positive knowledge and the character of positivist sociology. Analyzing Comte's social theory, Marcuse emphasizes that many of its traits are directly tied with general positivist concern with facts instead of illusions, with useful knowledge instead of contemplation, certainty instead of doubt, and with organization instead of critique. Thus, says Marcuse, "the new sociology is to tie itself to the facts of the existing social order, and though it will not reject the need for correction and improvement, it will exclude any move to overthrow or negate that order. As a result, the conceptual interest of the positive sociology is to be apologetic and justificatory" (Marcuse 1941: 341).

In this respect, according to Marcuse, Comte's theory differs from classical rationalism, which closely relates the idea of reason to the idea of freedom. He blames Comte's social theory for its methodological naturalism, conformism, stemming from the recognition of the immutable social laws, and "for the maintenance of prevailing authority and for the protection of vested interest from any and all revolutionary onset" (Marcuse 1941: 345).

It is quite natural for a leading representative of critical theory to blame Comte for his lack of social criticism. Yet Marcuse obviously misinterprets Comte when claiming that his ultimate goal was to fortify the given social order. Comte formulates a rather impressive (though not very attractive and not too realistic) program of social and political reforms. The fact that Comte was an adherent of private property makes him an advocate of the given social order only in the eyes of a Marxist for whom the change of property relations is a necessary precondition for any radical social change. Many other critical points Marcuse makes when criticizing Comte are based on his Marxist understanding of history and on a Marxist notion of social change. Thus he criticizes Comte for believing that the development of society is in itself not revolutionary but evolutionary, and for denying that the overall transformation of a given social system is possible (cf. Marcuse 1941: 352). Marcuse's criticism, however, would not convince those who do not share his Marxist beliefs. Its openly partisan character is reminiscent of Lenin's criticism of Mach.

One of the few things Marcuse praises in positivism is, characteristically, its collectivism, a firm perseverance to deal not with isolated individuals but with society, uniting them in a real comity which transcends national frontiers. He emphasizes that on this point Comte's sociology evidently surpasses Hegel's political philosophy. Yet, according to Marcuse, on other points the classics of rationalism outstrips Comte, because "the positivist repudiation of metaphysics was [...] coupled with a repudiation of man's claim to alter and reorganize his social institutions in accordance with his rational will" (Marcuse 1941: 344).

The Enlightenment project was based exactly on the idea of reforming social institutions in accordance with man's rational will. Marcuse does not see much similarity between positivism and the Enlightenment and rather stresses their differences. Yet in the *Dialectic of Enlightenment* written by two other prominent members of the Frankfurt school— Horkheimer and Adorno—only six years after *Reason and Revolution* was published, the relation between both movements is treated otherwise. Horkheimer and Adorno have no doubt that positivism is the continuation of the Enlightenment project. In their words, "Enlightenment, understood in the widest sense as the advance of thought, has always aimed at liberating human beings from fear and installing them as masters. [...] Enlightenment's program was the disenchantment of the world. It wanted to dispel myths, to overthrow fantasy with knowledge" (Horkheimer and Adorno 2002 [1947]: 1).

Positivism had the same goals. Elaborating a point already made by Horkheimer in "The Latest Attack on Metaphysics," Horkheimer and Adorno claim that positivism is the culmination in the development of the tendency present already in the eighteenth century seeking to reduce scientific knowledge to a purely instrumental one. That tendency is regarded by them as extremely dangerous because it enables the use of reason to pursue any goals including those inimical to the cause of freedom, equality, and justice.

Of course, the reason that the Enlightenment philosophers spoke of had not only an instrumental but also a substantive value component, tied to aforementioned goals. Yet the value component was shrinking with time. During the last two centuries, reason, under the influence of science, has lost its ties with religion and metaphysics and with them its substantive value content. All norms look insipid in comparison with the highest authority—science. And because modern science treats itself as an instrument, reason becomes a mere tool of power. By favoring utility at the cost of insight and comprehension, positivism subordinates the ideals of objectivity to those of effectiveness and of a drive to self-preservation and domination. And by strict adherence to the principle of the fact-value dichotomy, positivists become, as Horkheimer and Adorno claim, *de facto* supporters of the oppressors of the majority.

Of course, logical positivists blatantly rejected such accusations. They surely were no supporters of totalitarian or authoritarian dictators, or of the *haute bourgeoisie*. They had to concede the blame that value-neutral knowledge can be used in favor of evil, as well as good goals. But in this respect the social sciences were no different from the natural sciences. This conclusion comes as natural to those who believe in the unity of science.

Much later, in the 1960s, representatives of the Frankfurt School, in particular Adorno and Habermas, got involved in a discussion that came to be known as *Positivismustreit* (Positivism dispute). It sprung out at a conference of the German Society of Sociology held in Tübingen in 1961 and lasted eight years (cf. Adorno et al. 1976 [1969]). Even though it continued the dispute of the 1940s between the Frankfurt and Vienna Schools, this time the opponents of the Frankfurt School were not logical positivists, but critical rationalists—Karl Popper and Hans Albert (in 1950 the Institute for Social Research has renewed in its hometown the work carried out in New York in 1934–49). Popper and Albert were not eager to consider themselves positivists, but their views were so similar to those of the latter, that their opponents had no doubts that they were arguing against logical positivists.

The most prominent second-generation member of the Frankfurt School, Jürgen Habermas, presents his views on positivism in his well-known book *Knowledge and Human Interests* (1968). One of its most important chapters bears the heading "Comte and Mach: The Intentions of Early Positivism" (Habermas uses the term "early positivism" with a different meaning than I do). As is to be expected from the most celebrated living German, and probably European, philosopher, his criticism of positivism is more profound than that of other members of the Frankfurt School.

It is a delicate issue whether to treat Habermas as a neo-Marxist, because his distance from orthodox Marxism is greater than that of the first-generation members of the Frankfurt School. Kant's and the neo-Kantians' influence on Habermas is comparable to that of Marx and neo-Marxists, and he was influenced by many other thinkers—from Max Weber to Hans-Georg Gadamer. Yet he never denied his affiliation with Marxism, and even stressed it in the titles of some of his books such as the *Toward a Reconstruction of Historical Materialism* (1976). He has not tried to retract it during my conversations with him at the Goethe University in Frankfurt. Thus it is difficult to find a pressing reason to break with a tradition which considers the *entire* Frankfurt School to be part of modern Marxism. The term "critical theory," often used to describe the views of the School and its later followers, more expressly reflects certain features of these views, while the Marxist sources of the very theory raise no doubts. Of special importance for us is the fact that Habermas continues the critique of the Enlightenment project commenced by Horkheimer and Adorno and places his critique of positivism in the same broader context.

For Habermas, the affinity between Marx's methodological and epistemological orientation and that characteristic to positivism is beyond doubt. He emphasizes that for Marx, natural science is the principal paradigm of knowledge, and that Marx regards concern for power and control as the main interest which steers scientific knowledge. Being overly dependent on positivism which denies that a subject of cognition takes part in construction of the world he apprehends, Marx fails to approach science critically.

Habermas acknowledges that early positivism was interested in the liberation of humanity and was driven by a desire to destroy all dogmas. Its interest in science was fully justified at the time when it was believed that science is the means to redeem humanity. Moreover, Habermas strongly believes that scientific knowledge is a necessary condition of liberation. Yet, according to him, it is not a sufficient one.

By emphasizing the instrumental function of scientific knowledge, positivism facilitates the growth of a technocratic mentality. Thus, the critique of positivism is a precondition for establishing philosophy as a critical theory and a critical activity. For a proper comprehension of positivism, it is important to grasp its transformation from the philosophy of history that it was in Comte's time to the epistemology and philosophy of science that it became later. Habermas emphasizes that, at first glance, the appearance of positivism in the form of a new philosophy of history looks paradoxical "for the scientist content of positivist doctrine, according to which legitimate knowledge is possible only in the system of the empirical sciences is obviously at odds with the form in which positivism originated. [...] Comte's law of three stages [...] has a logical form that does not correspond to the status of lawlike hypotheses in the empirical sciences. [...] This paradox disappears as soon as we discern the intention of early positivism: the pseudo-scientific propagation of the cognitive monopoly of science" (Habermas 1987 [1968]: 71). Yet at the end of the nineteenth century the status of scientific knowledge was already high enough and there was no need to legitimate it. The other problem loomed large, namely the problem of the objectivity of knowledge produced by the sciences.

One could and should doubt Habermas here. At least positivists were concerned not with the problem of objectivity of scientific results, but the problem of their intersubjectivity. However, according to Habermas, Mach addresses not the second, but the first problem. Under such reading of Mach's goals, Habermas can only conclude that Mach's effort to base the objectivity of scientific knowledge on facts given to us in sense perception was unsuccessful, since facts given in our personal experience are subjective, and not intersubjective. Mach could retort that facts are neutral, and not subjective. Yet in Habermas's opinion Mach's attempt at ontologization of facts is not a way out of the maze because corresponding assumptions cannot be justified within the framework of the procedures of empirical sciences.

Habermas acknowledges that physicalism, which has replaced phenomenalism in the mid-thirties as the principal orientation of logical positivism, has eliminated some of the difficulties that Mach and phenomenalism faced. But that creates new problems. The possibility of objective knowledge cannot be explained without an explanation of how intersubjective comprehension in the process of communication is possible. Logical positivists are unable, according to Habermas, to explain the nature of everyday language and that of intersubjective consensus in general. Thus they are unable to grasp the real status of scientific knowledge.

Habermas claims that logical positivists methodically objectify reality, and so fail to discern the *a priori* element of the structure of scientific knowledge, and do not understand that this knowledge is the result of interaction between the knowing subject and reality. The first attempts to overcome such "misleading ontologizing," characteristic not only of positivism, have been made, according to Habermas, by Charles Sanders Peirce and Wilhelm Dilthey. He regards his own theory of communicative action the most significant effort towards the final overcoming of such objectification conditioned by the monologic treatment of language which limits its function to that of technical control and precludes the proper understanding of the role of conventions and rules of the language game presupposed by science.

Habermas's arguments against objectification of scientific knowledge seem well founded. Yet I am not certain that one could blame logical positivists for such objectification. Carnap's notion of a linguistic framework is clear evidence that they have well understood the role of conventions, rules, and other *a priori* elements of language in cognitive processes.

To conclude this analysis of the relation between positivism and Marxism we have to state that these two trends are related in many more ways than is usually recognized. However, which of their differences should be considered most fundamental? It would hardly be their views on language, as Habermas claims. It seems that the most critical are their differences in the treatment of experience which, as we will see later, are also crucial for understanding the relation of positivism to pragmatism.

Marxism gives a much wider meaning to this term than positivism. For the former, this notion encompasses both individual and social, current, and historical experience. The emphasis, however, is given to social and historical experience.

Comte's positivism is little different from Marxism in this respect. However, for both Hume and Mill the notion of experience was reduced to individual experience. That feature is especially salient in Mach's philosophy. The early, phenomenalistic logical positivism carried on this tradition. In the physicalist phase of logical positivism, this notion has lost its purely individualistic character, yet had not become a manifestation of social life *par excellence* as was the case in Marxism. The physicalist notion of experience refers to society only implicitly. The intersubjectivity of scientific statements would be impossible without the communicative exchange between members of a scientific community. Yet in physicalism, society itself is hardly visible, hidden behind language as the means by which individuals communicate. Although some logical positivists, namely Edgar Zilsel and Philipp Frank, were interested in the historical dimension of experience—and especially that of scientific experimentation—their investigations made no substantial impact upon mainstream logical positivism.

Quite a momentous element in the Marxist notion of experience is political, especially revolutionary praxis. Critical theory focuses upon the value aspects of social practice. This element is completely absent from the positivist conception of experience developed in the twentieth century.

Positivism and Pragmatism

Positivism and pragmatism have many common features. Both share the conviction that the development of science is a decisive factor in the progress of humanity. Both share the opinion that the source of real knowledge is experience, and both treat the British empiricists as their forerunners. Both pragmatism and positivism (especially twentieth century positivism) are antidogmatic and claim that philosophy is a method rather than theory. The theories they nonetheless formulate resemble each other. Specifically, their theories of meaning exhibit quite a great degree of affinity. Pragmatism attaches great importance to the practical efficacy of our concepts and theories, but Auguste Comte has also claimed that the positive mind is interested only in what is useful and practical.

In short, pragmatism and positivism are rather close to each other. They are so close that their distinctiveness may be called into question. In fact, William James in *The Meaning of Truth* admitted that many people hold that "pragmatism is only a re-editing of positivism" (James 1978 [1909]: 266). James was not happy with this opinion, all the more since he felt less affinity towards positivism than the founder of pragmatism, Charles Sanders Peirce. Nevertheless, James does not deny the resemblance between both philosophies. According to him, pragmatism "harmonizes" with positivism, sharing with it "its disdain for verbal solutions, useless questions and metaphysical abstractions" (James 1995 [1907]: 21). By attaching to his *Pragmatism* the subtitle *A New Name for Some Old Ways of Thinking*, he inadvertently reinforced the view that pragmatism was not a new philosophy. Thus, two years later, in *The Meaning of Truth* he found it necessary to point out that treating pragmatism as modified positivism is a serious mistake. And he is right: this view is undoubtedly mistaken.

However, pragmatism has a great deal in common with positivism. They share almost the same roots and both represent a "progressivist" answer to the cultural and social challenges of the industrial era. Certain similarities could be observed between the intellectual environment of the creator of classical positivism Comte and that of Peirce. The formation of Comte's views was heavily influenced by his studies and work at the École Polytechnique, the most famous French research center, and for Peirce it was his work (1861–91) in the U.S. Coastal Survey (called the Coastal and Geodetic Survey since 1878), which was the most important institution of scientific research in the United States at that time. Peirce's scientific interests were no less broad than Comte's—he worked in the spheres of chemistry, geodesy, metrology, spectrology, experimental psychology, mathematics, and history of science.

However, the similarities of positivism and pragmatism should not overshadow their differences. Peirce disagreed with Comte on many things. In general, he regarded the scientific side of positive philosophy as its advantage and claimed that exactly because of that this philosophy has been of real service to the world (cf. Peirce 1984: 122). He had, nonetheless, some objections both to Comte's treatment of the methods of science and to his conception of its interdisciplinary structure.

Contrary to Comte and Mill, Peirce was a probabilist, bearing more resemblance in this respect to Carnap and, especially, Reichenbach, than to the positivists of the nineteenth century. Unlike Mill, Peirce believed that hypotheses do not differ in epistemic status from inductively inferred knowledge. According to him there's no crucial difference between accepting a hypothesis and rejecting it: one should always be aware of the degree of empirical support and not forget it, for it may change. As to Comte's classification of sciences, Peirce found it of limited value due to its inadequate treatment of logic and psychology. Yet he praised Comte and positivism for taking a critical stance towards metaphysics, or as he preferred to say, "ontological metaphysics."

> In this respect, pragmatism is a species of prope-positivism. But what distinguishes it from other species is, first, its retention of a purified philosophy; secondly, its full acceptance of the main body of our instinctive beliefs; and thirdly, its strenuous insistence upon the truth of scholastic realism. [...] So, instead of merely jeering at metaphysics, [...] the pragmaticist extracts from it a precious essence which will serve to give life and light to cosmology and physics. At the same time, the moral applications of the [positivist] doctrine are positive and potent; and there are many other uses of it not easily classed. (Peirce 1966 [1905]: 192)

Peirce's words deserve commentary. First of all, Comte and Mill would retort, without doubt, that our instinctive beliefs may be wrong and, moreover, that some of them actually are wrong. One of the functions of science is to subject them to scrutiny. The critical examination of our beliefs is a necessity because some of them may thwart the advance of science and slow social progress. Secondly, they would certainly agree that some purification of philosophy is rather desirable. However, they would note, as many others actually did, that some ideas of the author of the article "How to Make Our Ideas Clear" clearly lack the demanded clarity. And is the doctrine of scholastic realism really "a precious essence" of metaphysics?

We have called Peirce "the founder of pragmatism." This title was bestowed upon him by James, and nobody has ever seriously questioned Peirce's influence for the development of pragmatism. His role is significant indeed, however, it usually is overrated. In fact, pragmatism was a product of joint efforts, and in this respect it was not different from logical positivism. It would be unreasonable to call Moritz Schlick the founder of logical positivism, even though he had founded the Vienna Circle. Whereas Peirce wasn't even a formal leader of the group in the meetings of which pragmatist philosophy has crystallized. I certainly have in mind the so-called Metaphysical Club.

This Club or, rather, discussion circle, was founded at the initiative of James (and not Peirce) in Cambridge, Massachusetts, at the beginning of

1872. Its most important members were James, Oliver Wendell Holmes, Peirce, Chancey Wright, Nicholas St. John Green, and Joseph Bangs Warner. Just like in Vienna, most of the members were university professors, scientists, and lawyers. There were, however, certain differences: two theologians were members of the Metaphysical Club, and it was *dominated* by lawyers. In the beginning of 1870s there were six lawyers among the twelve members of the club. Holmes was, of course, the most prominent of them. Whereas in Vienna lawyers—represented by Felix Kaufmann—played a minor role, and, it seems, no theologians have ever attended the Thursday discussions of the Vienna Circle. They would have had nothing to do there. But in Puritan New England of the nineteenth century, religion was important, and it mattered to most members of the Metaphysical Club. James was deeply religious (his father was a follower of the Swedish mystic Emanuel Swedenborg), while Peirce also has declared himself a theist (cf. Fisch 1964).

The name "Metaphysical Club" reflects the oppositions of its members to agnosticism and emphasizes that philosophical, not merely scientific, issues are discussed in it. In his letter to Holmes, James has expressed an intent to discuss "the very tallest and broadest questions" (James 1920, vol. 1: 126). Yet some other members of the Metaphysical Club used its name mainly with a hint of irony. The different make-up of both groups, the different interests of their members, in addition to differences of time, place, and cultural environment, explain why these groups have developed different philosophies. However, the manner in which it was being done was similar: intense discussions, at the end of which it was impossible to determine who said what, and whose contribution was most substantial.

In spite of obvious differences, both groups were deeply interested in science and its social and cultural effects. Members of the Metaphysical Club have clearly comprehended the significance of positivism so popular in both Europe and America. Chauncey Wright, the oldest and probably most respected member of the club, a mathematician, psychologist, and philosopher, often referred to himself as a positivist (cf. Kuklick 1977: 63).

While drawing parallels between the two groups, it is worth noting that the Metaphysical Club has published a text resembling in status the Vienna Circle manifesto. Following several months of intense discussions, Peirce began the preparation of a *credo*, explaining the philosophical position of the club. The text was under consideration as early as 1872, but had been published, with modifications, only four years later—Peirce sought to take into account all critical remarks and ideas brought out by the members. The final version is well known—it is the aforementioned article by Peirce "How to Make Our Ideas Clear" (Peirce 1966 [1876]).

The Metaphysical Club (like the Vienna Circle) subsisted for a relatively short time. After Peirce's departure (to continue his work on the Coastal Survey), the deaths of Green and Wright, and Holmes's diminishing interest in philosophy, new members were admitted to the Club, but this failed to help. After three years, the Club was disbanded (cf. Fisch 1964: 19).[3]

Attempting to define his relation to positivism in an early unfinished paper (it was published fully only in 1984) dedicated specifically to the appraisal of the merits and shortcomings of positivism, Peirce claims that the weakest side of positivism is its stance on religious issues. He spares no efforts to show that positivism does not promise much to the passions which, according to him, are actually intimately bound and and innate to our nature. He sums them up in one phrase—"the love of life." According to Peirce, positivism gives us no hope by claiming that "life upon the globe is a phase, quite accidental, tending as far as we know to no permanent end, of no sort of use, except in producing a pleasant titillation now and then on the nerves of this or that wayfarer on this weary and purposeless journey. [...] There is no good in life but its occasional pleasures; these are mostly delusive, and as like as not will soon utterly pass away" (Peirce 1984: 126). Meanwhile, the promise of theism is, according to Peirce, radically different because its capital principle is that there is a purpose in the nature of things. Thus theism is capable of producing consolation, whereas positivism is not.

Not only Peirce but also other pragmatists emphasize "the love of life" and the importance of any noncognitive human activity in general. Being human involves undergoing affections, passions, instincts, volition, longing for love and success. In their view, positivism is rather dry and schol-

---

[3] Readers of this book who have read the remarkable volume by Louis Menand, *The Metaphysical Club: A Story of Ideas in America* (Menand 2001), might feel a bit perplexed. He maintains that the Metaphysical Club existed only for nine months in 1872. On the contrary, I am suggesting (on the basis of Fisch's article [Fisch 1964]) that it existed at least for three years. For lack of convincing evidence it is difficult to prove who is right. The club was a small and *informal* group of rather close intellectuals. Following Peirce's departure from Cambridge after (as Menand claims) the last club meeting in December 1872, some members of this group were likely still meeting each other. Pierce was neither the founder of the club, nor its most revered member. By the way, Menand mentions Fisch in the notes to his book but he does not mentions the article which I cite here (even though it partly contradicts his story).

arly, although neither Comte nor Schlick would have found such a view justified. In laying stress on the vitality of man and the richness of life, pragmatism echoes Nietzsche, a contemporary of Peirce and James. Contrary to him, however, they held that if we do not find God in our hearts, we should rather conclude that we are dead, not God. Peirce clearly regarded belief in God to be part of the main body of our instinctive beliefs.

Like Peirce, James never regarded the questions concerning God and moral truths to be purely intellectual ones. In general, pragmatism displayed quite strong anti-intellectualist tendencies, and this is a very important characteristic distinguishing it from positivism (the late Comte being the main exception). We use the word "anti-intellectualist" in the sense it was used by Morton White, who distinguishes anti-intellectuals and anti-intellectualists. The latter, unlike the former, are not hostile to intellectuals but press the claims of the heart against those of the head. In Peirce's version of pragmatism this trait is not conspicuous. Yet in James's pragmatist philosophy and especially in the essay "The Will to Believe" (1897), the reasons of the heart quite often trump the reasons of the mind. White, in all probability, is right regarding James as "a grandchild, if not a child of the romantic movement" (White 1969: 89). Notwithstanding all the homage he pays to science, James did not hold the opinion that reason, observation, and experimentation alone could establish the truth.

Renouncing the positivist attitude towards religion, pragmatists were not able to renounce in the same romantic fashion the positivist attitude towards metaphysics. If they dared to go that far, they would have had no right to present themselves as "prope-positivists," because the anti-metaphysical attitude is a constitutive element of positivist philosophy. Pragmatists wanted to be progressive, and that required them to take a critical stance towards the metaphysical tradition. Therefore, they were keen to emphasize their disdain for "metaphysical abstractions" and "useless questions." Yet their disdain was half-hearted, and their conception of metaphysics much more ambivalent than that of positivists. Pragmatists desperately tried to avoid both the Scylla of positivism and the Charybdis of metaphysics but their ship was caught on the rocks more than once.

In their awkward maneuvering, pragmatists tried to apply at least three different strategies. The first one was condemning metaphysics as lacking a firm ground, producing fierce disputes without chance to reach an agreement, and having no practical interest.

The second strategy was to treat metaphysics not as an odd discipline involved in abstract discussion but as a body of important truths and beliefs which are instinctive and/or established by way of passions, emotions, and feelings. At least in "The Will to Believe," James adheres, it seems, to the opinion that metaphysics is a creature of sentiment.

The third strategy is closely bound to James's version of the pragmatic theory of truth, which, it seems, lacks logical consistency. On the one hand, he says that "true ideas are those that we can assimilate, validate, corroborate and verify" (James 1995 [1907]: 77) and that "truth for us is simply a collective name for the verification process" (ibid., 84). However, in James's *Pragmatism* as well as in other works, an interpretation of truth sounding very much like the positivist interpretation goes hand in hand with a quite different one. According to the latter, "an idea is true so long it is profitable to our lives" (ibid., 30). Since in the broad sense that James attributes to the term "profitable," profitable ideas (e.g., those that provide consolation) could be not verifiable, there's no surprise that James derives a completely nonpositivistic conclusion: "*if theological ideas prove to have a value for concrete life, they will be true, for pragmatism, in the sense of being good for so much*" (ibid., 97) And since, according to James, metaphysical ideas also have a value for concrete life, he can recognize them as true, despite that they are not empirically verifiable.

James elucidates this position by means of discussing the idea of the Absolute. He finds this idea "majestic" and yielding religious comfort to a class of minds. And as long as it yields such comfort it is true. To a reproach that he unjustifiably identifies good (that is profitable) ideas with true ideas, James replies: "truth is *one species of good*, and not, as is usually supposed, a category distinct from good" (James 1995 [1907]: 99).

My aim is not to subject pragmatist theory of truth to a detailed analysis. Yet I want to draw attention to one circumstance that is rarely noticed in the literature. The deficiencies of the pragmatist theory of truth are linked to the fact that the pragmatists rely upon an extremely broad concept of experience. Already in the works of Peirce it was employed in a broader sense than it had been used by positivists, and James expanded it further still, enriching it with all kinds of emotional and religious experience. On the other hand, by means of interpreting experience as an active encounter with the world, they brought their conception of experience closer to Marx's conception of *praxis*. Merging individual and social experience, emotions and *praxis*, James substantially modified, and in fact, blurred the notion which in empiricist philosophy had a relatively narrow but clear-cut sense. Positivists never accepted such a radical reinterpretation of this fundamental notion. Disagreement regarding the scope of the notion of experience forms the basis of most other disagreements between positivists and pragmatists.

After John Dewey assumed the role of the leader of pragmatism in the 1920s, more or less at the same time when logical positivism appeared in the philosophical arena, relations between pragmatism and positivism entered a new stage. Their contacts became much more intensive and the development of the initial links between the two movements culminated in the start of real cooperation.

The first decades of the twentieth century were marked in America by rising interest among philosophers in making philosophy more scientific. This interest was felt both within and outside the pragmatist movement. Some reports on the new developments on the European continent attracted considerable attention. An article, "Logical Positivism: A New Movement in European Philosophy," published in 1930 in the *Journal of Philosophy* by Albert Blumberg from Johns Hopkins University and Herbert Feigl, then at Harvard (both were former students of Schlick), is especially worthy to note. Sometime later, in 1936, John Sommerville and, independently, Ernest Nagel recognized significant connections between pragmatism and logical positivism, both in their efforts to clarify meanings of our ideas and in their social attitudes. In December of 1935 Rudolf Carnap moved to America, where Feigl had been since 1930. Due to political circumstances in Europe after Hitler's coming to power, many logical positivists, we recall, immigrated to the new continent.

In large part because of the growth of pragmatism, the soil for logical positivism in America was rather fertile. To some philosophers it seemed so favorable that they envisaged the possibility of creating out of the two trends one unified movement. Charles W. Morris was the greatest proponent of this idea. According to him,

> both in method of treatment and in obtaining access to an elaborated body of material in the formal and physical sciences, pragmatism has much to gain from its European cousin. In return, pragmatism can offer its store of socially and biologically oriented analysis of such concepts as "mind," "consciousness," "self," "truth," "symbol" and can perhaps aid logical positivism in doing justice to the full range of interests which have generally characterized the activity of philosophers by saving it from the scholastic spinning of webs which a too narrow concern with the logical analysis of a restricted set of meanings might tend to encourage. (Morris 1937: 23)

He has proposed the term "scientific empiricism" for the philosophy uniting both "cousins" and, maybe, some other related developments.

In the late 1930s this idea did not seem extravagant. Many of the differences between positivism and pragmatism which we have examined abated. On the one hand, logical positivism was much more cautious in its pronouncements regarding religion than was Comte. It had no plans to create a new religion to replace Christianity. On the other hand, pragmatists of the thirties were much less dependent on religion than Peirce and James. Dewey had almost no interest in religion.

Both pragmatism and positivism have slightly changed their attitudes towards metaphysics as well. The efforts of nineteenth century positivists to diminish its significance have been relatively successful. In the twentieth century, Americans exhibited little interest in transcendental speculations. Thus pragmatists of the 1930s were not especially interested in metaphysics either. On the other hand, by the end of the thirties, logical positivists no longer regarded metaphysics as meaningless. They, Carnap for instance, had to acknowledge that they had not managed to win their battle against metaphysics; they had not overridden it by means of the logical analysis of language. Therefore, logical positivists grew much more tolerant towards metaphysics by this time, and in this respect the position of the logical positivists came closer to that of the pragmatists. The social and political attitudes of both movements, with the moderate leftist reformer Dewey now in charge of pragmatism, became closer as well.

In the theory of meaning, logical positivists found more points of contact with pragmatists than did the positivists of nineteenth century. Initially pragmatism was first of all a theory of meaning based on Peirce's famous rule for attaining the highest grade of comprehension: "consider what effects, which might conceivably have practical bearings, we conceive the objects of our conception to have. Then our conception of these affects is the whole of our conception of the object" (Peirce 1966 [1876]: 124). Logical positivism itself, unlike classical positivism, was primarily a theory of meaning. Of course, it was interested in the meaning of concepts and propositions, not of "objects." Logical positivists preferred to talk about "observable consequences" instead of "practical bearings." Nevertheless, precisely the theory of meaning, or to put it more broadly, the theory of signs, including syntax, semantics, and pragmatics, was the field in which cooperation between pragmatists and logical positivists promised to be especially fruitful.

A very important stimulus for cooperation was the shared concern with the unity of science. Pragmatists, and especially Dewey, were supporting many ideas of the unity of science movement initiated by the logical positivists. Moreover, it was precisely in this area that the real cooperation between pragmatists and logical positivists began. Dewey took part in preparation of the *International Encyclopedia of Unified Science.* Not only did he write a study entitled "Theory of Valuation" published as separate volume of the *Encyclopedia* (cf. Dewey 1969 [1939b]), but he also joined forces with Neurath, Bohr, Russell, Carnap, and Morris in preparing the introduction to the *Encyclopedia*, where he examined the unity of science as a social problem (cf. Dewey 1969 [1939a]).

There were, of course, some differences between Dewey and the logical positivists in their understanding of how to achieve the unity of science. Aiming at the advance of the sciences, Dewey was more interested in fostering cooperation between scientists, which was important for practical reasons, than in establishing a unified world conception on the basis of a single, unified language of science. Clearly having the physicalism of logical positivists in mind, he found it necessary to emphasize that "the needed work of cooperation cannot be done mechanically or from without. [...] The attempt to secure unity by defining the terms of all the sciences in terms of some one science is doomed in advance to defeat" (Dewey 1969 [1939a]: 34).

Dewey held the opinion that the most important task in advancing the unity of science is to build bridges between the sciences. He emphasized that a distinction has to be made between science as an attitude and method, and science as a body of knowledge. Attitude and method come prior to facts and theories; therefore, Dewey stressed that he was much more interested in achieving a unity in attitude and method, and not in creating a unified scientific description of the world.

Cooperation between scientists and *their unity* was especially important for Dewey, who was interested in the full use of science's social potential. He saw powerful forces interested in the isolation of science from everyday life. Only a united scientific community would overcome this artificial isolation and thus increase its power and ability to solve the problems of everyday life. By reminding us that the scientific method is not confined to those who are called scientists and must be used in dealing with social problems, Dewey resembles Comte. An important difference between them, however, lies in Dewey's striving to secure the link between science and democracy. Comte, on the contrary, wanted to use science as an instrument of elitist rule.

Dewey calls his own theory of scientific inquiry instrumentalism, which he defines as "an attempt to establish a precise logical theory of concepts, of judgments and inferences" (Dewey 1984 [1925]: 14). Logical positivists could define *their* theory of inquiry in exactly the same words, and say with Dewey that they aim at a theory of the general forms of conception and reasoning. However, Dewey's ideas presented in his *Logic:*

The Theory of Inquiry (1938) differ substantially from the logic of science developed by the logical positivists. Some of the differences between them result directly from different views about scientific inquiry. Dewey conceived it as "the controlled or directed transformation of an indeterminate situation into one that is so determinate in its constituent distinctions and relations as to convert the elements of the original situation into a unified whole" (Dewey 1938: 104). For the logical positivists such understanding seemed quite vague.

Although in his *Logic* Dewey retreated from radical holism, which he took initially from Hegel, and assimilated many pluralist elements of British empiricism, he still was not interested in formulating requirements which a scientific statement should satisfy, or methods to be used in establishing the degree of their empirical support. In this he diverges from logical positivists.

More elements of his conception of inquiry set Dewey apart from logical positivists. Dewey explicitly declared that he wanted to establish principles of inquiry on the basis of the study of actual methods of inquiry used by experimental sciences. As for the logical positivists, they were more interested in inferring the principles of inquiry from principles of logic and logical semantics rather than from the practice of inquiry. Another difference was no less important: although logical positivists emphasized the intersubjectivity of the results of scientific inquiry, they regarded it essentially as an individual endeavor. For Dewey, inquiry is a collective enterprise presupposing that a community of researchers is able to cooperate. And much earlier than Quine, Dewey held that no scientific statement may be regarded as absolutely certain. Inquiry, according to Dewey, is a self-correcting procedure, and any knowledge claim must be assessed critically.

In the words of Morton White, "Dewey has spent a good part of his life hunting and shooting at dualisms: body-mind, theory-practice, percept-concept, value-science, learning-doing, sensation-thought, external-internal" (White 1969: 121). Logical positivists, on the contrary, were very proud of all dichotomic distinctions they were able to make (or thought that they had made). Of course, the avoidance of dualisms makes Dewey pay its price: his ideas become insufficiently clear and precise.

We briefly examined the pragmatist, or rather, Peirce's and James's, notion of experience. It is worth extending this examination now. It played a key role in Dewey's philosophy. If logical positivists spoke about experience, even though they spoke about it surprisingly rarely, especially after closing their discussion on protocol sentences in the early thirties, they regarded it only as an unproblematic source of empirical facts, and were not much interested in the role experiment plays in physics and other sciences. They projected all activities of inquiry onto the plane of the language of science and were interested mainly in concepts, sentences, and their logical relations. Thus, as was already mentioned, their notion of experience was rather narrow. The adherents of pragmatism, like Marxists, were extending this notion, but in a somewhat different direction, by emphasizing the role of multifaceted relations between a person and the reality he creates.

For Dewey, experimentation is a very important form of acquiring knowledge, but experience means much more for him. For him experience is a form of a person's existence. A human being is a being who acts, enjoys, and suffers. Experience is its direct contact with nature and with other human beings. Thus the experience Dewey speaks about is a *whole* person's experience. The everyday life of a person is always Dewey's reference point. A very important trait of this life is its openness to the future and its active creative character. In this, Dewey emphasizes that his point differs from that of empiricism: "Whereas for empiricism, in a world already constructed and determinate, reason or general thought has no other meaning than that of summing up particular cases, in a world where the future is not mere word, where theories, general notions, and rational ideas have consequences for action, reason necessary has a constructive function" (Dewey 1984 [1925]: 12–13).

A distinctive feature of Dewey's conception of experience, setting it apart from that of logical positivism, is the dimension of values. Dewey underlines that "desires, affections, preferences, needs and interests […] exist in human experience; they are characteristics of it" (Dewey 1984 [1929]: 36). Values are not unchangeable objects of a higher realm. They exist in the world in which we live, they are part of our experience, they direct it.

In his article in the volume of *The Library of Living Philosophers* on Dewey, Reichenbach acknowledges that "in restoring the world of everyday life as the basis of knowledge, […] Dewey […] is establishing the sphere of values, of human desires and aims, on the same basis and in analogous form as the system of knowledge" (Reichenbach 1939: 163). For Dewey there is no fundamental difference between the sphere of value and the sphere of knowledge. Dewey blends ethics and social theory by establishing a direct relation between valuation and knowledge of the consequences of the decisions made in the past. According to Dewey, facts may constitute an empirical basis for valuations, which are by no means only purely emotive attitudes. Logical positivists, as is well known, held contrasting views.

Robert B. Westbrook, the author of a biographical book *John Dewey and American Democracy* (1991), treats Reichenbach's article as a step of logical positivists towards a union with the leader of American pragmatism (see Westbrook 1991: 498). This raises no doubt. Yet it is worth adding that this union was important not only for logical positivists, but for Dewey himself. The positions of pragmatism in America in the late 1930s were not as strong as in 1910s or 1920s. The influence of logical positivism in American universities was growing, and most young American philosophers considered it interesting, precise, and more modern than pragmatism. During the following decade it actually became the dominant intellectual force in American philosophy. Logical positivists, as newcomers, were interested in alliance with pragmatists, but the declining leader of American pragmatism hoped to gain a lot from the union with the young lion.

Both parties were more concerned with *exhibiting* good relations rather than with achieving real affinity of their views. However, the interaction of logical positivism and pragmatism has given birth to a new trend in American philosophy. It is sometimes called logical or analytical pragmatism. Later we will meet its most prominent representative—Willard V. O. Quine.

Positivism and Critical Rationalism

Among the critics of positivism, Karl Popper is a remarkable figure. More than anyone else Popper was eager to emphasize that he was an irreconcilable opponent of positivism in all its forms. Beginning with the *Logic of Scientific Discovery*, the German original of which appeared under the title *Logik der Forschung* in 1935, he claimed that positivism had fundamental flaws and was an essentially fallacious philosophy. In his later works dealing with political philosophy he criticized positivism for many real or imaginary sins, especially historicism, yet the main target of his attacks in the *Logic of Scientific Discovery* and in other works concerned with philosophy of science and epistemology, is positivism's *inductivism*. According to Popper, inductivism is the basis of the positivist philosophy of science, and this basis crumbles under his scrutiny.

The most radical form the results of this scrutiny acquired in his *Conjectures and Refutations* (1962) was as follows: "Induction, i.e. inference based on many observations, is a myth. It is neither a psychological fact, nor a fact of ordinary life, nor one of scientific procedure" (Popper 1962:

53). In the *Logic of Scientific Discovery* he was more moderate and claimed that it is impossible to decide whether or not scientists and researchers actually use induction (the principle of induction). In the *Logic* he presented himself as an opponent of the *naturalistic* methodology according to which it is a study of the actual behavior of scientists. Claiming that methodology should not be taken for empirical science, Popper has asserted that he does not believe "that it is possible to decide, by using the methods of empirical science, such controversial questions as whether science actually uses a principle of induction or not" (Popper 1972 [1935]: 52). Yet at the same time he stated that a reconstruction of scientific procedure without any reference to induction is much more coherent then the inductivist one, and does not face the insoluble problem of justification of induction.

Bearing in mind how much attention Mill and Carnap devoted to the analysis of inductive reasoning, which according to them is indispensable for any research based upon empirical evidence, Popper's claim that there is a very close link between positivism and inductivism seems quite convincing. Popper maintains that positivist philosophy of science is incapable of understanding the role that bold conjectures, criticism, and falsification (refutation) play in science.

According to Popper, the logic of scientific inquiry comprises the following elements and procedures. First, knowing the facts of a certain field of science and aiming to explain them, bold hypotheses should be proposed. They are considered to be bold if they are sufficiently remote from the facts (i.e., are sufficiently general) and allow one to derive empirically testable conclusions about the facts yet to be observed. Second, empirically testable conclusions are deductively inferred from those hypotheses. Third, incompatible facts, i.e., contradicting the hypothesis, are sought. Fourth, once such facts are discovered, the hypothesis is rejected, since according to the laws of logic, if a conclusion deductively inferred from a premise is false, then the premise is also false. Fifth, once the hypothesis is proved false, i.e., after it is falsified, it is refuted. Sixth, a new bold hypothesis explaining the facts (including those newly discovered) is proposed, and it is viewed just as critically as the previous one, requiring all efforts to falsify it.

According to Popper, *inductivism, verificationism, and positivism* are three facets of the same faulty methodology, just as *deductivism, falsificationism, and negativism* are three sides of the correct, Popperian, one. The general statements of science are not conclusions of inductive inferences, but rational conjectures. From them we deductively derive testable conclusions. Treating verification as a *proof of truth* of a statement (although logical positivists understood it only as *determination of a truth-value* of a statement, i.e., a test of that statement), Popper claims that it is impossible to verify scientific statements about the world. We cannot prove that they are true, since general scientific statements are not mere descriptions of observed facts; their content is broader. However, we can prove them to be false. The characteristic feature of a scientific statement is the possibility to falsify it. That is precisely what we should aim at. A scientist must subject all scientific statements, even those he himself formulates, to rigorous and unrelenting tests.

Popper calls his own philosophy of science critical rationalism, although it is not based on the view, characteristic of classical rationalists like Descartes or Leibniz, that reason is the source and foundation of all knowledge. He wants to emphasize that it is contrary to positivist empiricism. And since empiricists take induction to be the principal method of science, Popper devotes special attention to criticism of induction in his philosophy of science.

In order to see whether this criticism is well-founded, we should bear in mind one of its features—Popper provides no clear definition of the induction which he assails. While eluding precise definition of this notion, he is satisfied to note that according to inductivists "science starts from observation and proceeds by induction, to generalizations, and ultimately to theories" (Popper 1962: 154). The induction he speaks about is a "more or less simple procedure of inference" (ibid., 331). According to him inductivists believe that, acting cautiously, it is possible to ensure the soundness of "nondemonstrative" inferences on the basis of "positive instances" (Popper 1972 [1935]: 20), whereas caution requires to be confined to "cautious" generalizations of facts, not moving too far from them. When speaking about induction in the aforementioned and other writings, Popper often invokes a textbook example of an inference about white swans—that upon observing white swans in several instances one concludes that all swans are white—and this fact suggests that he understands induction as enumerative induction, that is, as a simple generalization.

Yet logical positivists have never restricted themselves to such a narrow understanding of induction. They were speaking about hypotheses, but they never thought that hypotheses were simple generalizations of facts. They took any nondeductive inference to be inductive. The difference between inductive and deductive inference is that the truth of inductive premises does not guarantee the truth of the conclusion.

Among the representatives of nineteenth and twentieth century positivism there are, however, important differences regarding the nature of inductive inference. The notion of induction Mill was most interested in, i.e., enumerative and eliminative induction, may be called *genetic*. Such induction is used in the *context of discovery*. Mill was concerned about how on the basis of inductive rules one could *infer* from the facts their causes. The notion of induction that logical positivists had in mind may be labeled *methodological*, and they were interested in the *context of justification*. They were concerned with establishing a logical link between already formulated hypotheses (conjectures, in Popper's terminology) and the relevant empirical data.

According to Popper, the only link between them is the deductive one: observational statements are deductively inferred from the hypothesis. Logical positivists had no doubts about the importance of this link either. Yet they also emphasized the existence of another, inductive link. Inductive inference meant for them, however, not the inference of a hypothesis from the facts but one of the two things: either (1) the establishment of the degree of empirical support for a hypothesis; or (2) tentative, probational acceptance of a hypothesis on the basis of empirical data and recognition of its empirical superiority over other competing hypotheses.

Knowing perfectly well Popper's arguments against induction (and Popper was only reiterating attacks on induction made by Duhem, not to mention earlier anti-inductivists), logical positivists persevered in emphasizing that the induction in the aforementioned sense is an absolutely indispensable part of scientific procedure. Or, to put in another way, they were convinced that reconstructions of the procedures of empirical science which do not take induction into account were blatantly inadequate.

Popper, of course, admits that scientific hypotheses can be not only rejected, but sometimes, following the failure of persistent attempts to refute them, can be tentatively accepted. Otherwise it would be impossible to speak about scientific *knowledge* at all. On the basis of accepted hypotheses new ones are proposed, theories are constructed from them, and so science develops. But contrary to Popper's claims, acceptance of a hypothesis, however tentative, is certainly an instance of (methodological) induction. It has to be regarded as an inductive procedure, because in making decisions regarding the acceptance of a hypothesis after its empirical testing, we are moving from empirical evidence to hypothesis, and in stepping from the particular to the general we are performing not a deductive, but an inductive movement. Popper simply did not want to admit that *induction is any nondeductive inference*. Yet logical positivists referred precisely to this definition of induction. It is also accepted by the majority of philosophers of science.

Since an empirical science which does not take into account the results of testing hypotheses (and taking them into account does not always mean rejecting a hypothesis) is inconceivable, induction is vital for science. Popper's main argument against induction has been that it is impossible to justify it. We have already seen that by induction logical positivists meant something different from the kind of inference which from singular statements about observed white swans leads to the conclusion that all swans are white. It was the question of justification of this kind of inferences that Popper devotes his attention, and he concludes that they cannot be justified. Does his conclusion also apply to the methodological notion of induction?

Before answering this question it is worth reconsidering Popper's arguments. First of all he points out that those who are seeking to justify inductive inference must establish a principle of induction, on the basis of which "we could put inductive inferences into a logically acceptable form" (Popper 1972 [1935]: 28). Then he maintains that this principle cannot be a purely logical truth. Thus it must be a synthetic statement, a universal synthetic statement, as a matter of fact. Yet if we try to infer it from facts, we face the same problems which required the introduction of this principle:

> My own view is that the various difficulties of inductive logic here sketched are insurmountable. So also, I fear, are those inherent in the doctrine, so widely current today, that inductive inference, although not "strictly valid," can attain some degree of "reliability" or of "probability." According to this doctrine, inductive inferences are "probable inferences." [...] For if a certain degree of probability is to be assigned to statements based on inductive inference, then this will have to be justified by invoking a new principle of induction, appropriately modified. And this new principle in its turn will have to be justified, and so on. Nothing is gained, moreover, if the principle of induction, in its turn, is taken not as "true" but only as "probable." In short, like every other form of inductive logic, the logic of probable inference, or "probability logic," leads either to an infinite regress, or to the doctrine of apriorism. (Popper 1972 [1935]: 29–30)

This was written at the time when Reichenbach had already started developing his *probability logic* and before Carnap outlined his original project of probabilistic *inductive logic*. Their notion of induction was different, I repeat, from the Popperian notion. Still, it seems that their undertakings face serious problems. Logical positivists, first of all Carnap, cherished the hope to be able to assign *with certainty* the exact degrees of probability to hypotheses. This certainty was to be the watered-down form of the ideal of certainty revered by philosophy and science for ages. It seems, however, that we must reconcile ourselves with the thought that not only statements of the empirical sciences, but even our judgments concerning their probability cannot be certain, rock solid. This means that we can never know with certainty whether a statement of empirical science is true or false, or what its probability is.

Popper, however, does *not* make this claim. He is convinced that the hypotheses of empirical science can be appraised as *false* and ultimately rejected, while it is impossible to prove the *truth* of scientific hypotheses because it is impossible to justify the principle of induction. But why is it necessary to reject the falsified theory? Apparently, falsification of a theory shows that it is untrustworthy. That means, however, that we base our judgment on the assumption that a theory falsified by empirical facts at present will be falsified by empirical facts in the future. But this assumption is quite analogous to the assumption of inductivism: a theory confirmed by facts at present will be confirmed by them in the future as well. As Ayer noted in *The Problem of Knowledge*, it would be inconsistent to believe that a falsified hypothesis is universally valid, yet there is no inconsistency in believing that it is more likely that a falsified hypothesis will be valid in the future. It is imaginable that falsification is like chicken pox, which gives immunity after the first infection (cf. Ayer 1966 [1956]: 74). Thus, at least in this respect, Popper's theory does not improve upon that of the logical positivists.

It is more important, however, that Popper is wrong in claiming that the falsity of a hypothesis can be proven with *absolute certainty*. Deducing conclusions from a hypothesis, especially when it bears a high degree of generality, is often a long and complicated process. In this process we infer one statement from another, and then the third from the former, and often only after several deductive inferences do we arrive at a conclusion that is directly empirically testable. In this process of deduction one has to rely on various premises, the truth of which may always raise doubts. For, as Popper himself emphasizes, the truth of synthetic statements can never be strictly proved. Therefore, from the methodological point of view, an observable empirical fact that contravenes a deductive entailment of a hypothesis proves only that at least one of the statements used in the process of deductive inference is false. *Either* a hypothesis *or* some auxiliary premise might be false. Moreover, the very description of the fact contradicting the hypothesis may turn out to be false—an observer may be mistaken for some reason. Thus our claim that the hypothesis is false is but a new hypothesis. This means that not only the truth, but also the falsity of a hypothesis can never be proved with *absolute* certainty. And so in accepting a hypothesis after testing it empirically, as well as in rejecting it, we make a judgment that may have to be revised in future.

Popper always claimed that the differences between him and logical positivists were huge because their standpoints regarding the character of scientific knowledge and the ways to elaborate the philosophy of science were diametrically opposed. Agreeing with this claim is difficult. It is enough to compare his *Logic of Scientific Discovery* (1935) with Carnap's *Testability and Meaning* (1936/37), which best represents the point of view of logical positivism of the period, to assure ourselves that Popper and Carnap held very close beliefs on the most important issues.

First of all, both believed that the philosophical investigation they are interested in is logical and methodological analysis. Carnap declared it many times, and in the *Logic of Scientific Discovery* Popper defines his task as to give "a logical analysis of the method of empirical science" (Popper 1972 [1935]: 27).

Second, both investigate the context of justification, not of discovery. The title of the English translation of *Logik der Forschung* as *Logic of Scientific Discovery* published in 1959, is flagrantly, even ludicrously misleading because at the outset of the book the author who titled it *Logic of Inquiry* says that "there is no such thing as a logical method of having new ideas, or a logical reconstruction of this process" (Popper 1972 [1935]: 32) assenting to Einstein's view that no logical path leads a scientist to the universal laws of science, and adding a few lines below that every discovery contains an irrational element.

Third, both Popper and Carnap sought some criterion allowing us to draw a dividing line between science and metaphysics and thus to identify empirical science as an object of logical analysis. Though Popper regarded the criterion of verification as inadequate, they were both addressing the so-called demarcation problem, i.e., the problem of separating empirical science from metaphysics.

Fourth, both treated scientific knowledge as a logical system of statements satisfying strict requirements and governed by rigorous laws of logic.

Fifth, they both investigated relations between a theory and relevant facts, not between different theories replacing each other.

Sixth, both used the principle of the dichotomy of analytical and synthetic statements without any restrictions.

Seventh, both treated methodology (philosophy of science) as a normative discipline.

Eighth, both had no doubt that the statements of singular facts form the empirical foundation of a theory, and it has to be tested on their basis.

Ninth, both regarded all synthetic statements (including statements of a singular fact) to be hypothetical and not absolutely certain.

To summarize, Popper, at least in the mid-1930s, was very close to logical positivism. He maintained contacts with members of the Vienna Circle, and the fact that German original of *Logic of Scientific Discovery* was first published in the series of *Schriften zur Wissenschaftlichen Weltauffasung*, edited by members of the circle (namely Frank and Schlick) together with Carnap's, Neurath's, Frank's, and Schlick's books, is quite revealing. Moreover, even much later Popper was often treated as a representative of positivism as well. We have already mentioned that in the famous positivism dispute (*Positivismusstreit*), in 1961 the German Sociological Society chose Popper to represent positivism. A few years later at the conference on sociology in Heidelberg Habermas sparred with the second major representative of critical rationalism and close associate of Popper, Hans Albert, who was also treated as a positivist.

It is noteworthy that the main target of Popper's criticism in *Logik der Forschung* was phenomenalist positivism of the end of the 1920s. The development of the ideas of the Vienna Circle at the beginning of the 1930s was so dynamic that when Popper's book left the printer's office, his criticism was already somewhat obsolete. At that time physicalist analysis has replaced in the Circle the earlier, phenomenalist one. Popper himself recorded this: in a footnote added after reading the proofs in 1934, he indicated that his criticism of positivism did not concern Carnap's *Logische Syntax der Sprache* which had appeared the same year. In the mid-thirties there was also no difference between Popper and the logical positivists regarding the issue of objectivity of scientific knowledge which Popper began to treat as the crux of their disagreements in *Objective Knowledge: An Evolutionary Approach* published in 1972. Popper's thesis presented in *Logic of Scientific Discovery* that "the objectivity of scientific statements lies in the fact that they can be inter-subjectively tested" (Popper 1972 [1935]: 44) fully corresponded to the views of Carnap and his colleagues.

One of the differences between Popper and logical positivists he stresses in 1959 in the preface to the English translation of *The Logic of Scientific Discovery* concerns his attitude to the logical analysis of language which suffered some changes. Logical positivists and early Popper regarded logical analysis as the principal method of philosophy. In the preface he presents a different opinion. "I do not deny," wrote Popper, "that something which may be called 'logical analysis' can play a role in this process of clarifying and scrutinizing our problems and proposed solutions; and I do not assert that the methods of 'logical analysis' or 'language analysis' are necessarily useless. My thesis is, rather, that these methods are far from being the only ones which a philosopher can use with advantage, and that they are in no way characteristic of philosophy" (Popper 1972 [1959]: 16).

In 1959 Popper indicates that, in particular, he is critical of the construction of the artificial models of scientific language. He claims that proponents of this method are unable to construct a full-scale working model of a language of real science, say physics, thus "we find them engaged in the construction of intricate working models in miniature—of a vast system of minute gadgets" (Popper 1972 [1959]: 20). According to Popper, by doing this they miss the most important problems of epistemology, and, especially, that of the growth of knowledge, that of the relation between theories that are overthrowing each other. One could partly agree with this opinion. However, it must be noticed, that Popper himself does not deal with this problem in his *Logic of Scientific Discovery*—it is not examined in the main text of a book written twenty-five years earlier. In it he scrutinizes theories and their empirical basis, falsifiability, testability, simplicity, and probability, but he is not bothered by the dynamics of science.

Incidentally, Popper, having discussed the problems of the growth of scientific knowledge in his collection of papers *Conjectures and Refutations* (1962), later devotes more attention to a different issue. In 1967, at the Third International Congress of Logic, Methodology, and Philosophy of Science held in Amsterdam, he promoted the now well-known doctrine of the *third world*, which he compared with Plato's world of ideas and Hegel's world of the objective spirit. Popper delineates the first world of physical objects or physical states, the second world of mental states, and the third world of the objective content of thought. While speaking about the latter he means the contents of the scientific problems and theories that are supposed to be independent of what certain thinkers in some particular epoch think about these problems and theories.

Popper treated this doctrine as a challenge to the "philosophy of belief," which was content to examine knowledge only in the subjective sense. Among representatives of this philosophy, Popper mentioned both Hume and Russell, and it seems, he was also inclined to impute this doctrine to logical positivists, although the philosophy of science of logical positivism from the mid-1930s was no longer subjectivist: its representatives were not interested, as well as Popper, in the cognizing subject and his actions, but concentrated their efforts on examining the structure and relations of purely logical cognitive constructions: concepts, statements, theories.

In the late 1960s Popper decided that the theory of science needs an extralogical foundation and must be based on realist epistemology and the metaphysics of three worlds. He came to treat objects and phenomena of the physical world as objectively existing referents of terms in which scientific propositions are cast. This provided the prerequisites necessary for the full objectification of the content of scientific theories, problems, and their solutions by way of setting apart objective and subjective knowledge. Logical positivists deliberately refused to draw clear-cut lines between the mental, the physical, and the logical, since they were convinced that the claim about the existence of three different worlds is clearly metaphysical.

Thus we can conclude that the differences between Popper's and logical positivists' standpoints, minor initially, tended to grow and become more pronounced. Yet that does not prove, as Popper claimed, that they were giving *opposite* methodological advice to practicing researchers. First of all, Popper was very keen to repeat that he called for a search for negative, falsifying facts whereas logical positivists suggested that empirical data be sought which confirmed theories. Yet logical positivists never suggested such a procedure. It is worth stressing once again that logical positivists understood verification as the procedure of determining the truth-value—one or another—of a statement or a theory, and not as a proof of its truth. They were convinced that the scientific theory or hypothesis must be assessed on the basis of *all available* evidence, whether it confirms or disconfirms the hypothesis. On the other hand, Popper suggested that bold, improbable hypotheses must be given preference over cautious and very probable ones. Yet a hypothesis with low *a priori* probability may acquire, after severe but successful testing, a quite high *a posteriori* probability. Thus *both* Popper and Carnap may lend preference exactly to the same hypothesis out of several competing ones.

Imre Lakatos, himself a prominent philosopher of science, once compared Popper's and Carnap's views on the appraisal of scientific theories in rather witty terms. Explaining Popper's standpoint, Lakatos wrote: "Bold theories undergo *severe tests*. The 'severity' of tests is assessed by the difference between the likeliness of the positive outcome of the test in the light of our theory and the likeliness of its positive outcome in the light of some rival 'touchstone theory' (already articulated in the extent body of science, or only articulated in the proposal of the new theory)"

(Lakatos 1968: 379–80). Only in the case the previous theory is refuted, another theory is (tentatively) accepted. "This Popperian jungle," continued Lakatos, "contrasts starkly with Carnap's civilized society of theories. The latter is a peaceful welfare state of fallible but respectably aging theories, reliable [...] to different but always positive degrees, which are registered daily with pedantic precision in the office of an inductive judge. Murders are unknown—theories may be undermined but never refuted" (ibid., 380).

This comparison is revealing in at least one respect. The appraisal of a theory logical positivists were interested in, was in fact similar to that an impartial judge would make. Popper, however, was more interested in a different kind of appraisal. When demanding from scientists to be severe and critical, to always look for indictments, negative evidence, and seek falsification of their own theories as soon as possible, he was advocating the partiality and bias characteristic of a public prosecutor. But even the latter does not usually demand that the defendant prove his own guilt. Another point, however, is even more important. The later Popper ignored entirely the fact that a public prosecutor would be completely out of place in his third world. It is peculiar that Popper, who considered himself *objectivist*, was concerned with the *subjective* attitudes taken by researchers, while logical positivists were little interested in them.

Positivism and the Analytic Tradition

Critics of positivism were not lacking within the analytic tradition of philosophy, which should be characterized by the view that the analysis understood in one way or another is the principal or even the sole method of philosophy, and the clarification of language—the main task of philosophy. Logical positivism was a particularly important form of such philosophy. But this fact has not precluded some analytic philosophers to take a critical stance towards some of its tenets.

It is natural to start with Ludwig Wittgenstein, who, together with Bertrand Russell, and (before them) Gottlob Frege, is usually considered to be one of the founders of this tradition. (Sometimes this fame is shared by G. E. Moore for his analysis of ethical concepts.) Wittgenstein can justly be held as one of the two (next to Russell) most important representatives of logical atomism, and later, as the founder of another trend in analytic philosophy—ordinary language philosophy. The fact that in the mid-thirties he embarked on the project of developing ordinary language philosophy is clear evidence that he was unsatisfied with his earlier philosophy. And because the logical positivists drew on the *Tractatus Logico-Philosophicus*, it is rather natural to expect that he also had to take a critical stance towards their philosophy. However, we will find not much direct criticism of logical positivism in Wittgenstein's philosophical heritage. He did not like to engage in circumstantial critique of the conceptions of other philosophers in general. And his aphoristic style of writing was little suited to such critique.

Wittgenstein's views changed. Yet the differences between the earlier and the later Wittgenstein are not as great as they are sometimes presented. Notwithstanding the variety of its forms, analytic philosophy displays a considerable degree of unity, and the main pillar of this unity was the conviction of its founders that all philosophy is or should be a critique of language, and that the aim of such a critique is the clarification of language and the demonstration of how it is sometimes misused.

Viewed from this perspective, the main change occurring in Wittgenstein's philosophy in the 1930s is, it seems, not the repudiation of certain conceptions of the *Tractatus*, but the broadening of his conception of language and the widening of the spectrum of the ways by which it may be analyzed and criticized. The change of Wittgenstein's views is well expressed by a short note, almost put aside in parentheses in his *Remarks on the Philosophy of Psychology*: "(There just are many more language-games than are dreamt in the philosophy of Carnap and others.)" (Wittgenstein 1988 [1980], vol. 1: 920).

At first glance this remark seems rather irrelevant because nobody says that every possible game must be played. You play the game you know and like, and logical positivists played the game of science. However, in all probability, Wittgenstein is suggesting that according to logical positivists, the rules of this one game are the general rules of every game which may be played meaningfully. This was unacceptable for the later Wittgenstein. And the restriction of the means which may be used to clarify language to rational reconstruction was even more unacceptable for him. By replacing some fragment of a scientific language by its rational reconstruction, you only stiffen an already rigid language. You move away from, not towards the real use of language. You are not playing; you confine yourself in the prison you yourself constructed.

It seems that representatives of ordinary language philosophy, and first of all John L. Austin, did not have much to add to Wittgenstein's appraisal of logical positivism. In the 1950s and early 1960s they carried out their delicate, refined, and a bit fragmented analyses and were not especially interested in the works of logical positivists. When Peter Strawson, coming from a different background, started looking for ways to integrate logical and ordinary language analyses in order to create logical metaphysics, he also did not expect to gain much from the criticism of the philosophy of science of logical positivists.

The most damaging critique of logical positivism came not from the British but from the American analytic philosophers. American analytic philosophers, unlike their colleagues in Britain, were almost unaffected by the analysis of ordinary language. The influence of Norman Malcolm, Alice Ambrose, Charles L. Stevenson, and other disciples of the later Wittgenstein was relatively insignificant in the U.S.A. The minds of analytic philosophers in America were primarily shaped by logical positivism. Harvard University became an important base for logical positivists in the United States. The Fifth International Congress for the Unity of Science was held there in September 1939, at a time when World War II was already starting in Europe. Some historians of philosophy consider Cambridge, Massachusetts, where Harvard University is located, to be an analogue of Vienna in the New World. A noted physicist and philosopher Gerald Holton published a substantial text with an eloquent title "From the Vienna Circle to Harvard Square: The Americanization of a European World Conception" (Holton 1993). A little earlier one of the most famous French supporters of logical positivism, Pierre Jacob, edited and published a collection of papers by logical positivists with a similar title—*De Vienne à Cambridge. L'héritage du positivisme logique de 1950 à nos jours* (1980).

Harvard has played a really important role in propogating the ideas of logical positivism in the United States. Carnap delivered lectures there for a time. World-famous philosophers Willard Van Orman Quine and Hilary Putnam were professors at Harvard for many years. Quine, as I already mentioned, was collaborating with the Vienna Circle; Putnam, who was younger and began lecturing at Harvard later than Quine, had been a doctoral student of Hans Reichenbach at the University of California, Los Angeles, and he took Carnap to be his most important teacher.

Quine was not mentioned by accident. For many he's not as much an adherent of logical positivism as its adversary. We shall return to this issue later. However, it was his paper published in *The Philosophical Review* in 1951 that made a tremendous impact on the fortune of logical positivism. The paper I have in mind is, of course, "Two Dogmas of Empiricism."

Quine begins his article with a few sentences summarizing its content that are worth citing:

Modern empiricism has been conditioned in large part by two dogmas. One is a belief in some fundamental cleavage between truths which are *analytic*, or grounded in meanings independently of matters of fact, and truths which are *synthetic*, or grounded in fact. The other dogma is *reductionism*, the belief that each meaningful statement is equivalent to some logical construct upon terms which refer to immediate experience. Both dogmas, I shall argue, are ill-founded. One effect of abandoning them is, as we shall see, a blurring of the supposed boundary between speculative metaphysics and natural science. Another effect is a shift toward pragmatism. (Quine 1980 [1951]: 20)

When arguing his points, Quine proceeds in the following way. First of all he emphasizes that the theory of meaning must be separated from the theory of reference. After that is done, there is no need to consider in the theory of meaning "obscure intermediary entities," i.e., meanings; all that is important is the synonymy of linguistic forms (i.e., sameness of meanings) and analyticity of statements. In this respect his opinion does not differ from Carnap, who also tried to avoid any talk about "meanings," concentrating his efforts on the examination of meaningfulness and analyticity. However, Quine is unhappy with Carnap's explication of analyticity in terms of state-descriptions: analytic is the statement that is true under any state of a system, i.e., when it comes out true under every state-description.

Quine emphasizes that the real problem in defining analyticity is not the logical truths which remain true under all interpretations of their non-logical components. The real problem is the second class of analytic statements represented by "No bachelor is married." Statements of this kind may be transformed into logical truths—such as "No unmarried man is married"—by substituting synonyms for synonyms. Carnap's explanation of analyticity in terms of state-descriptions presupposes that the atomic statements of a language out of which state-descriptions are constructed, are mutually independent. Yet the statements "John is a bachelor" and "John is married" are not mutually independent. Therefore, though the conceptual apparatus of Carnap's state-descriptions is helpful, according to Quine, in clarifying problems of probability and induction, or maybe even that of logical truth, it does not clear up the problem of analyticity when synonymy is involved.

Quine rejects the claim that analytic statements of the second class may be reduced to logical truths by definition. By defining "bachelor" as "an unmarried man" on the basis of an observed synonymy, the lexicographer presupposes the notion of synonymy which has still to be explained. The same relation between definition and synonymy holds in formal languages. Thus, Quine concludes, "definition—except in the extreme case of the explicitly conventional introduction of new notions—hinges on prior relations of synonymy" (Quine 1980 [1951]: 27).

Other attempts to draw a dividing line between analytic and synthetic statements on the basis of, say, the notion of interchangeability or semantic rules, are not successful according to Quine. Hence he ends his examination of the problem of analyticity with the following words:

> It is obvious that truth in general depends on both language and extralinguistic facts. [...] Thus one is tempted to suppose in general that the truth of a statement is somewhat analyzable into a linguistic component and a factual component. Given this supposition, it next seems reasonable that in some statements the factual component should be null; and these are the analytic statements. But, for all its a priori reasonableness, a boundary between analytic and synthetic statements has not been drawn. That there is such a distinction to be drawn at all is an unempirical dogma of empiricists, a metaphysical article of faith. (Quine 1980 [1951]: 36–37).

After exposing to criticism the first dogma of modern empiricism (read: logical positivism), Quine turns his mind to the second one, that of reductionism. According to it, each meaningful statement is a construct from terms which refer to immediate experience.

Quine claims that in logical positivism the principle of reductionism found its expression in the verification theory of meaning. According to him, this theory traces its parentage to Peirce. At the same time he emphasizes that (radical) reductionism well antedates the verification theory of meaning. "Thus," Quine notes, "Locke and Hume held that every idea must either originate directly in sense experience or else be compounded of ideas thus originating; and taking a hint from Tooke [an eighteenth-century philologist] we might rephrase this doctrine in semantical jargon by saying that a term, to be significant at all, must either be a name of a sense datum or a compound of such names or an abbreviation of such a compound" (Quine 1980 [1951]: 38).

According to Quine, this is an unnecessary restriction. It is more reasonable to conceive of the doctrine of radical reductionism as a requirement that each of our statements as a whole be translatable into sense-datum language. Quine points out that Carnap attempted to carry out the program of radical reductionism in the *Aufbau*. However, Carnap did not succeed in demonstrating how statements about physical objects may be translated into statements about immediate experience and later gave up this idea. Yet, according to Quine, Carnap never renounced the dogma of more tenuous reductionism according to which "to each statement, or each synthetic statement there is associated a unique range of possible sensory events such that the occurrence of any of them would add to the likelihood of truth of the statement, and that there is associated also another unique range of possible sensory events whose occurrence would detract from that likelihood" (Quine 1980 [1951]: 40–41). Thus, Carnap purportedly thought that it's possible to confirm or disconfirm every statement individually, in isolation from others.

When criticizing the first dogma, that of a sharp division between analytic and synthetic statements, Quine claimed that empiricists did not show how it is possible to draw a line between them. It is worth noting that speaking about the second dogma, Quine is even more cautious because there is no doubt that Carnap did show how to associate observation statements with some statements we want to test and thus confirm or disconfirm. Quine's holistic position which he modestly calls his "counter-suggestion," is that "our statements about the external world face the tribunal of sense experience not individually but only as a corporate body" (Quine 1980 [1951]: 41).

Is Quine's criticism sound? It seems that many analytic philosophers lean towards Quine's point of view, and almost no philosopher outside the analytic tradition has any doubts that Quine is right. But is he right indeed?

While trying to answer this question, it has to be stressed that Carnap never claimed that it is possible to give an exact definition of analyticity applicable to the natural language. The meanings of the words of natural language are not precise; therefore, it is impossible to determine when two words have exactly the same meaning. Thus the major part of Quine's paper in which he discusses the problem of analyticity of natural language statements is not relevant as a criticism of Carnap's views. There is no disagreement between Carnap and Quine on this issue.

What Carnap did claim was the following: when the *new artificial language system* is being laid down, both its syntactical and semantical rules must be chosen and precisely formulated. Hence Quine's argument becomes relevant as a critique of Carnap only when Quine begins to discuss the problem of the analyticity of sentences of language systems, i.e., languages with codified rules. Quine does not deny that it is possible to single out—in one way or another—some sentences of some language system and call them "analytic." He claims, however, that this procedure does not help him to grasp the meaning of the term "analytic."

A few months after Quine's article was printed, in February 1952, Carnap wrote a retort to Quine. This text was not published until 1990 when it appeared in the volume *Dear Carnap, Dear Van: The Quine-*

Carnap Correspondence and Related Work (Quine and Carnap 1990), that is, twenty years after the death of the author. I do not know the real reasons why Carnap did not publish it, although the manuscript—written in shorthand—contains a marginal note: "(Perhaps a separate paper)." One could only guess that he intended to present his main counterarguments in Schilpp's volume. In 1952 the volume on Carnap was already under preparation although for various reasons it was published only in 1963 (Schilpp 1963). This led most members of the philosophical community to hold that Carnap had no answer to Quine's arguments.

In fact, Carnap had an answer. The text written in 1952 presents Carnap's immediate reaction to Quine's criticism and is of special interest to us. It has been given a title "Quine on Analyticity" by Richard Creath, the editor of the book *Dear Carnap, Dear Van*. In this text Carnap lends special importance to the fact that from Quine's article it is difficult to understand when Quine speaks about analyticity as *explicandum* and when as *explicatum*. This indeterminacy stems from Quine's failure to strictly distinguish natural languages from the artificial, codified languages, that is, languages with strictly specified rules.

Exactly for that reason Carnap qualifies as ambiguous one of the most important claims made in "Two Dogmas," namely that "semantical rules determining the analytic statements of an artificial language are of interest only insofar as we already understand the notion of analyticity; they are of no help in gaining this understanding" (Quine 1980 [1951]: 36). Carnap's rejoinder is fairly predictable: "We have an understanding of the notion of analyticity, in practice clear enough for application in many cases, but not exact enough for other cases or for theoretical purposes. The semantical rules give us an exact concept; we accept it as an explicatum if we find by comparison with the explicandum that it is sufficiently in accord with this" (Carnap 1990: 431).

Like Duhem (whom he mentions) and Dewey (whom he does not mention) prior to him, Quine claims that no statements of science are sacrosanct: even laws of logic may be revised. At first glance, Carnap's point of view is rather similar. One form of logic may be replaced by another, and Carnap directly refers to the possibility of replacing the usual form of logic by an intuitionistic or a three-valued one (cf. Carnap 1990: 431). Carnap held this point of view from the mid-thirties: the logical principle of tolerance was put forward in *The Logical Syntax of Language* (cf. Carnap 1937 [1934]: 51–52).

On the *explicandum* level "analytic" means approximately "true by virtue of meanings." But as Carnap notes, by changing the logical structure of language we may change the *meanings* of the signs. The same sequence of words or symbols may be analytic in one language and synthetic in another. Yet the difference between analytic and synthetic statements is, according to Carnap, "a difference internal to two kinds of statements inside a given language structure; it has nothing to do with the transition from one language to another" (Carnap 1990: 431). This elucidation enables Carnap to conclude: "Since the truth of an analytic sentence depends on the meaning, and is determined by language rules and not by observed facts, then an analytic sentence is indeed 'unrevisable' in another sense: it remains true and analytic as long as the language rules are not changed. The attribution of truth-values to synthetic sentences changes continually, induced by new observations, even during a period in which the logical structure of language remains unchanged. A revision of this sort is not possible for the analytic sentences" (Carnap 1990: 432).

The debate between Carnap and Quine on analyticity is continued on the pages of Schilpp's volume on Carnap. Because of delays in the publication of this volume, Quine lost his patience and published the text "Carnap and Logical Truth" in 1960 in *Synthese*; it was reprinted in 1963 in Schilpp's *The Philosophy of Rudolf Carnap* (see Quine 1963 [1960]).

In this text Quine widens his criticism of Carnap by attacking the doctrine of logical truth according to which "logical truths are true by virtue purely of the intended meanings, or intended usage of the logical words" (Quine 1963 [1960]: 388). He finds this doctrine unacceptable because of the obscurity of "true by virtue of language." That means that "analyticity" is also an obscure notion. "In any event," notes Quine, "we at present lack any tenable general suggestion, either rough and practical or remotely theoretical, as to what it is to be an analytic sentence. All we have are purported illustrations, and claims that the truths of elementary logic, with or without the rest of mathematics, should be counted in" (ibid., 403). He does not think that in "Meaning Postulates" (see Carnap 1956 [1952]) Carnap managed to strengthen his position on analyticity. Quine ends his article with a passage which has since become famous: "The lore of our fathers is a fabric of sentences. In our hands it develops and changes, through more or less arbitrary and deliberate revisions and additions of our own, more or less directly occasioned by the continuing stimulation of our sense organs. It is pale gray lore, black with facts and white with convention. But I have found no substantial reasons for concluding that there are any quite black threads in it, or any white ones" (Quine 1963 [1960]: 406).

In his reply to Quine, Carnap repeats his complaint of 1952 that he has not been able to determine the meaning of Quine's formulations "with sufficient clarity." Carnap emphasizes once again that Quine does not seem to be disputing his explication of analyticity in terms of semantical rules, since it is absolutely exact. However, Quine thought that customary explanations of analyticity are too vague and incomprehensible, and hence required an empirical criterion for analyticity, even though he has not required an empirical criterion for truth. Such a criterion would allegedly allow determining the content of the concept of analyticity that is to be explicated.

Carnap's response is as follows. The statement that sentence S is analytic in language L for a person X is an empirical hypothesis which may be confirmed or disconfirmed by observations of the verbal behavior of X (see Carnap 1963b: 919). Carnap clearly has in mind that if a person X never abandons some of his statements despite his changing experience, he takes those statements to be analytic.

However, Quine did not find this reply of Carnap convincing. In his *The Roots of Reference*, Quine emphasizes that each of us learns to use and understand sentences, and *sometimes* by learning a sentence a person learns to assent to it. Thus for a person who has learnt a language under certain conditions the sentence "A dog is an animal" would be analytic. However people learn in different ways and different order. Therefore different persons would hold different sentences to be unconditionally true. It is impossible to precisely identify analytic sentences, or to discern them from the sentences one might call nearly analytic that are unconditionally true for many but not for all speakers. The notion of analyticity, according to Quine, has not a logical, as Carnap affirmed, but social character.

Quine's and Carnap's debate over analyticity is, likely, the most famous dispute in the history of analytic philosophy. It has provoked vast literature. Some scholars ingeniously defended Carnap's point of view, claiming that Quine's argument is of doubtful value, because it is perfectly possible that the notion of analyticity belongs to the family of linguistic notions which are not definable in terms of nonlinguistic notions (cf. Grice and Strawson 1956). Others maintained that, in fact, Quine's arguments were directed against two different notions. However, most scholars have supported Quine's view. Thus Jaakko Hintikka has claimed that "the Carnapian idea that the meanings (and *a forteriori* references) of our words and expressions are predetermined by set of meaning rules (whatever they are or may be) is unrealistic. What Quine has completely correctly tried to get at is a much more realistic picture of how semantics of our actual language really works" (Hintikka 1991: 212).

I think that Hintikka is right. But Carnap never analyzed the semantics of natural language. It is very likely that he would have responded to the just quoted words of Hintikka in a way familiar from his response to Quine. Carnap would have said that he is concerned with formalized languages, and only in such languages a clear-cut line between analytic and synthetic sentences may be drawn. The questions as to "how semantics of our actual language really works" and "how we must construct formalized languages" are completely distinct. Therefore, unsurprisingly, the answer to each question is different.

It seems advisable to leave here the inexhaustible subject of analyticity and to go on to another one. Quine's criticism of the second "dogma" of empiricism, that of reductionism, seems to be weaker than that of the first, and may be countered more easily. The point is that in the *Aufbau*, Carnap really sought to reduce physicalist statements about the external world to phenomenalist statements about sense-data. He failed. Later, however, when he embarked on the project of constructing a logical theory of confirmation, he was not interested anymore in reducing scientific hypotheses about the physical world to sense-data language, or in Quine's words, to "sensory events." In Carnap's inductive logic, both hypothesis and evidence confirming it are formulated in the same (physical) language. Therefore the relation between hypothesis and evidence has nothing to do with the reduction of the language in which we speak about the external world to the language of sense-data.

I do not think that the connection between the verification theory of meaning and the individual confirmability of scientific statements is as close and direct as Quine claims. Logical positivists were engaged in two major projects: one of establishing an adequate criterion of empirical significance, and second, of constructing a logical theory of confirmation. There is a genetic but not a logical connection between them. The idea of developing a logical theory of confirmation was put forward when it became clear that scientific statements having empirical content cannot be verified directly and conclusively. Yet when embarking on the second project, Carnap did not abandon the first. In the course of its development, the criterion of empirical significance was transformed substantially. According to its latest version proposed by Carnap, statements of theoretical language are *not individually confirmable*, but that does not mean that they are void of empirical significance (cf. Carnap 1956).

Before leaving the subject of the two dogmas altogether it is worth making a few general remarks on Quine and positivism. It is a commonplace to regard him as its opponent. Yet this is not right. When criticizing

Carnap's conception of analyticity, Quine had no intentions to destroy or weaken positivism. He was unhappy that while insisting on sharp formulations, Carnap himself was not sharp enough when using the word "analyticity." It is precisely the *positivist* attitude that the use by a philosopher of a term having no clear-cut meaning is illegitimate. Writing "Two Dogmas" Quine was following this rule. He was criticizing it not from the outside but from the inside. Instead of destroying positivism, he wanted to improve it by eliminating an ambiguous term from its vocabulary. He did not think that the term is necessary or even useful. In "Reply to Herbert H. Bonnert" in Quine's volume of *The Library of Living Philosophers* he claims: "I see little use for it [analyticity] in the epistemology or methodology of science" (Quine 1986: 95).

During one of our several long conversations in 1996, Quine told me, a little unexpectedly, that the importance of "Two Dogmas" is *overrated* by philosophers. As far as I understood, Quine had in mind not only the relevance of his later publications on the subject but also the interpretation of the "Two Dogmas" as an attack on positivism.

Two nonformal but important honorary titles have been conferred on Quine. Hilary Putnam has called him "The Greatest Logical Positivist" (cf. Putnam 1992 [1990]), and Ernest Gellner dubbed him "The Last Pragmatist" (cf. Gellner 1975). When I asked Quine which one of these characterizations is more truthful, he responded without hesitation that his philosophical position is much closer to that of logical positivism, though the *greatest* logical positivist was, obviously, Carnap. Moreover, he said he was not certain that pragmatism's philosophical position could be clearly and precisely defined at all. Quine, it seems, was convinced that the naturalism present already in "Two Dogmas" and expounded in *Epistemology Naturalized* (Quine 1969) and later works remains true to the fundamental principles of logical positivism and, especially, to the creed that philosophy must become science.

It is quite natural that the distinction between analytic and synthetic sentences has been an important problem for *analytic* philosophy, its adherents and close thinkers. However, beyond analytic philosophy another issue that was investigated by its followers had been even more important: the problem of the fact-value dichotomy, or the distinction between descriptive and normative sentences. It still plays an enormous role in philosophy, social sciences, and everyday life. Culturally it is perhaps the most significant part of the positivist heritage.

The idea of fact-value dichotomy was defended not only by the members of the Vienna Circle and their philosophical allies. It was supported by other prominent thinkers, such as Max Weber. However, it is a positivistic idea that played an important role in the development of positivism as early as the era of Hume.

The relation between facts and values is important, as was already mentioned, not only for philosophers. It has much wider cultural resonance because, as Putnam rightly notes, one particular answer to the question of the fact-value relation, namely "the answer that fact and value are totally disjoint realms, that the dichotomy "statement of fact *or* value judgment" is an absolute one, has assumed the status of a cultural institution" (Putnam 1994 [1981]: 127).

Putnam is, without doubt, the most famous critic of this dichotomy among the philosophers shaped by the analytic tradition. He spares no pains to show that this dichotomy so cherished by logical positivists lacks any rational basis. However, he is forced to concede that at least for the time being no efforts of philosophers who hold that this dichotomy is unfounded can change the cultural situation, i.e., the view, clearly dominant in the contemporary Western world, that facts and values are distinct. This conviction guides the everyday lives of millions of people, including those who have never heard that this dichotomy is the foundation of positivistic thinking.

Although Putnam discusses the fact-value dichotomy in many of his papers, his main argument line took shape already in a chapter "Fact and Value" of his likely best-known book *Reason, Truth, and History*, first published in 1981 and reprinted almost every year since then. Putnam modestly acknowledges in it that the argument is not fully his own invention. "The strategy of my argument," Putnam claims, "is not going to be a new one. I'm going to rehabilitate a somewhat discredited move in the debate about fact and value, namely the move that consists in arguing that the distinction is at very least hopelessly fuzzy because factual statements themselves, and the practice of scientific inquiry upon which we rely what is and what is not a fact, presupposes values" (Putnam 1994 [1981]: 128).

As is well-known, many proponents of the fact-value dichotomy concede that science presupposes values indeed, but claim that these values are *cognitive* and have nothing in common with *ethical* values. Thus Putnam's task, as I see it, is to show that ethical and cognitive values are not only closely interconnected, but inseparable.

Aiming at this, Putnam asks what criteria of rational acceptability are used in science when constructing a scientific representation of the world. His answer does not differ essentially from the answers other philosophers tackling the problem have given: we want our representation to be coherent, comprehensive, functionally simple, and instrumentally efficacious. But he is not content with this statement and asks why we want this kind of representation. Putnam's answer to *this* question is the following: "The reason […] is that having this sort of representation is part of our idea of *human cognitive flourishing*, and hence part of our idea of total human flourishing, of Eudaemonia" (Putnam 1994 [1981]: 134).

Putnam's argument at least in part depends on the rejection of metaphysical realism, and, accordingly, the spectator theory of knowing. From the perspective of metaphysical realism, as Putnam puts it, "the world consists of some fixed totality of mind-independent objects. There is exactly one true and complete description of 'the way the world is'" (Putnam 1994 [1981]: 49). Therefore, a metaphysical realist could argue that we are interested in criteria of rational acceptability only because these are instruments to get our notional world to "match" the world in itself. In the world, as metaphysical realist sees it, facts do not depend on values. However, together with quite a number of contemporary philosophers, Putnam rejects metaphysical realism, or, as it is sometimes called, the *God's Eye point of view*. In an earlier chapter of his book he emphasizes that different descriptions of the world reflect different human purposes and interests. Therefore we can neither know nor imagine the God's Eye point of view (cf. ibid., 50).

To the perspective of metaphysical realism Putnam counterpoises the perspective that he calls the *internalist* one "because it is characteristic of this view to hold that *what objects does the world consist of?* is a question that it only makes sense to ask *within* a theory or description" (Putnam 1994 [1981]: 49). Now from the internalist point of view, we build a theoretical picture of the "empirical world" using as an instrument our criteria of rational acceptability. The interplay of criteria and facts accounts for the dependence of "real world" upon our values.

This clarification still leaves open the question of the relation of cognitive values to ethical ones. Putnam reminds us that in ancient Greece wisdom was an ethical value; only later Judaism and Christianity narrowed the notion of ethical value because being wise was not *directly* related to salvation. But this is rather a digression. His main point is to stress that "coherent" or "simple," like "kind" and "beautiful," are used as terms of praise. Putnam claims that the search for truth and rationality like any other human activity is guided by the general idea of the good.

Putnam notes that *subjective* values are often juxtaposed with *objective* facts. But his discussion of metaphysical realism shows that facts are not "ready-made" and their articulation depends upon our needs and interests.

On the other hand, such values as "coherence" do not stand only for our subjective attitudes. It is especially worth emphasizing that quite often we use the same expressions both for description of a fact and for a value judgment. Pronouncements like "John is inconsiderate" or "Mary is stubborn" may express a precise description or moral appraisal.

> When we think of facts and values as independent we typically think of facts as stated in some physicalistic or bureaucratic jargon, and the "values" as being stated in the most abstract value terms, e.g. "good," "bad." The independence of values from fact is harder to maintain when the facts themselves are of the order "inconsiderate," "thinks only about himself," "would do anything for money." [...] Today we tend to be too realistic about physics and too subjectivistic about ethics, and these are connected tendencies. It is *because* we are too realistic about physics, because we see physics (or some hypothetical future physics) as the One True Theory, and not simply as a rationally acceptable description suited for certain problems and purposes, that we tend to be subjectivistic about descriptions we cannot "reduce" to physics. (Putnam 1994 [1981]: 139–43)

Putnam's words are quite convincing. However, it is notoriously difficult to present knock-down arguments in philosophy and Putnam's arguments are no exception. They depend on some theoretical premises, and Putnam is perfectly aware of that. I have in mind primarily his conception of "internal realism" which, it seems, needs more elaboration. The same may be said about Putnam's version of "moral realism" which he juxtaposes with the emotivism of logical positivists. One of the characteristic features of Putnam's conception of moral realism is the holistic notion of the good. He would protest to calling it "Putnam's notion of the good." He would rather claim that he *discovered* it in the same sense that scientists discovered galaxies or elementary particles, and did not *create* it. Putnam concedes that the idea of good is not an eternal one—ancient Greeks had one, and Christians have another. But then he faces a standard reproach: if the notion of good is culturally conditioned, why can't it be personally conditioned as well? In other words, if different cultures may have their notions of the good, why can't individuals? A moral realist cannot answer this question without accepting some kind of teleology or theology. Putnam's answers to some of Giovanna Borradori's questions in the book *The American Philosopher* seem to indirectly confirm this guess. In his interview, Putnam agrees with the point of view that anyone establishing for himself what is right and wrong is like God because he transcends "human limits." And Putnam adds, "I think that man is the worst god there is" (Borradori 1994 [1991]: 65).

Probably this is right. However good or bad a person may be, he is doubtless capable of using his mind instrumentally, and sometimes he uses it very skillfully for the execution of abhorrent plans. A nefarious intellect, say that of an efficient terrorist, may be instrumentally good but morally bad. It follows that the values implicit in science do not make an integral part of our idea of good.

Thus, Putnam's criticism of the fact-value dichotomy is not as conclusive as he would like to think.

Positivism, Kuhn, and Postmodernism

The attitude of the philosophical community towards positivism, and logical positivism in particular, was heavily influenced not only by direct criticism of its fundamental principles, but also by the development of a philosophy of science that appeared to be alternative to positivism. In a book published in 1962, its author, who was initially considered to be a historian of physics and who later attained recognition as a distinguished philosopher of science, aimed to explain how science develops, and, more precisely, how and why scientific revolutions occur. Logical positivism, the author seemingly thought, was not interested in the subject; therefore he did not feel obliged to take a direct critical stance towards its theories and conceptions. However, the book I am referring to, Thomas Kuhn's *The Structure of Scientific Revolutions*, performed a fateful role in the history of logical positivism. Its contribution to the decline of popularity of this philosophy was tremendous. It turned the majority of young philosophers of science from the kind of analysis of knowledge logical positivists practiced to a rather different kind of research. The so-called historical school of philosophy of science emerged, and by the 1970s it was already prevailing over the logical school inspired by the ideas and exemplars of logical positivism. Besides Kuhn, it is worth noting at least Imre Lakatos as one of its representatives.

The basic idea of *The Structure of Scientific Revolutions* is as follows: there are two phases in the development of science. The first is the *normal* phase, when scientists engage in "puzzle solving," i.e., in a rather routine activity on the basis and within the confines of an accepted *paradigm*—a certain way of viewing the world. The second is the *revolutionary* phase, when an old paradigm (say Aristotelian or Newtonian) is supplanted by a new one.

The most important elements of a paradigm are clearly defined theoretical assumptions, fundamental laws of science and the standards of their application in research. During the normal phase of scientific development, these components are not changed even when newly discovered empirical facts seemingly contradict the principles of an accepted paradigm. Scientific revolutions occur when accepted paradigms are changed for various reasons, only some of which may be classified as empirical.

Kuhn also emphasizes that different paradigms are incommensurable: they are different world-conceptions which cannot be compared with each other. To be more exact, a scientific statement formulated in terms of one paradigm cannot be translated (as a rule) into a statement of another paradigm. Thus empirical facts cannot provide sufficient grounds for deciding which paradigm is better, although one can rely on those facts when considering which of the hypotheses formulated *within one and the same paradigm* better corresponds to these facts.

The picture of science presented by Kuhn in a rather convincing and detailed way differed greatly, at least at first glance, from that of the logical positivists. The latter were interested in the *timeless* analysis of scientific knowledge; they were concerned with the *structure* of knowledge. Kuhn, to the contrary, engages in a different kind of inquiry, that of the *development* of knowledge. The sort of analysis that logical positivists were interested in was *logical* analysis, while Kuhn's analysis is rather an *historical* one, and he does not use any technical logical instruments whatsoever. The claims made by logical positivists had mainly a normative, prescriptive character. The criterion of empirical significance was a rule specifying the requirements sentences *must* meet in order to be considered as scientific sentences. The change of the word "rule" into "proposal" in mature logical positivism did not matter: logical positivists were interested in the *rational reconstruction* of concepts, theories, and arguments, but not in the empirical description of actual scientific practice. Kuhn, on the other hand, seemed to be interested only in the description and explanation of how science actually functions and develops, but not in prescribing how one must conduct scientific research.

The whole flock of fledgling philosophers of science faced in the sixties an alternative—either to continue Carnap's complicated and intricate work on the logic of science, or to engage in research aimed at understanding how the great scientific revolutions which result in changes of the scientific conception of the world take place. The prevailing majority opted for the latter.

This line of research not only seemed more exciting, but it did not require systematic, cumbersome, and sometimes dull logical training. Carnap was excessively technical and his works seemed somewhat dry to most readers. Kuhn, on the contrary, was very adept at *telling* his story—it seemed that not too grueling preparation was required to continue his work. Thus there is nothing strange about the fact that soon after the publication of *The Structure of Scientific Revolutions* Kuhn became a serious authority to a significant part of younger philosophers of science. Carnap, unfortunately, died in 1970.

The most important of Carnap's postwar books on the philosophy of science is the *Logical Foundations of Probability*. I do not know exactly how many students of philosophy have read this voluminous, although not very complicated from a logical point of view, opus. Yet I dare estimate that Kuhn's slim volume had *at least* ten times more readers than Carnap's. Thus Kuhn's "victory" over Carnap was rather easy: he commanded overwhelming forces.

But were they competitors after all? One could maintain that they competed at least for the number of readers, although even that wouldn't be right: Carnap, it seems, was not very interested in his own popularity. But did they compete as philosophers? Did Kuhn say (not to mention—prove) anything antithetical to Carnap's philosophical views? I have *serious* doubts. But before giving my reasons, let me mention several historical perplexities.

The first, 1962 edition (the second, revised edition followed in 1970) of Kuhn's famous book was published in a series initiated by logical positivists, namely in the "International Encyclopedia of Unified Science." In it, Kuhn claimed that great scientific revolutions embrace *almost all* of science. Thus Kuhn's work was rather amenable to the ideas of the unified science movement, although it was a little ambiguous about the possibility of radical changes taking place in different sciences at different times. Nevertheless, the willingness of logical positivists to cooperate with their *enemy*—as some authors regard Kuhn today—is a bit puzzling. Yet much more perplexing for Kuhn's adherents must be another little-known fact: Carnap's *positive* attitude towards the ideas presented in *The Structure of Scientific Revolutions* (cf. Reisch 1991).

These strange facts may be explained, however, by supposing that Carnap apprehended the closeness of his and Kuhn's views, or to be more exact, the possibility of translating the most important of Kuhn's notions into his, Carnap's, language. We can rely on the works by Reisch (1991), Friedman (1991; 1992; 1993), Earman (1993), and, especially, Irzik and Grünberg (1995) in claiming this.

That it is possible to express Kuhn's ideas in Carnap's language becomes apparent upon recalling Carnap's views on linguistic frameworks and scientific theories. A linguistic framework might be characterized as a set of rules governing the ways in which we are talking about objects in which we are interested. It is important to remember that according to the (logical) principle of tolerance, we choose our linguistic frameworks freely on the basis of conventional, pragmatic considerations. Moreover, we are free to change them when we feel that it is advisable.

Every scientific theory is formulated in one or another linguistic framework. The later Carnap, having divided the language of scientific theory into observation language and theoretical language, stresses that theoretical postulates containing only theoretical descriptive terms must be distinguished from correspondence rules that connect theoretical and observational terms. Theoretical terms acquire their meaning, according to Carnap, not only through correspondence rules but through theoretical postulates (laws of a theory formulated in purely theoretical terms) as well, because not every theoretical term is connected via these correspondence rules with empirical language directly. Carnap emphasizes that the very meaning of the theoretical terms is dependent on theoretical postulates. Thus Irzik and Grünberg are perfectly right in claiming that the later Carnap is a semantic holist subscribing to the doctrine "that the theoretical postulates of a theory contribute to the meaning of theoretical terms occurring in them and that a change in the theoretical postulates results in a change in meaning" (Irzik and Grünberg 1995: 289). At least beginning in the late 1930s, Carnap does not require that each theoretical term be individually defined in purely empirical terms. One cannot define the term "electron," for instance, in purely empirical terms.

From Carnap's semantic holism (which is shared by Quine and Kuhn) follows the thesis that scientific theories are incommensurable, or—which is the same—that some sentences of one theory are not translatable into sentences of another. Carnap's semantic holism finds its clearest expression in his publications of the 1950s and 1960s. Yet already in 1936, in "Truth and Confirmation" Carnap stressed that many statements of modern physics cannot be translated into statements of classical physics because these statements presuppose different forms of language, or as he would later say, different linguistic frameworks (cf. Carnap 1949 [1936]: 126).

Irzik and Grünberg provide convincing arguments proving that for Carnap not only theoretical, but also empirical scientific statements were incommensurable, untranslatable from one framework to another. At least in the physical language preferred by Carnap from the mid-1930s, even such seemingly purely empirical concepts as "red" refer to a property of an object defined in terms of wavelength which have meaning only in a certain recognized physical theory. Thus, in a mature Carnapian philosophy of science developed from 1936, it is emphasized that the boundary between the empirical and theoretical terms is essentially conventional, and whichever way it is drawn, the meaning of empirical (observational) terms is theory dependent and revisable. This was pointed out already by Oberdan (1990).

Now we may come back to the problem of translating Kuhn's main points into the language of Carnap. Carnap would say that scientific revolutions occur when the rules of the linguistic framework of a theory or its theoretical postulates undergo change. The normal phase in the development of science takes place when the meaning rules and postulates remain intact, but the truth-values of statements which are not fixed by them are changed or added. It is also clear that the addition of some new correspondence rules does not involve drastic change. Calling changes of meaning rules and postulates changes of the first kind, and changes in truth-values and the addition of new correspondence rules changes of the second kind, Carnap wrote: "A change of the first kind constitutes a radical alteration, sometimes a revolution, and it occurs only at certain historically decisive points in the development of science. On the other hand, changes of the second kind occur every minute. A change of the first kind constitutes, strictly speaking, a transition from a language L_n to a new language L_{n+1}" (Carnap 1963b: 921). I would like to emphasize that these words were *written* presumably before Kuhn's *Structure of Scientific Revolutions* appeared (as I have mentioned earlier, the publication of Schilpp's volume on Carnap was unduly delayed). Moreover, Carnap already spoke about "radical revolutions in the system of science" much earlier (cf. Carnap 1956: 51).

Thus, according to Carnap, scientific revolutions occur when one scientific language is replaced by a new one. And we can talk about the normal phase of the development of science when researchers rely on the accepted framework of a scientific language and do not transform it, that is, as Kuhn would say, they do not transform its fundamental assumptions and laws. There may be many different pragmatic reasons for changing them. The discovery of some new empirical facts is not the main reason to change the language, because the simpler way to accommodate them is to change the truth-values of the sentences not determined by its rules. And from what has been said above, it must be clear that different theories may be incommensurable in the same sense that Kuhn's paradigms are.

Thus we must conclude that Kuhn said little about the fundamental aspects of scientific development that Carnap wouldn't have known, stated, or anticipated in his published works. This conclusion contradicts the widespread opinion that a shift from positivist and logical, to an historical and first of all Kuhnian philosophy of science, has itself been a kind of *revolution* in philosophy of science. Of course, Kuhn and his followers introduced some new ideas, began analyzing new problems, but it would be *completely* wrong to claim that logical positivism construed the growth of scientific knowledge solely as an accumulation of new empirical facts and cautious revision of hypotheses.

During the glory years of historical philosophy of science, perhaps in the 1970s, one of its representatives publicly asked a rhetorical question about logical empiricism's philosophy of science: how was it possible that competent researchers maintained such strange (read—positivist) views concerning the nature of scientific inquiry. It is an irony of fate that nowadays the opposite question looms large: "Why has the logical positivist movement been misunderstood so badly?" (Irzik and Grünberg 1995: 305).

The simplest answer to the latter question is as follows: logical positivism has not been studied deeply enough. Its rivals managed to change the way of doing philosophy of science, because what they took to be the new paradigm in philosophy of science seemed more promising and less technical. For Kuhn never claimed that a scientific revolution occurs only after it is proved that the old paradigm was false. Both Kuhn and Carnap could explain the changes in philosophy of science in the same pragmatic terms in which they could explain changes taking place in other scientific disciplines.

Apropos, the scale of these changes today no longer appears to be as vast as it seemed in the 1970s or 1980s. Even though positivism had weakened, the historical school had hardly become absolutely dominant. Such trend of a contemporary philosophy of science as the new experimentalism is closer, at least in spirit, to Carnap that to Kuhn. Also the tradition of probabilistic appraisal of knowledge, developed by Carnap, related to the application of the so called Bayesian method, has many adepts today (see, e.g., Chalmers 1999).

It is worth touching upon one more issue connected to the relation between Carnapian and Kuhnian philosophy of science. There are not only connections, but also important differences between them. Carnap's philosophy of science, as that of the positivism in general, is normative. Although positivism has attempted to create a scientific philosophy, while claiming that (empirical) science ought to limit itself with the description, explanation and prediction of facts, it has not applied this requirement to itself. Though not always with sufficient clarity, it realized that positive philosophy had not only the right, but also an obligation to formulate the *criteria and requirements* for science. Science must strictly subscribe to the fact-value distinction, but positive philosophy that formulates this principle cannot apply it to itself. In this respect, it is rather a *metatheory* or metascience, determining what science *shall* be like. Meanwhile Kuhn's philosophy of science does not aspire to be normative. It seeks to describe and explain the procedures actually used by scientists, the facts of scientific practice. It aspires to be one of the sciences, not a meta-science.

In order to correctly understand the nature of positivism, one must understand that the statements of its representatives about science are meta-theoretical statements. Metatheory is the theory of a theory. For example, the statement "Every physical theory is a hypothesis: despite being empirically confirmed many times, it may nevertheless contradict new observational or experimental results" is obviously a metatheoretical statement. It is about physical theories, but it is not itself a statement of physics. The criterion of verifiability is also of metatheoretical character. That is precisely why I am convinced that positivist philosophy of science can be treated as metatheory.

It is true that the notion of metatheory is well-defined only in meta-mathematics and metalogics. An epistemologist cannot define it as precisely. However, treating logical positivism as metatheory allows a better understanding why its adherents were formulating normative value statements while at the same time demanding that only factual statements be formulated in scientific theories. For even their claims such as "The ground of scientific knowledge is experience" were essentially normative in character, since what is meant is that scientific knowledge *must* rely on experience. The character of a metatheory may differ from that of the theories it is about. Therefore, it is not surprising that the seven principles of positivism formulated in the introduction to this book are not, strictly speaking, statements of facts. For what they speak about is what scientific knowledge and also scientific theories should be like.

Kuhn or Lakatos did not show and did not even have the intention to show that positivism "was false." The 1960s, however, witnessed the outset of developments in philosophy that were much more ambitious. They were wider in scope and aimed not specifically at positivist philosophy of science, or positivism in general, but at almost entire modern philosophy. *Modernism* as such came under attack. The different, supposedly radically divergent from modernist, way of apprehending and understanding the world has become famous as *postmodernism*.

Postmodernism attempted to assimilate Kuhn's ideas by claiming that Kuhn and some other representatives of the historical philosophy of science were clear-cut postmodernists. I do not think that Kuhn or Lakatos can be considered postmodernists. However, answering the question whether one or other thinker can be called a postmodernist is complicated by the vagueness of the term "postmodernism." This term and its derivatives today can be found in various texts and are used in different contexts. Postmodernism is undoubtedly still popular and fashionable. Sadly, fashionable ideas and concepts rarely are well-defined, clear, and precise. Clarity and precision would perhaps make them difficult to handle and manipulate. The notion of postmodernism is no exception to this rule.

This notion is ambivalent and somewhat incoherent for several reasons. Some of those are worth noting. First of all, it was used primarily to describe some new developments in the arts, especially in architecture. Later its meaning was extended to indicate the direction of recent cultural changes in general, and only subsequently was the term used as the name of a philosophical movement closely connected with the aforementioned developments in the arts but having (or at least avidly seeking) its own intellectual roots. Secondly, postmodernist authors—Jean-François Lyotard, Jean Baudrillard, Jacques Derrida, and others are often not even considered philosophers, especially in the Anglo-Saxon countries.

In the Anglophone world, especially the U.S.A. and the United Kingdom, postmodernists are regarded as cultural or literary critics rather than "proper" philosophers. Postmodernists certainly do look like impostors, or at least like outsiders, marginal figures against the background of philosophy mainly shaped by the analytic tradition. Both in the United States and the United Kingdom, university departments of English show, as a rule, much more interest in postmodernism than do departments of philosophy. The latter rarely offer courses on postmodernism. I did not find any books by Lyotard, Baudrillard, Michel Foucault, or Derrida on the philosophy shelves of the Harvard Book Store, which is a respected and independent institution. Those books were located under the category "Cultural and critical theory." As a matter of fact, Richard Rorty is considered a philosopher, but in the United States he is usually treated not as a postmodernist, but as a "romantic pragmatist" (see Nevo 1995).

Thirdly, postmodernists in one way or another appeal to the ideas of many quite different philosophers—Marx, Peirce, Nietzsche, Dewey, Wittgenstein, Freud, Heidegger. Thinkers drawing their ideas from such diverse sources quite naturally face rather troublesome problems while unfolding their views, which in principle cannot be put into a clear, harmonious, and logically coherent whole. Fourthly, postmodernists themselves do not always desire to be called philosophers. Fifthly, even those among them, who consider themselves philosophers, frown upon being connected into one single school, trend or movement. In the context of fragmented postmodern reality, they take significant differences among themselves to be quite natural.

In fact, postmodernists deny not only the possibility but also the desirability of an integral postmodernist philosophy, although there are some positive, however poorly defined, principles that many postmodernists share. The main principle of postmodernism—its opposition to modernism—is an obviously negative one (just like the main principle of positivism—its opposition to metaphysics). As to the question "what is modernism?," postmodernists give different answers. Although one may think that the determination of chronological boundaries of modernism is a simpler task than the definition of its guiding principles, one encounters a wide disarray of opinions as to the beginning of the era of modernity. Some postmodernists date it as early as with Socrates and Plato, others associate it with Descartes.

There is a much greater degree of consensus regarding the end of modernity and the dawn of postmodernity. As far as I know, nobody has claimed that postmodernism originated earlier than 1875. The majority of postmodernists agree that it appeared only in the twentieth century, most likely in the 1960s. Quite often the exact year of the thunderous rupture between two eras is given—1968. One event postmodernists have in mind is, of course, the *Student Revolt*. Actually, postmodernism grew out of the *protest* against entrenched values and practices of the bourgeois society, the new *ancien régime*.

Almost all the thinkers (at least the French and German ones) regarded as representative of postmodernism had rather close ties to Marxism, therefore the failure in 1968, both in Paris and (for completely different reasons) in Prague, where the last hopes to create socialism "with a human face" collapsed, was perceived by them as an epoch-making event that demands a change of theoretical attitude. That does not mean that after 1968 they abandoned Marxism altogether. Most of them clearly rely upon Antonio Gramsci's—probably the most famous Marxist (after Lenin) of the twentieth century and the founder of the Italian Communist Party—conception of *hegemony* in which the orthodox Marxist doctrine of class-hegemony is transformed by exalting not only economic and political, but also other kinds of hegemony and oppression. The espousal of all possible sources and forms of domination and exploitation from sexual to intellectual is very characteristic of postmodernism and clearly indicates its Marxist origin.

Although postmodernism as a philosophy is extremely, if not impossibly diverse, and its different representatives sometimes stress radically different issues, there is something like a standard definition of postmodernism which is widely, although not universally, referred to. It was laid down by Jean-François Lyotard in his book *The Postmodern Condition: A Report on Knowledge* published in 1979. This book performed an important role in consolidating—to a certain degree—the postmodernist movement. In it Lyotard defines the notion "*postmodern*" in negative terms, juxtaposing it with the notion "*modern*," which he regards primarily as a characterization of a certain state of science. At the same time, changes taking place in science and transferring it from a modern to a postmodern state are tied by him to altering rules of the game in other domains of culture. These changes are explained in terms of the crisis of *narratives*.

> Science has always been in conflict with narratives. Judged by the yardstick of science, the majority of them prove to be fables. But to the extent that science does not restrict itself to stating useful regularities and seeks the truth, it is obliged to legitimate the rules of its own game. It then produces a discourse of legitimation with respect to its own status, a discourse called philosophy. I will use the term *modern* to designate any science that legitimates itself with reference to a metadiscourse of this kind making an explicit appeal to some grand narrative, such as the dialectic of Spirit, the hermeneutics of meaning, the emancipation of the rational or working subject, or the creation of wealth. For example, the rule of consensus between the sender and addressee of a statement with truth-value is deemed acceptable if it is cast in terms of possible unanimity between rational minds: this is the Enlightenment narrative, in which the hero of knowledge works toward a good ethico-political end—universal peace. [...] Simplifying to the extreme, I define *postmodern* as incredulity towards metanarratives. (Lyotard 1993 [1979]: 71–72)

This incredulity is based on the belief that there are many language games leading to knowledge, not just a single one. The game characteristic of modern science was based on principles of commensurability, determinacy, and efficiency. The latter principle, that of efficiency, was especially important for the legitimation both of scientific truth and of social institutions.

Lyotard thinks that postmodern knowledge is not obliged to seek consensus through discussion. Consensus is incompatible with the variety and heterogeneity of language games. Moreover, consensus, even regarded as a purely regulative principle, produces intellectual terror. Every postmodernist agrees with Michel Foucault that knowledge is power (they understand this statement differently from Bacon) which at present exceeds traditional kinds of power—economic, political, or military. Hence they consider intellectual terror as the most insidious and baleful sort of oppression.

Contrary to knowledge based on consensus, "post-modern knowledge," Lyotard claims, "is not simply a tool of the authorities; it refines our sensibility to differences and reinforces our ability to tolerate the incommensurable. Its principle is not the expert's homology, but the inventor's paralogy" (Lyotard 1993 [1979]: 73). While speaking about the homologousness of experts, Lyotard means that their opinions have to be alike, whereas the opinions of postmodern "researchers" may vary and they are permitted to commit unintentional logical fallacies, which make their arguments invalid and their conclusions false (paralogism is precisely one of such unintentional fallacies). The words of Lyotard are echoed by Baudrillard in a different context: he denounces "the delirious dream of reunifying the world under a unitary principle" and the "mythic imperative of rationality" (Baudrillard 1988 [1976]: 141).

Kuhn's concept of a paradigm is widely used in postmodernist writings, and they regard him as one of the main exponents of postmodernist thought. The use of the concept of paradigm reflects the belief that the mind's nature is essentially interpretative. In principle, postmodernists should regard different paradigms as being of equal standing. But because postmodernism constitutes itself by opposing the single paradigm of modernity which is denounced as the worst conceivable, it produces in fact a (meta)paradigm of the postmodern mind bearing the main traits of different postmodernist paradigms. Historians of ideas have, it seems, no difficulty in grasping this *Postmodern Paradigm*. Admitting that "the postmodern paradigm is by its nature fundamentally subversive of all paradigms, for at its core is the awareness of reality as being at once multiple, local and temporal, and without demonstrable foundation," Richard Tarnas, nevertheless, is able to indicate some widely shared principles comprising the backbone of this paradigm. According to him:

> There is an appreciation of the plasticity and constant change of reality and knowledge, a stress on the priority of concrete experience over fixed abstract principles, and a conviction that no single a priori thought system should govern belief or investigation. It is recognized that human knowledge is subjectively determined by a multitude of factors; that objective essences, or things-in-themselves are neither accessible nor positable; and that the value of all truths and assumptions must be continually subjected to direct testing. The critical search for truth is constrained to be tolerant of ambiguity and pluralism, and its outcome will necessarily be knowledge that is relative and fallible rather than absolute and certain. (Tarnas 1993 [1991]: 401; 395–96)

I will call postmodernism characterized by these words *moderate post-modernism*. A radical form of postmodernism rejects any attempt to seek cognitive unity, coherence, comprehensiveness, and even simple under-standing based on shared meanings. *Radical postmodernism* is, without any doubt, a new metanarrative denying the existence of universal, inter-subjective truths. Yet by claiming that there are no universal truths, it postulates at least one such truth, and thus is self-refuting in the same sense radical skepticism is. Hence, from an epistemological point of view, it is uninteresting.

Despite their different opinions regarding the relation of modernism and postmodernism, or regarding the role of consensus in scientific discourse, neither Lyotard nor Rorty advocate radical postmodernism. However, Jacques Derrida's deconstructivism is dangerously close to it. Uncon-strained expansion of the practice of deconstruction, decentering, deferral, dissemination, differance, demystification, and dispersion *may* indeed lead eventually to the disappearance not only of the last traces of knowledge itself, but even of the weakest wish to *know* anything whatsoever.

Recriminations of a similar kind cannot be held against moderate post-modernism. What it is opposed to is not knowledge as such, but a specific form of knowledge and specific principles of construction, justification, in-terpretation, and use of knowledge. Postmodernism quite often counterposes itself to the Enlightenment and its conception of human knowledge. How-ever, when postmodernists attack some specific epistemological principles unacceptable to them, they usually speak not of Immanuel Kant or Étienne Bonnot de Condillac, but about positivism. Positivism is regarded by them as a new version of the Enlightenment project (see Laclau 1993 [1988]: 332). They take the same stance on this issue as Horkheimer and Adorno. And they may be right, although it might be worth reminding the reader of the remarks and reservations made in the beginning of this book (see the final section of Chapter 1, "The Idea of Progress in the French Enlightenment").

Being relatively modern and influential, positivism has become a fa-vorite target of postmodernist attacks. Positivism (with a peculiar excep-tion of the *skeptic* Hume) is regarded by them as an embodiment of every-thing that is the worst in the Spirit of the Enlightenment: its emphasis on wholeness, unity, unequivocalness of knowledge, its views on science as means to achieve social progress, and of course, its *foundationalism*—the belief in the existence of the ultimate grounds of knowledge that ensures its firmness and truthfulness. Foundationalism is regarded by postmodern-ists as the ultimate source of all oppression, including the one under the banner of social progress.

There is no doubt that in many of its incarnations, postmodernism is quite different from positivism. One of the most influential theorists of postmodernism, Ihab Hassan, correctly claims that "as an artistic, philosophical, and social phenomenon, postmodernism veers toward open, playful, optative, provisional (open in time as well as in structure or space), disjunctive, or indeterminate forms, a discourse of ironies and fragments, a 'white ideology' of absences and fractures, a desire of diffractions and invocation of complex, articulate silences" (Hassan 1993 [1987]: 283). This is surely not characteristic of positivism. However, a closer look into the epistemological principles of (moderate) postmodernism reveals that many principles regarded by postmodernism as constitutive of its own philosophy are shared by it with positivism.

The exposure of "the obsolescence of the metanarrative apparatus of legitimation" and the rise of postmodernism is tied by Lyotard directly to the crisis of metaphysical philosophy (cf. Lyotard 1993 [1979]: 72). Postmodernism is overtly antimetaphysical. But so is positivism. At least in this important respect, postmodernism rather continues the positivist tradition than denies it. Postmodernism and positivism, logical positivism especially, share some other epistemological features, including those that are *allegedly* antipositivist. Both share relativism and fallibilism. Postmodernism equates knowledge with power, but in this respect it does not seem especially original. Of course, a postmodernist would claim that, whereas positivists regard knowledge as a source of power, he treats it as power itself, and that he is concerned not with the technological issues of the use of science, but with the social ones. However, I am convinced that the basic postmodernist concepts describing the complex, reciprocal relation between power and knowledge, such as Foucault's notion of "regime of truth" are expressible in positivist terms.

The claim that postmodernism, contrary to modernism and thus to positivism also, is *antifoundationalist* is already a tired and tattered cliché. This claim has at least two different versions depending on the interpretation of the term *antifoundationalism*. According to the first version, the characteristics "foundational" and "antifoundational" are treated as attributes of epistemology. Thus postmodernist epistemology is anti-foundational because it denies that knowledge rests on indubitable foundations. Taking a critical stance towards positivism, postmodernism implies that positivists believe in such foundations: they must be indubitable because of their *purely* experiential character. This allegation is simply not true. Some positivist thinkers, e.g., Schlick, indeed have searched for the firm foundations of knowledge, but neither Comte nor (from the early 1930s) Carnap, Reichenbach, or Ayer supported the foundationalist point of view.

In the second version postmodernism is claimed as being anti-foundational because it renounces "the view that casts philosophy in the role of founding discourse vis-à-vis social criticism. That 'modern' conception gives a way to a 'postmodern' one in which criticism floats free of any universalist theoretic ground" (Fraser and Nicholson 1993 [1988]: 416–17). It seems that these authors convey rather exactly the meaning of Lyotard's words that postmodernism rejects grand narratives (or metanarratives).

Positivism indeed carried on the great Enlightenment metanarrative of progress, reason, and liberation from dogmas, and I do not think that it should be ashamed of this. Moreover, I have serious doubts that postmodernist social criticism "floats free of any universalist theoretical ground." Postmodernist antifoundational epistemology in conjunction with the thesis that intellectual rigorism amounts to oppression provide a perfect universalist theoretical ground for social criticism that they practice. Without such a ground postmodernism would lose any philosophical interest.

Postmodernism counterposes itself to positivism (and modernism in general) by renouncing its alleged optimism and progressivism. Positivism is truly a rather optimistic and progressivist philosophy. But it is enough to remember the claim of Peirce, that for positivists human life is but a tiresome and purposeless journey (since it is not life towards God), in order to understand that, in the opinion of the founder of pragmatism, positivism displays lack of optimism rather than its excess.

On the other hand, it is not difficult to see that postmodernism is not a pessimistic philosophy at all. Most often it presents itself as a movement of intellectual and social liberation. As I have already mentioned, most postmodernists, especially in France, are simply obsessed with the idea of struggle against the oppression allegedly flourishing in Western democracies. Fighting for the rights of the oppressed, jobless, illiterate, immigrants, downtrodden, and minorities, they follow Mill, Marx, Lincoln, or Freud, and continue the battle for the human rights, that were fought for by the modernists for centuries. Postmodernists seek to improve the human condition, to foster liberty and solidarity, hence they may be perfectly regarded as (a bit anarchical) progressivists.

In postmodernist literature one may even find an amazing analogy to Comte's law of the three stages. In *Contingency, Irony, and Solidarity* regarded by many adherents of postmodernism as something like their Holy Writ, Richard Rorty writes:

I can crudely sum up the story which historians like Blumenberg tell by saying that once upon a time we felt a need to worship something which lay beyond the visible world. Beginning in the seventeenth century we tried to substitute a love of truth for a love of God, treating the world described by science as a quasi divinity. Beginning at the end of the eighteenth century we tried to substitute a love of ourselves for a love of scientific truth, a worship of our own deep spiritual or poetic nature, treated as one more quasi divinity.

The line of thought common to Blumenberg, Nietzsche, Freud, and Davidson suggests that we try to get to the point where we no longer worship *anything*, where we treat everything—our language, our conscience, our community—as a product of time and chance. (Rorty 1995 [1989]: 22)

Positivism was always infused and informed with contingency (let us recall the words of Peirce once again) and solidarity. Irony, perhaps, was in short supply, though sometimes it also sparkled: Carnap's witty characterization of metaphysicians as poor musicians, who are unable to mix sounds and thus combine ideas, comes to mind. And I have no doubt that Carnap would have looked at postmodernists and their efforts to present themselves as revolutionaries (for which time in the history of philosophy?) with a really great dose of irony. Especially when reading Rorty's words that these "revolutionaries," these "strong philosophers" at the end of the twentieth century "are interested in dissolving inherited problems rather than solving them" (Rorty 1995 [1989]: 20), the words that almost verbatim repeat what Wittgenstein or Carnap said in the beginning of the century.

There is no reason to regard postmodernism as a revolution in philosophy which overcame all modernist philosophies, including positivism. Ernesto Laclau is right when he claims that "postmodernity does not imply *change* in the values of Enlightenment modernity but rather a particular weakening of their absolutist character" (Laclau 1988: 332). I would add to this that positivism, and especially logical positivism, weakened them to a significant degree by means of undermining the belief in absolute truths and demonstrable values, and not much *reasonable* work in this area has been left to postmodernists.

Having explained the influence of positivism upon postmodernism, and their common traits, the reader should not be left with the wrong impression that there are no significant differences between them. They will be discussed in the next chapter in the context of the controversy of positivism and postpositivism (postmodernism is one of its forms).

CHAPTER 6

The Impact of the Positive Mind Outside Philosophy

Positivism's Impact on the Natural and Social Sciences

The main, direct aim of positivism was to guide scientific research. Positivism presented specific ideals of scientific research and a clear standpoint on what should and what should not be recognized as scientific knowledge. Science had to be detached from metaphysics, and metaphysical elements had to be removed from science. The underdeveloped (from the positivist point of view) sciences had to be advanced, the language of science had to become clearer, more precise and more empirical, and the degree of unity of science had to be increased.

Positivists had rather clear views about what science should be like, what goals it must pursue, what methods it should employ, what kind of problems it should solve, what the relations between different scientific disciplines must be, what logical standards apply to statements science should produce, what the structure of scientific theory must be. Thus it is perfectly legitimate, in fact inevitable, to pose the question: did positivists succeed in influencing the development of science (or the sciences) in a way that it would comply with their requirements?

Answering this seemingly simple question is a complicated task, at the very least because there have been different forms of positivism. There were different ways in which different positivists in different epochs wanted to *help* science fulfill its real—as they understood it—role and to foster the quickest possible advance of science. Hume aimed to develop a real science of man. Comte sought to complete the already existing system of natural sciences by supplementing it with a social science. Mill elaborated the methodology of scientific inquiry which various researchers could use and which could guarantee that they reach agreement on the results of their inquiry. Mach strove to purify science from metaphysical and anthropomorphic elements and, even more energetically, from valuations. Logical positivists worked to establish general and rather strict requirements applicable to a language of science, its concepts, propositions, and theories.

A thorough examination of positivism's influence outside philosophy in one chapter of a medium-sized book is a quite formidable task. For beyond the boundaries of philosophy there are not just special sciences. Even positivism's impact on science is a rather cumbersome topic. That is witnessed by the vast amount of books published on the subject. Many books are titled *Positivism and X* where *X* may be substituted by the name of almost any discipline of science (especially social). I will mention just a few of them: *Positivism and Sociology* (Giddens 1974), *Positivism and Sociology: Explaining Social Life* (Halfpenny 1992 [1982]), *Positivist Sociology and Its Critics* in three volumes (Halfpenny and McMylor 1994), *Behaviorism and Logical Positivism: A Reassessment of the Alliance* (Smith 1986), *Positivism in Psychology: Historical and Contemporary Problems* (Tolman 2009 [1991]). In addition to these are plenty more books, which, having assessed the influence and significance of this philosophy, suggest going "beyond positivism" and seeing what follows: *Beyond Positivism: Economic Methodology in the Twentieth Century* (Caldwell 1994 [1982]), *Beyond Positivism: Critical Reflections on International Relations* (Sjolander and Cox 1994), *International Theory: Positivism and Beyond* (Smith, Booth, and Zalewski 1996).

In recent years there have been many attempts to step "beyond positivism." However, it seems that the influence of positivism remains strong. An examination of one of the issues of *Psychological Review* reveals that "six of seven articles contained in it demonstrate an adherence to a positivist understanding of science without, of course, ever mentioning positivism, or any other philosophy of science" (Tolman 2009 [1992]: 2). Thus positivism has left its trace in science, although it could be more pronounced in some disciplines than in others.

Positivism, Mathematics, and Physics

When examining the foundations, methods, and language of science, most positivists took physics as a model science. According to them, by the eighteenth century physics was already a developed discipline based on sound methodological principles. Although less exact, it had at least one advantage over mathematics, which was regarded by philosophers as the model science in earlier times: unlike mathematics, physics is systematically based on observation and experience, and for positivists with their empirical orientation this was very important. The empirical foundations of mathematics were not *obvious*, and positivists had no universal agreement about them. Yet from the time of Galileo and Newton, physicists widely used a mathematical apparatus for descriptions and analysis of the results of observations and experiments, as well as for the formulation of laws of physics. Thus it is natural to start with a brief discussion of the influence of positivism upon mathematics.

Most positivists regarded mathematics not as a proper science based on facts with its own empirical object, but rather as a tool science makes use of: the language of science. Primarily they were interested in the possibilities of its application in other sciences and only secondarily they were concerned with its foundations and development. This does not mean that positivists themselves made no contribution to the progress of mathematics.

Perhaps the most important mathematical discipline which made a great advance in the twentieth century and found numerous applications in the natural and social sciences was the theory of probability. One of the most prominent researchers in this field was Richard von Mises. In his *Positivism* (1939), which begins with words that it is not merely a book *about* positivism, but a book written by a *positivist*, he emphasized that precisely empiricist and positivist ideas had led him to the new definition of probability in terms of *limit of frequency* on the basis of which he achieved some important mathematical results. Carnap's and Reichenbach's inquiries into the foundational problems of the theory of probability, being less important from a purely mathematical point of view, furthered, after all, the development of mathematical statistics and game theory (they are both based on the theory of probability).

Positivists contributed much more to the development of physics. Their works enriched the understanding of the character and value of physics, its methods and theories. Much of Mach's, Duhem's, Schlick's, Reichenbach's, and Frank's investigations in physics and the border area between physics and other sciences received much applause from the leading physicists of the twentieth century, including such authorities as Einstein and Bohr. Yet likely more important is to note that through the careful examination of the nature of physical knowledge and especially of the new, epoch-making physical theories, positivists were able to clarify the foundations of physics, the relation of theory to observation, the ways of testing and appraising hypotheses and building physical theories. In many respects we now see physical knowledge through positivist eyes.

When talking about positivism's influence on physics, the central figure is Mach, especially his philosophical criticism of the conceptual foundations of Newtonian mechanics. His ideas on the observability of physical phenomena and the measurability of physical parameters not only made a lasting impact on the general understanding of physics, but were also conducive to many very well-known, revolutionary achievements including Einstein's theory of relativity. Mach's methodological ideas influenced many twentieth century physicists. Duhem's hypothetism elaborated by logical positivists also guided many physicists in their framing of physical theories.

Nevertheless, the foundations of classical physics were already laid by Galileo and Newton, and neither Hume, Comte, nor Mill were able to improve them. However, when the time for a new scientific revolution came at the turn of the twentieth century, empiriocriticist, conventionalist, and hypothetist ideas played a certain role in the process of shaping new physics. As already mentioned above, without Mach's criticism of the Newtonian conception of absolute space and time, the theory of relativity hardly could have been created. But in quantum physics Max Planck or Erwin Schrödinger quite often proceeded against the recommendations made by Mach: they were not cautious enough and sometimes pursued rather crazy theoretical ideas.

Those ideas were much more acceptable to logical positivists, who were much more pro-theoretical than Mach. Niels Bohr was even one of the contributors to their *Encyclopedia of Unified Science.* Yet, generally speaking, in physics and in most of the physical sciences, logical positivists usually played the role of the owl of Minerva. They were helpful in making clear for physicists what the latter were doing and what was the meaning and significance of their discoveries and theories. Yet they were mainly *interpreting* developments for which they were not *directly* responsible, rather than *transforming* the science called physics. They fostered the formation of a probabilistic picture of the world which is, in all likelihood, their greatest success in the field, but even here they were following the lines of development which commenced with the formation of statistical physics quite a few years before the Vienna Circle took shape.

One may conclude that the scope of positivism's impact on physics was limited and that physics had a greater impact on positivism. When explaining what the scientific method really is, they were usually analyzing the method of contemporary physics, explicating and reinterpreting it. They regarded the method of physics as an exemplar other sciences must follow. From the thirties onwards logical positivists subscribed to physicalism. Not only did they think that other sciences should follow the method of physics, but also, if possible, adopt its language. The latter requirement was in accord with the spirit of the *unity of science movement*. Almost every scientist who belonged to this movement believed that physics is an exemplary science.

Positivism had no significant direct impact on chemistry, perhaps except for promoting the application of the methods of physics in it. As for biology, however, the situation was a little different and is worthy at least a brief discussion. At the time when Comte was writing the *Course*, he considered physics and chemistry to be already in the positive phase of their development, but biology was not yet a positive science. Some biologists were still employing such concepts as *entelechy* or *vis vitalis*. Ties between different departments of biology were rather obscure, links with chemistry and physics extremely vague, and Darwin's *Origin of Species* was still to be written. Comte played a significant role in bringing biology closer from a methodological standpoint to physics and chemistry. One of the obvious fruits of his effort is the view prevailing in contemporary biology that both its theoretical foundation and (partial) methodological basis is physics and chemistry. A majority of biologists accept that view today, although probably not many of them know that it was Comte's idea or at least an idea advanced and popularized precisely by him. Later, in the twentieth century the members of the unity of science movement had no doubts about the unity of physics, chemistry, and biology. Only on the basis of such views could molecular biology or contemporary genetic engineering be created and the human genome deciphered.

Positivism's influence upon Darwinism is no less important. Doubtless, the most important concept in Darwinian theory is evolution. Positivism made the field fertile for Darwinism. Although ideas of evolution and progress were not solely the fruit of positivist reflections, they were the central concepts of classical (social) positivism. Jean Baptist Lamarck's speculative theory of biological evolution, as is well known, did not win far-ranging acceptance from his contemporaries. At the same time the idea of social evolution put forward by positivists gained wide popularity and at the very least facilitated the wide recognition of Darwin's theory of evolution.

Positivism's Effect on Psychology

According to the opinion widespread both among psychologists and philosophers, the main school in psychology on which positivism made a major impact is behaviorism. Some psychologists and historians of psychology believe that behaviorism was the most important single psychological movement of the twentieth century. And even those who disagree are convinced that positivism was responsible for the emergence and development of behaviorism. In fact there is a link between logical positivism and behaviorism, as well as between logical positivism and cognitive psychology.

Both logical positivism and behaviorism were manifestly antimetaphysical movements. Also both were famous for their radical views. Just like logical positivists, behaviorists thought that most problems considered to be important in previous times were mere pseudo-problems. They were emphatically not interested in subjective, mental phenomena and unanimously insisted that scientific claims including claims made by psychologists must be intersubjective and empirically testable. Psychologists who adopted such an attitude concluded that psychology must limit itself to the investigation of human (or animal) *behavior*, with the aim to predict and control it. Thus it is inappropriate to conduct investigations relying on any suspicious (intersubjectively not testable) hypotheses about the mental processes or mechanisms allegedly causing that behavior.

There is ample evidence that positivism and behaviorism were two closely related intellectual movements. Two famous behaviorists, namely Edward C. Tolman and Clark L. Hull, took part in the project of the *International Encyclopedia of Unified Science*. The most brilliant and famous of behaviorists, B. F. Skinner, discussed issues of common interest with Carnap at Harvard and made Feigl one of his closest friends at Minnesota. Moreover, the heydays of both movements—from the thirties until the sixties—coincide almost perfectly. According to Thomas H. Leahey (his influential *History of Psychology* was published in 1980) and many other authors who wrote on the subject, logical positivism exerted a substantial influence on behaviorism and especially on its belief in the possibility of an objective psychological theory. The standard account boils down to the claim that behaviorism was based on a methodology imported from logical positivism.

However, it is only partially correct. As Laurence D. Smith convincingly shows in his detailed and well-documented examination of the relationship of both movements (Smith 1986), behaviorism received more influence not from *logical* positivism, as it is often thought, but from empiriocriticism. In other words, Mach, rather than Carnap or Feigl, profoundly influenced behaviorism. If we consider that none of the major representatives of logical positivism was a psychologist, while Mach in fact was, Smith's conclusion does not seem unfounded. Remember that behaviorism was first promulgated by John B. Watson in 1913 (Watson

1913), and gained wide recognition after the publication of his *Behaviorism* in 1924 (Watson 1967 [1924]). Since logical positivism was only starting to develop in the 1920s, it would be naive to think that it was already capable of exerting an influence upon the new school in American psychology.

Yet Mach was very popular in America at that time. He was highly regarded by the best known American psychologist and philosopher of the period, William James, who (as well as Pragmatism in general) greatly influenced Watson. The latter borrowed many of Mach's methodological views from James. Smith presents ample evidence that Mach's influence on Skinner was even stronger than on Watson. Skinner was strongly influenced not only by Mach but by Comte as well. He adopted Comte's view that introspective psychology has no scientific value because psychology based on the method of self-observation cannot produce a corpus of objective (intersubjective) knowledge.

The view that introspection is scientifically worthless served as the basis on which behaviorists grounded the formation of the new direction of research. Watson convincingly argued that mental events or states cannot be the object of psychology as science, and hence psychology must investigate not mental processes but human behavior. Of course, some of the notions employed by traditional psychology in relation to mental states, such as hunger or fear, may be interpreted in a scientific way as patterns of behavior, and these patterns may be fruitfully investigated because they may be empirically observed and examined. From a behaviorist standpoint, animal psychology is a good model for a science of human behavior which, like animal psychology, must be based on the examination of how humans *respond* to definite *stimuli*.

One must admit, however, that the term *behavior* is not overtly clear. Watson was convinced that behavior must be defined predominantly in the terms of measurable physiological parameters, such as pulse, blood pressure, contraction of muscles. Yet most behaviorists are concerned with more general patterns of behavior, say the fight or flight response. Skinner adopted precisely such a notion of behavior.

Although Skinner thoroughly studied works by various positivists, it was Mach who, as Skinner himself acknowledged, influenced him the most (cf. Smith 1986: 373). Like Mach, Skinner emphasized that science is an extension of biological knowledge and is a practice which serves self-preservation. Even though science is practiced *within* the confines of culture and not of nature, it has no value without its contribution to the survival of the species. "Survival is the only value according to which a culture is eventually to be judged, and any practice that furthers survival has survival value by definition" (Skinner 1971: 136). As Smith indicates, Skinner claims like Mach earlier, "that the purely descriptive approach possesses greater efficiency than the hypothetico-deductive approach, that the use of unnecessary terms violates that ultimate simplicity of formulation that it is reasonable to demand of a scientific system and that the criteria for judging a system are supplied principally by the usefulness and economy of the system with respect to the data at hand" (Smith 1986: 270).

Like Mach, Skinner had less interest in theories than observable correlations and regularities of phenomena. Although he was not as radical as Mach and did not believe that, *in principle*, theories are superfluous he clearly preferred observed functional dependencies to theoretical postulates. For this reason he was critical of the view that one could and should formulate rigorous methodological rules and develop a formalized logic of science characteristic of logical positivists.

Here I must emphasize that although logical positivists were interested in behaviorism, they did not consider it to be the only prospective trend of psychological research. They recognized that other modes of inquiry are also valuable, and that psychologists, no less than physicists, are free to use *theoretical* notions and produce bold theories. In this sense, their attitude towards Freud is typical. (Most would name Freud as the most influential psychologist of the twentieth century, rather than Skinner.) According to Popper, Freud's theories, just like those of Marx, are alien to a really scientific spirit—they lack empirical content and are unfalsifiable. Yet logical positivists and, first of all Carnap, showed great interest in Freud's inquiries and treated him as a true scientist who furthered the advance of psychology. Carnap did not think that psychoanalysis betrays the positivist ideals of science. Nor, it seems, are they betrayed by the rapidly developing cognitive psychology and cognitive science, although this direction of psychological research, despite its connections with behaviorism, comes close to the mentalist tradition, which Skinner took to be unscientific. Cognitive science through its application of the methods of formalization and computer simulation, is positive in spirit as well. Thus we may maintain, together with Tolman, that "positivism is alive and well, indeed is thriving in psychology today even in what are normally understood as the most antipositivist tendencies of its recent history" (Tolman 2009 [1992]: 5).

A Positive Economics

Among the social sciences economics is doubtless the most influential and the one which has achieved the most tangible results. At the same time (and probably, by no coincidence) economics is the science which has adopted most positivist ideals and standards, and which raises verifiable hypotheses and widely applies quantitative methods. The spread of these ideals and standards was furthered by the Nobel Prize winner and the leader of the famous Chicago school of economics, Milton Friedman. He expounded his chief methodological ideas in the essay "The Methodology of Positive Economics," which had a tremendous influence on most economists of the second half of the twentieth century.

When talking about positive economics, Friedman has in mind the economics which describes and explains actual economic processes. Such economics is contrasted with the normative, which is concerned with what has to be done in economy and how the set goals should be pursued. But there is also another meaning, which Friedman associated with the name of positive economics—it is economics which follows the ideals of positive science as it was understood by positivists. Friedman's criteria for positive economics were such that any logical positivist would support them without reservation.

In short, positive economics is, or can be an "objective" science, in precisely the same sense as any of the physical sciences (Friedman 1953: 4–5). Of course, economics deals with the interactions and interrelations of human beings, and the investigator is himself part of the subject matter being investigated. But, according to Friedman, these circumstances do not result in any fundamental differences between the physical sciences and economics. Normative economics has the right to exist, but it is dependent on positive economics. Any decision in economic policy or mere evaluation must rest on a prediction about the consequences of doing one thing rather than another.

Thus the ultimate goal of the science of economics is to yield predictions and explanations of phenomena, and hence economic theories have to be judged by their predictive and explanatory power. Friedman also notes another important, although less clearly defined criterion for assessment of a scientific theory—simplicity. According to this criterion, if two theories have the same explanatory power, the one which adopts less variables and basic statements is considered to be superior.

Following the positivist philosophy of science, Friedman makes a strict distinction between two notions of testing hypotheses: by comparing observed phenomena with the *implications* of hypotheses and by appraising the conformity of their *assumptions* to empirical reality. Friedman claimed that it is wrong to think that a scientific hypothesis can be accepted as valid only when their assumptions conform to reality. Hypotheses in economic science rely on assumptions that are rarely truthful. The best examples of such assumptions are the concepts of "perfect competition" or "perfect market"—in reality neither exists, but the explanatory models used by economists rely on these idealized notions. According to Friedman, unrealistic assumptions by no means are a defect of a scientific theory. He writes:

> Truly important and significant hypotheses will be found to have "assumptions" that are wildly inaccurate descriptive representations of reality, and, in general, the more significant the theory, the more unrealistic the assumptions (in this sense). The reason is simple. A hypothesis is important if it abstracts the common and crucial elements from the mass of complex and detailed circumstances surrounding the phenomena to be explained and permits valid predictions on the basis of it alone. To be important, therefore, a hypothesis must be descriptively false in its assumptions. (Friedman 1953: 14–15)

Friedman behaves like a true follower of Carnap by treating the crucial assumptions of a theory as the key elements of the abstract model from which empirically testable consequences must be derived. Friedman's views on the structure of economic theory is also similar to Carnap's.

The influence of Friedman's methodological ideas for the development of economics was substantial, although not too original in the context of the positivist notion of science. It is no accident that he (together with John Maynard Keynes) was one of the two most famous economists of the twentieth century. By the way, the methodological views of Keynes and Friedman were much less different than their theoretical views. The controversy between positivists and antipositivists, although less pronounced than in other social sciences, exists. But so far most economists adopt the notion of economics as a positive science. Positivist ideas ensured that economics has, compared with other social sciences, a substantial degree of unity, which guaranteed its rapid progress.

Of course, in the contemporary *methodology* of economics, positivist ideas are approached more critically than previously. Yet there is no clear indication that the antipositivist attitudes would significantly affect *practicing* economists. In this respect, the situation in economics is not different from that in other social sciences: positivism is a subject of attacks launched by some methodologists of various persuasions, but the effect of these attacks on the scientific practice of the leading representatives of a discipline is rather limited. The science of economics can be taken as an example of positivism's success in the social sciences.

Positivism's Influence on Sociology

Positivism's influence on the development of sociology deserves special attention at least because Comte regarded it as the most important science. Although far from every sociologist would agree that Comte had *founded* this science, he certainly made a lasting impact on the discipline. One can be critical of Comte's own sociological conceptions, but one must at least acknowledge that the next most serious candidate to the title of the father of sociology—the famous French *fin-de-siècle* sociologist Émile Durkheim—drew heavily on Comte. This French researcher accomplished a lot in establishing sociology as an academic discipline and founded the first European department of sociology at the University of Bordeaux in 1895. Although Durkheim was somewhat critical of several aspects of Comte's teaching, he explicitly acknowledged his debt to Comte. He "adopted all of Comte's major themes—empiricism, sociologism, naturalism, scientism and social reformism" and "added to Comte's abstract philosophical themes another, quite independent tradition, that of statistics" (Halfpenny 1982: 23–24).

Even though in his *Course* Comte clearly expressed the general precept to use mathematics in all sciences—including sociology—as widely as possible, he himself rarely used it. He was more interested in the universal history of mankind and grandiose schemes of social reforms than in a detailed analysis of "minor" problems such as unemployment or suicide rates. Durkheim was more interested in the problems of the latter kind, and taking example from Adolphe Quetelet and other social statisticians, began widely applying mathematics in social science. The entire field of sociology followed Durkheim in this respect. Logical positivism, with its emphasis on mathematical analysis of functional relations between phenomena, only strengthened the statistical orientation of empirical sociology. It mainly follows positivist guidelines.

Another name that must be mentioned when discussing the positivist influence on sociology is Paul Lazarsfeld, who established sociology as an academic discipline in the United States, and thus fulfilled a role comparable to that of Durkheim in France (although the Department of History and Sociology was established in the University of Kansas already in 1891). Having begun his career as a sociologist in Austria, and having continued it in the U.S. from the 1930s, Lazarsfeld described himself as a positivist and had some indirect links with the Vienna Circle (cf. Stadler 1997: 382–83). The Bureau of Applied Social Research at Columbia University that he established in 1937 became a global center of sociological research. Many famous sociologists of positivistic views worked there, such as Robert K. Merton.

What was the influence of positivist ideas on the other eminent sociologist of that era—Max Weber? His relation to positivism is more vexed than that of Durkheim or Lazarsfeld. He rejected many ideas cherished by Comte and his followers. Yet Weber was instrumental in lending support and scientific credibility to what is likely the most significant positivist notion—the idea that social science must be value-free. Mach also swayed Weber, who adopted this stance. According to Weber, allowing normative judgments in social research turns social science into a part of social policy. Value judgments may become an *object* of scientific inquiry with the aim, say, to establish dominant values of different social groups and their fluctuation. However, a *scientist* cannot claim that the values of one group are higher than those of another. Within the confines of science, we can speak about values but we cannot appeal to value judgments, nor can we lean upon them when building our theories. Of course, our values might influence both what issues we choose to examine and the appraisal of the results of an investigation as important or unimportant. Weber is perfectly aware of this, yet his conviction that scientists *as scientists* do not make value judgments is firm and unwavering. And although this point has been made earlier by other thinkers, Weber contributed most to making it a proper part of sociologist's understanding of their work.

It would be difficult to evaluate how many of the sociologists still share the view that their research should be value-free. Having conducted sociological research myself some time ago, I can confirm that it is prevailing in the area of empirical research. Yet the influence of positivism on theoretical sociology is, in general, substantially weaker than on empirical inquiry. Assessing positivism's influence on theoretical sociology, one must emphasize that positivists aimed to merge it with the practice of empirical sociological inquiry and to separate it from social philosophy as much as possible. If we separate *sociology* from *social philosophy*, whose representatives are often inclined to label their speculative constructions as *social theory*, we will have to admit that positivism also still profoundly influences theoretical sociology.

Yet the clear-cut demarcation line that positivists attempted to draw between theoretical sociology, based on the results of empirical inquiries, and more speculative social philosophy is supported by far from all social theorists. Positivists would refrain from calling most of them sociologists, but these theorists usually disagree with such an opinion. Thus it is unsurprising that precisely they are responsible for most criticism of positivist principles in sociology—they were simply obliged to do that in order to defend their status as sociologists. Not all social theorists looked to philosophy for inspiration (e.g., psychoanalytic sociology tried to establish itself as a discipline independent from philosophy), but most postpositivist or antipositivist sociologists turn towards one or another nonpositivist philosophical movement. Critical sociology, phenomenological sociology, symbolic interactionism, and postmodernism—all these sociologies, or, more precisely, trends of social theory, criticized positivism because they rested on different philosophical views and assumptions. Most sociological schools opposing the positivist schools dropped the criterion of empirical testability as too positivistic, or modified it to such an extent that it became inoperative. And without the aid of a strict criterion of empirical testability it is very hard to decide which social theorist is right and which is wrong.

According to Anthony Giddens, the "positivist attitude" in sociology may be said to comprise the following connected suppositions:

1. That the *methodological* procedures of natural sciences may be directly adapted to sociology. According to this standpoint, the phenomena of human subjectivity, of volition and will, do not offer any particular barriers to the treatment of social conduct as an "object" on a par with objects in the natural world. Positivism here implies a particular stance concerning the sociologist as *observer* of social "reality."

2. That the *outcome* or end-result of sociological investigations can be formulated in terms parallel to those of natural science: that is to say, that the goal of sociological analysis can and must be to formulate "laws" or "law-like" generalizations of the same kind as those which have been established in relation to natural reality. Positivism here involves a definite view of the sociologist as *analyst* or "interpreter" of his subject matter.

3. That sociology has a *technical* character, providing knowledge which is purely "instrumental" in form; in other words, that the findings of sociological research do not carry any logically given implications for practical policy or for pursuit of values. Sociology, like natural science, is "neutral" in respect of values. From this aspect, positivism presumes a specific standpoint upon the sociologists as practically involved in the social order. (Giddens 1974: 3–4)

Certainly, not all sociologists endorse these suppositions. Still, Giddens notes that the two most famous sociologists of the twentieth century—Durkheim and Weber—and the very important sociological traditions associated with their names were heavily influenced by positivism.

The first, according to Giddens, is obviously positivistic and the second attempts to synthesize the subjectivism of the German historical school (and neo-Kantian philosophy) with the views "which were quite definitely positivistic" (Giddens 1974: 5).

The schools of Durkheim and Weber for the most part of the twentieth century were dominant in *theoretical* sociology. Yet in the last decades positivism went out of vogue in the field. Discussions in *social theory* concentrate now on issues rather distant to the positive mind. Social theorists nowadays discuss topics quite alien to positivism, such as "interpretation of subjective meanings," "direct understanding," or *Lebenswelt*. For a sociologist involved in empirical research these discussions look at best philosophical and not directly relevant to their investigations. Yet quite a few *theoreticians* do not regard the philosophical character of the discussions going on in contemporary sociology as a deficiency which must be overcome, for, let me repeat, they hesitate to merge theoretical sociology with empirical research, and, on the other hand, to separate it from social philosophy. They would completely agree with the opinion expressed by Peter Winch in his influential book *The Idea of Social Science and Its Relation to Philosophy* that "the central problem of sociology, that of giving an account of the nature of social phenomena in general, itself belongs to philosophy" (Winch 1971 [1958]: 43). This view is shared by many of those theoretical sociologists who, like Alfred Schutz, would find it somewhat difficult to decide whether their deliberations on social phenomena belong to phenomenological sociology or, rather, to sociological phenomenology, i.e., philosophy.

Positivism's Impact on Political Science

Positivism's impact on political science is in many respects reminiscent of positivism's impact on sociology, since it directly depends on how we define political science and its relation with political and social philosophy, ethics, and social theory. Just like in sociology, those political theorists who deny the desirability of demarcating political *science* from other disciplines (first of all from philosophy) are, to put it mildly, not very enthusiastic supporters of positivism. They emphasize that positivism is just one among many philosophies, hence its influence on political science should not be overestimated. Their colleagues who disagree blame them for disregarding the boundaries between political science and other disciplines, which threatens the status of political science as an autonomous and academically established science.

However, considering the above, it must be stressed that there is one significant difference between political science and sociology: political scientists and theorists generally feel much less sympathy towards most nonpositivist philosophical trends, e.g., phenomenology, which are more highly valued by some sociologists. Although the phenomenological influence on sociology has been considerable in the last decades, it is much less tangible in political science. Antipositivist philosophical ideas became more prominent in political theory only from the end of 1980s, and they are mostly connected with critical theory and postmodernism. Political scientists of a traditional, i.e., positivist, methodological outlook usually consider postpositivists as marginal: those who produce much criticism (towards positivist methodology and epistemology, and the theoretical views of scholars accepting them), but are not able to suggest how particular problems of political science should be resolved. Those political scientists who aspire to seem more modern opt rather for constructivism (which will be discussed in the next section).

On the basis of my personal experience of working in political science, I risk claiming that positivism affected this discipline no less than it affected economics. The mood prevailing among those political scientists who do *not* regard their discipline as being a part of philosophy is clearly positivistic. These scientists are involved mainly in examining the behavior of various political actors starting with individuals and ending with states and international organizations. Because of their involvement in politics in which law, protocol, and other formalities play a more important role than in social life in general, these actors and their actions are more easily recognizable and definable than social processes and their principal participants.

Political actors act quite often on a particular political scene, and in this respect they are even more visible than actors in an economy. When investigating their behavior, it is possible to formulate hypotheses concerning it in rather well-defined terms, to put them to empirical tests, to collect and examine empirical (often statistical) data and thus to do research following the positivist scientific agenda. The same could be said about the study of the forms of political organization of society (determined, at least partly, by the constitution of a country). On the other hand, on the basis of data concerning patterns of political behavior prevailing in different countries, it is quite difficult to construct a general political theory, i.e., to achieve something similar to what has been done in economics.

Positivism had an especially strong impact on the study of international relations. Its influence is clearest in the two most widely adopted paradigms: realism (neorealism) and liberalism (neoliberalism). Their representatives do not doubt that international relations must be analyzed as an interaction of clearly identifiable states (and maybe other international actors), and its results should be explained on the basis of empirically discernible differences of the participants of such interaction. Characteristically, the neorealism of Kenneth Waltz is even more positivistic than the classical realism of Hans Morgenthau and Henry Kissinger. Waltz abandons any theoretical elements that could be considered as metaphysical or normative (e.g., Morgenthau's deliberations on the nature of humans or states and his instructions for politicians). Waltz's theory is more abstracted from history, is built deductively rather than inductively (of which logical positivists would most likely approve), its concepts are defined as thoroughly as possible. Thus, although methodological discussions in the field of international relations continue most practitioners are far from believing that the theory of international relations has overstepped the clear line that positivism has drawn between what is and what is not a science.

Besides international relations, positivism's influence is also strong in other fields of political science. For manifest proof, one only needs to open a few issues of the *American Political Science Review*, a quarterly journal of the American Political Science Association, which is regarded the most prestigious in the field. Authors of almost every paper adhere to methodological principles of logical positivism (which is, of course, conditioned by the journal's editorial policy).

Positivism—The Postpositivism Debate. Constructivism

In summary of what was revealed in the previous section, we can affirm that positivism had a significant, although different, impact upon various sciences and fields of research. This impact was especially strong upon the social sciences, many of which became autonomous disciplines, i.e., disciplines separate from philosophy, only in the nineteenth century. These sciences, at least psychology, sociology and political science, formed as independent sciences at the time when Auguste Comte and John Stuart Mill—the main founders of classical positivism—had already presented their views about what form social science or the social sciences should take, and upon what methodological foundations. Since they were the most popular philosophers of the nineteenth century—not Hegel, Marx, Kierkegaard, or Nietzsche—their ideas found followers among the thinkers and researchers who were developing social sciences mostly under the influence of positivist ideas. Comte, as we have already noted, was one of the founders of sociology. And Mill was one of the most famous economists of the nineteenth century. Thus, their authority was tremendous. They were both relying on the principles of methodological naturalism and were convinced that the development of social sciences should be based upon the methods applied in natural sciences, and first of all on observation.

Positivism had a huge impact not only upon the various sciences and scientific disciplines, but also on the notion of science itself—we mostly comprehend science from the point of view of positivists and consider a thorough empirical testing of hypotheses to be a standard procedure of scientific investigation.

Positivism still decisively influences the practice of empirical research in the social sciences, although at the theoretical level it competes with other philosophical conceptions. But even in theoretical discussions, positivist positions are not weak, and some recent methodological trends bear the marks of positivism and seek to bestow an even more scientific character on the examination of social problems. Rational choice theory and formal sociology successfully employ most of the methodological recommendations of (logical) positivism.

The ideal of unified science so prized by the positivists has not been achieved. However, a rather coherent system of sciences has been created. The classification of sciences suggested by Comte (or rather its part from mathematics to biology) constitutes the foundation of contemporary understanding of interrelations among the sciences. Today no one doubts that mathematics, physics, chemistry, and biology are closely connected. Since Comte, positivism has played the most significant role in realization of the idea of social science based on empirical methods. No social science yet exists that relies on the common and universally recognized methodological principles of which positivists dreamed, but they have achieved another important goal: the dominant ranks in the social sciences are taken by schools which rely mainly on the scientific methodology positivists valued.

Positivism also established many other elements of contemporary understanding of science and its benefits. Few would doubt today that science must have (at least potential) practical value. The claim that the object of scientific investigation is the world of phenomena presently seems to be almost too obvious to mention. The same could be said about the distinction between facts and values (between statements of fact and value judgments) so strenuously defended by the positivists. The view that norms cannot be derived from facts was rarely accepted at the beginning of the nineteenth century. In our century it is almost an axiom.

Hence, the general conclusion is this: positivism had a tremendous influence on science. This influence clearly exceeds the influence on science by any other philosophical trend. This came as a result of two circumstances. First, no other philosophy was interested in science to a comparable degree as positivism. Second, no other philosophy has so consistently strived to be scientific. However, in some sciences practically every researcher relies on the positivist notion of science and its methods, while in others this is not the case.

The natural sciences in the broad sense, i.e., physical, biomedical, and technological sciences, are almost all developed under the principles of positivist epistemology and methodology, although those principles themselves were formulated partly on the basis of the analysis of scientific praxis established in those sciences, first of all in physics. The situation is different with the humanities and the social sciences.

The preeminent place among the humanities is taken by philosophy and history. Positivism's impact on philosophy was thoroughly discussed in the previous chapter, and here we will not discuss it further. Meanwhile, a few words about history are worth mentioning. Many significant historical works in the nineteenth, twentieth, and the twenty-first centuries were produced on the basis of positivistic views. The determination of historical facts on the general basis of archive material and methods of archeology, chemistry, physics, and astronomy thorough critical analysis of the reliability of historical sources, the systematization and generalization of historical data, the formulation of empirically testable hypotheses and strict testing of such hypotheses are all typical traits of the application of positivistic methods in the science of history. But, of course, I am not claiming that all historians adhere to positivist attitudes. Some of them were considerably influenced by entirely different schools of thought. From the 1980s the influence of postmodernism is clearly felt in history as well as in other humanities and social sciences.

However, as far as I know, unlike in the social sciences, stormy disputes between positivists and postpositivists do not take place in the humanities. Thus I intend to concentrate below precisely on the continuing discussions between positivists and postpositivists in the social sciences because of their significance for the farther advancement of our social knowledge.

I use the term "postpositivism" in the broad sense by which it is also used in such discussions (from the 1960s until the 1980s the term was mainly used in slightly different and narrower sense to name the views of Thomas Kuhn and of an entire historical school of philosophy of science). Such discussions are focused not only on purely methodological issues, but also on epistemological and ontological questions. However, they are usually called *methodological*, in order to separate them from *theoretical* discussions about more particular, interesting from the theoretical point of view, but nonmethodological and nonphilosophical issues. In methodological discussions in the social sciences, postpositivism is usually understood as critical theory or postmodernism. I suppose that the reader already has some understanding of critical theory and postmodernism and of some of their disparities. When discussing those schools of thought in this section I will, however, not pay much attention to their differences because they are attacking positivism on similar lines. And I will mostly disregard the differences between various social sciences as well because the methodological entanglements of most of them or, at least, of those which interest me most (and of which I have some firsthand knowledge), i.e., sociology and political science, are quite similar.

The position of the participants of these discussions can be characterized by certain views that they usually group as (1) ontological, (2) epistemological, and (3) methodological. Ontology, let's recall, is the philosophical theory of being that attempts to answer the question what exists, what sort of entities of fundamentally different types exist, and what are their relations—which are primary and which are secondary (not in the logical, but in supposedly the more fundamental, though less easily definable, ontological sense). Speculative metaphysics, which is concerned with what exists *over and above or beyond* experience, could be treated precisely as ontology. For instance, Plato as ontologist (although this term originated later) claims that only ideas conceivable by the intellect *really* exist, while the world of sensory things and phenomena is but a pale shadow of the world of ideas, and thus a *mere appearance*, not reality. It is worth noting that in answering an ontological question—what exists—Plato invokes epistemological notions that describe *knowledge* and understanding: "conceivable by the intellect," "perceivable by the senses." In fact, it is impossible to draw a clear line between ontology and epistemology (as well as, by the way, between epistemology and methodology).

Before explaining what are the ontological views of certain postpositivists, let us pose the following question: what are the ontological views of the positivists? This question puts us into a difficult position, at least at first sight, for positivists were fiercely critical of metaphysics. For them Plato's claim that only ideas really exist was neither true, nor false—they took it to be cognitively meaningless. Schlick spent a lot of time explaining the difference between positivism and realism, and Carnap took the question about the reality of ideas and things to be as meaningless as the question about the reality of numbers. It is true that Mach claimed that the world is constituted by neutral elements, which, depending on the issue under consideration, could be interpreted as qualities of physical objects or as sensations, but logical positivists were suspicious about such conception, for it resembled *metaphysics too much.* They were not willing to be entangled in the senseless, in their opinion, ontological disputes, for instance, the dispute between materialism, which claimed that only matter really exists (or that it is primary), and idealism, which claimed that only thoughts or ideas are real (or primary).

Thus, strictly speaking, positivism has no ontology of its own. What ontological views are then disputed by the social scientists who consider themselves postpositivists? What they have in mind while speaking about the positivist ontology is often not positivism, but materialism. But more often they change the philosophical meaning of the term "ontology" and define it, for instance, as a "theory connected with the things, properties and events that exist in the world; what exists to be investigated" (Daddow 2009: 238).

Understood in this way, ontology means the fundamental assumptions of different *theoretical* paradigms that Kuhn spoke about, for they include the assumptions about "what exists to be investigated." Among the physicists on the verge of the nineteenth and twentieth centuries who shared positivist views, some thought that atoms exist, and some have denied that. In the theory of international relations, realists (neorealists) relying on positivist views, might claim that the only (independent) international actors are states, while the liberals (neoliberals) sharing the same methodological views would claim that there also are nongovernmental international actors playing important political and economical roles. Thus, positivists who adhere to different theoretical paradigms within the same branch of science may rely on different ontologies, in the sense of the term as it is used by representatives of special sciences in their methodological disputes, but not in the sense of the term used by philosophers. Even in the former sense there is no basis for talking about a unified positivist ontological position, i.e., about positivist ontology.

Still, there is one question raised in ontological discussions within the social sciences that may also interest a philosopher: is the character of social reality studied in those sciences the same as that of natural reality? Representatives of the social sciences take divergent positions with regard to this question.

Most substantially they agree with materialists that the *natural* world is material. But the positions regarding the social world diverge. Some, and Marxists first among them, claim that the basis of social life is material production, economy. Upon the relations of production, i.e., human relations in the sphere of production, exchange, and distribution (of material goods), depend their social and political relations, institutions, and ideas. By claiming that entire social development, including the development of our ideas, is dependent on economic development, they adhere to substantially materialist views.

Other representatives of social sciences think that the social world is different from the natural world, because in the former ideas are *primary*. Those ideas are realized, materialized in the course of human activities; they determine our attitude towards the world. The position of these social scientists may be considered idealist.

Is it possible to talk about the uniform positivist position with regard to this question? My answer is negative. Comte's view on this issue was that the development of human reason (ideas) determines the entire social development, as well as social and political order. Mill's position on the issue is very close to that of Comte. However, logical positivists, who were not hiding their sympathy for Marxism—especially Neurath, Carnap, and Reichenbach—thought that material production is the foundation of social life, and its development determines social development. They favored Marx's point that people make their own history, but not under self-selected circumstances. Thus, in this respect, there's also no basis for talking about a uniform positivist ontology.

Now we can proceed to the analysis of arguments against positivism that are called *epistemological* by the postpositivists. Epistemology as the philosophical theory of knowledge is an inquiry into what can be known, where our knowledge comes from, and whether it can be considered true.

The answers to these questions given by the researchers of positivists views are quite clear and in certain regards quite simple. According to them, we can only gain knowledge of phenomena and their regular relations. Our knowledge of the world comes from experience. The basis of this knowledge are facts, which should be strictly distinguished from values; valuations or normative statements cannot be derived from statements of facts. The question whether under such view knowledge can be considered true is slightly more complicated and we will discuss it later.

However, the epistemological debates among social scientists are centered around different questions. Probably the most important question is the following: does the investigator of the social world investigate this world from the outside, like the investigator of the natural world, or, on the contrary, does he always analyze this world from within?

The position of the researchers of positivist attitudes, regardless of the significance in social life that they ascribe to ideas, is this: the social world is different from the natural world, but from the cognitive point of view, the difference in the relation of an investigator to one world or another is not large enough to make the investigator incapable of observing, describing, and explaining social phenomena in a manner resembling the observation, description, and explanation of natural phenomena. The investigator of the natural world is not only social but also a natural creature, but it does not preclude him seeing the natural world from the outside (with respect to phenomena under examination). The investigator of the social world is not only natural, but also a social creature, but it does not preclude him seeing the examined social phenomenon from the outside (or as if from the outside). He approaches the phenomena under investigation literally from the outside when dealing with the phenomena that took place in the past or take place in other cultures, civilizations, countries, or regions. However, even when he investigates, say, the voting behavior of his cocitizens, despite being a voter himself, or when investigating their moral values, although he has some values of his own, as a *researcher*, he can (and ought to) dissociate from his own political views or moral values and take the stance of impartial scientist, who approaches his phenomena *as if* from the outside. As a *researcher* he has no right to dispose with the sociological questionnaires when most of the respondents express the political or moral views that are unacceptable to him as a person, and claim that society supports the views and attitudes of his own (or of the party he belongs to) and thus to attempt to influence the public opinion and/or the results of the forthcoming election. Of course, the values, attitudes, and ideas of the investigator of the social world will influence what problems he will be interested in, what questions he will ask his respondents, but other researchers with different values, attitudes, and ideas will opt for different issues, so our intersubjective, shared social knowledge acquired through the effort of various researchers will be more comprehensive and less subjective.

The opinion of sociologists and political scientists adhering to postpositivist views is the opposite: the investigator is not capable of approaching the social reality from the outside; he can only see it from within. Human relations depend on what human beings think about each other. Institutions are the outcome of realization of certain ideas: for instance, the parliament embodies the idea of political representation. Our actions depend on our opinions. Moreover, those opinions expressed in various ways constitute a very important part of our social life. Postopositivists emphasize that the *investigator* of this social life is heavily influenced by social norms and values, rules and conventions, ideals and aspirations. As a social individual, the investigator is bound by multiple relations with society, and he is not capable of breaking those relations, dissociating from them, and declaring that *he as investigator* will from now on approach reality impartially from outside. For he cannot stop living in the society and just investigate it. Thus, his cognitive attitudes are dependent on his social attitudes.

Looking at the social world from within the researcher may *understand* it, but he cannot (and should not) *explain* it, in the sense in which positivists comprehend explanation. From the positivist point of view, in order to explain some phenomenon we must be able to derive *particular* statements about that *individual* phenomenon from the *general* laws and theories of science. Understanding, on the other hand, requires the ability to empathize with the other person, understand the motives of their actions, aims and values, instead of calmly observing and explaining their behavior.

The dichotomy of explanation and understanding is older than the contemporary debate between positivism and postpositivism. From the point of view of methodological individualism, which is different from the point of view of methodological collectivism, holism, or structuralism, the cornerstone of social sciences is the investigation of social actions of individuals. According to Max Weber, it is characteristic of social action that individual ascribes *subjective* meaning to it. The humanities and social sciences which seek to reveal this meaning are different from the natural sciences, which explain phenomena and don't need to deal with such meanings. The differences between the natural sciences and the sciences of the spirit (*Geisteswissenschaften*) were emphasized by Wilhelm Dilthey in the nineteenth century (cf. Dilthey 1986 [1883]). The *hermeneutical* view that in order to understand the text we need not only to master the grammar and semantics of the language it is written in, but also to be able to grasp the author's intentions and thoughts, was adapted by Dilthey for the understanding of human actions.

According to the hermeneutical approach, in order to understand the action from within (and that is the only way to understand it), one must know two things. "One is that the investigator needs to know the rules, conventions, and context governing the action—the meaning of the action regarded as a move in a socially defined 'game.' The other is that the investigator needs to know what the agent intended by and in performing the action—why this agent played this move in the 'game'" (Hollis and Smith 1990: 72).

In the 1960s within many social sciences the discussion took place between the so-called *traditionalists* and *behaviorists*. The former claimed that understanding in the aforementioned sense is an especially important (although not necessarily the main) goal of social sciences and that it can only be achieved on the basis of history, psychology, and morals. An investigator of international relations, for instance, in order to understand the decisions of the statesman must be able to grasp his intentions and motives, as well as the moral dilemmas gnawing at him (connected with, say, the use of violence). Only then will he be able to provide recommendations for the statesman on how to act in this or that situation. Meanwhile, behaviorists preferred that social sciences be restricted to the investigation of empirically observable behavior of individuals (or their groups), without relying on hardly testable or completely untestable assumptions about the motives and goals of individuals. Behaviorism was a radical form of positivism that emerged in psychology on the basis of Mach's ideas that encouraged researchers to confine themselves to the description of the object of investigation and to avoid theorizing. The restrictions of the *problems*, *language*, and *methods* of research proposed by behaviorists were viewed critically by logical positivists, although in the early 1930s Carnap had claimed that all the statements of psychology, including those couched in phenomenal, sensory language, may be translated into physical language, used to describe the behavior of individuals (cf. Carnap 1959 [1932/33]). But behaviorism had its influence upon the methodology of social sciences and, among other things, contributed to the fact that the fundamental positivist view—the principle of the fact-value distinction—was more consistently applied within the field. It is worth recalling that this principle was adopted by the entire sociological school of Max Weber, though the latter emphasized the epistemological differences between the social and natural sciences.

Some time ago, postpositivists who have launched a wide frontal attack on positivism in social sciences (although at different times in different sciences) concentrated their attention on the criticism of this fundamental attitude. The representatives of critical theory as well as postmodernists emphasize that the goal of social theory is not a mere *understanding* of human social actions, but also a *critique* of social reality, thus social theorists, contrary to the views of positivists and Weber, not only have a right, but also an obligation to *evaluate* social reality. Thus they must formulate not only factual, but also value statements. For if a social theorist wants to change reality which does not fit the ideals of justice, freedom, equality, democracy, and others, he must clearly name its vices. Critique is a very important, probably the most important, task of a social theorist. Understanding his task this way, one must abandon the positivist notion of a theory as an empirically testable system of statements which have factual content and are connected by strict logical bonds. The researchers adhering to positivist views, obviously, have no intention to discard this notion.

Now we can move to the discussion of the last of the aforementioned epistemological questions, i.e., to the question about truth. From the positivist point of view asking the question "what is truth?" is as pointless as asking the question "what is meaning?" It suffices to be able to determine which of our statements are meaningful, and which are true. The truth of the analytical statements of logic and mathematics is determined in one way—by strict proof, inference from axioms, while the truth of synthetic, factual, scientific statements is determined differently—by empirical testing. In traditional epistemology, the question whether our knowledge is true is often understood as the question about the relation between subject and object, and the only knowledge that is true is knowledge possessed by the subject which is objective, i.e., corresponds with the object. But logical positivists did not raise the general philosophical question about the relation of subject and object, although they were not against *calling* the investigated phenomenon "object" without loading this term with any specific philosophical meaning.

Accordingly, positivist philosophers avoid talking about *objective knowledge*, although the researchers adhering to positivist views who are less demanding towards their language (or who follow Popper rather than Carnap), often talk about it in methodological discussions. As I have already mentioned several times, logical positivists talked about the *intersubjectivity of knowledge*, not its objectivity. Intersubjectivity presupposed the possibility of exchanging the mutually understood information (knowledge) and possibility for any researcher to test this knowledge (provided he has due training and equipment). The view of logical positivists could be summarized as follows: scientific knowledge must be intersubjectively verifiable, i.e., testable.

They were reluctant to speak about the objectivity of knowledge, for if we wanted to answer the question whether it is objective, we would have to consider the philosophical question of the relation between the subject and object and to find out whether an object exists independently of a subject. Since it is impossible to empirically test the answer to this question, the question is devoid of cognitive value. Because of their antimetaphysical views, logical positivists avoided not only ontological questions, but also those epistemological questions which were so closely connected with the ontological that they could not be regarded as questions for philosophy of science, i.e., questions about the ways of gaining and testing scientific knowledge, its foundations, structure, and development.

When talking about the truth of knowledge, they would not forget to remind us that all our knowledge of the world (including empirical statements describing facts) is hypothetical. That should neither prevent us from recognizing it, nor to rely on such knowledge when developing a theory, but we must clearly understand that our cognitive decisions may have to be changed based on new evidence.

The least misunderstandings lie in the dispute between positivists and postpositivists within the field of methodology in the narrow sense. Researchers who adhere to positivist views claim that in order to investigate one or another realm of phenomena, one must, while relying on partially known but possibly incomplete and imprecise facts, formulate hypotheses about the connection of those facts or explain those connections. Once a hypothesis is raised (on the basis of analogy or other insight), it must be tested. A clear hypothesis allows one to decide what facts would verify or falsify it, i.e., what sort of facts are significant, relevant. Thus our hypothesis influences the direction and character of empirical research. Empirical facts may be formulated also in terms of the previously accepted theories (e.g., when we use an ammeter to find out that the strength of a current in a circuit is 5 mA, we formulate the *measurement data* in terms of a certain physical *theory*). After the hypothesis was tested it is either accepted, rejected, or modified, although the investigator must clearly understand that in the future he may need to change his opinion about the completion of testing of a particular hypothesis, as well as about the acceptance, rejection, or modification of it.

It is not always possible to test a hypothesis individually, although even in the course of the individual tests of a hypothesis one must often rely on various additional assumptions, e.g., that the examined phenomenon is affected by no other factors except those described in the hypothesis. Sometimes one needs to perform empirical tests of systems of statements connected by rather complex logical relations. Such systems are called theories and they may contain statements formulated exclusively in new theoretical terms (these terms may be used to describe abstract, idealized, directly empirically unobservable objects, or to create theoretical schemes and models). But even the statements that are formulated only in newly introduced theoretical terms must be in one way or another connected with less abstract and empirically testable propositions.

How do postpositivists respond to this? First, they emphasize that the positivist scheme of scientific research ignores the differences between the natural reality and the social reality. The researchers who adhere to positivist principles thus misunderstand the goals of social sciences. The usual and traditional ideas that serve for the domination of the powerful must be deconstructed and their flaws revealed. Therefore the social sciences cannot rely on the positivist distinction of facts and values.

It is also unnecessary to raise and test hypotheses, since social theorists have more important goals to pursue: they must first of all not be impartial observers, but active members of society, defending certain values. Moreover, there are no facts at all which could verify or falsify these hypotheses—what is considered a fact is a mere interpretation, which depends on our theoretical views and which, in turn, are influenced by our values. The knowledge of the social reality cannot exceed its interpretation and critique.

These are the main arguments that postpositivists (the representatives of the critical theory and postmodernists) direct against positivistic methodology. But can these arguments be called methodological? Do postpositivists have any methodology of their own that would be comparable with the positivist one?

One may talk about certain postpositivist methods, e.g., genealogy or deconstruction, but only with significant reservations. Most postpositivists, and postmodernists in particular, would not like to call them methods, for they associate the very term "method" with what they consider to be repulsive positivism, and thus they avoid defining these methods more precisely.

Genealogy and deconstruction are ways to analyze discourse therefore in order to understand the meaning of those terms, some things must be said about the word "discourse" so often used by postmodernists.

The term "discourse" has many meanings, and, in fact, this multiplicity of meanings makes it particularly favored by postmodernists who often avoid clarity, precision, and univocality. Discourse is language (spoken or written, including any text whatsoever), speech, conversation, argument, discussion, or communication of its participants. One can talk about the contemporary political discourse or discourse of political science. In the discipline called discourse analysis, any text longer than a sentence is taken to be a discourse. Yet postmodernists usually understand discourse as the ongoing discussion taking place in some particular field or discipline for a long period of time (years or even centuries in some cases), which is not necessarily direct, and is not understood as mere verbal activity. Consider an architect who designed and erected a building of a new architectural style—thus he joined the architectural discourse while some other participants of this discourse may be engaged only in theoretical considerations of what contemporary architecture should be like. Speeches of politicians, their declarations, laws and the projects of the laws, diplomatic notes, and international agreements are all parts of political discourse. Even their decisions, e.g., to start a war, or *the war itself* can be treated as elements of political discourse. In this very broad meaning, discourse is a totality of social (verbal and nonverbal) praxis, which forms and changes the social relations according to the meanings of certain signs. However, when actually analyzing discourse, postmodernists devote most time to texts, and the most radical of them take the entire social world to be a text.

We can now proceed to genealogy. Michel Foucault, who is often invoked by postmodernists, treats genealogy as the investigation of the historic practice which creates disciplines in order to reveal ruptures of discourse, i.e., the termination of considerations of certain topics, their elimination and marginalization. Probably the most important aim of such an investigation is to figure out what was forgotten and ignored by the discipline. Genealogy in this sense does not mean investigation of origin or kinship, although this was precisely the sense of the term already used by ancient Greeks.

The term "deconstruction," on the other hand, is new. It was introduced by Jacques Derrida. Although he has insisted that its general definition is neither available nor possible, since each and every instance of deconstruction is unique, this concept may be roughly described as follows: deconstruction is an analytical and critical way of reading a text, which enables the identification and dismounting revelation of the logic of binary oppositions that is inherent in the text. The examples of such binary oppositions are: life/death, good/evil, friend/enemy, familiar/strange, peace/war, West/East.

The first (central) element of each opposition is ascribed a positive content, the second—negative. The first is the dominant, and the second (which may be called complementary) is considered marginal. The multiplicity of binary oppositions is characteristic to Western philosophy and thus thinking (including scientific).

The principal goal of deconstruction, as a postmodern, antitraditional way of thinking, is to deconstruct the traditional understanding of the world and demonstrate that there are no stable meanings ascribed to the

elements of binary oppositions, and thus that the usual, standard oppositions are unfounded. For instance, the analysis of the opposition between
the West and the East may attempt to show that we are of excessively
good opinion about the West and Christianity, and of excessively bad
opinion about the East and Islam. Westerners traditionally saw the East as
an exotic, but uncivilized world, though in fact Western culture rests on
many discoveries and inventions from the East—from monotheism to the
alphabet invented by the Semitic nations of the Eastern coast of the Medi-
terranean Sea. And how about Chinese paper, gunpowder, and the compass? Or the inventions in ancient India of crucible steel, the use of zero,
or cataract surgery? Hence the level of civilization (using this word in the
ancient Roman sense, and not in the Arnold Toynbee's or Samuel Hunt-
ington's sense) of the Eastern world for a long time exceeded that of the
West in many fields.

Postmodernists provided many, and will probably provide even more,
insights and arguments worthy of attention. But let us return to the question whether they have their own methodology. Do they have any clear
and strict *criteria* for accepting or recognizing scientific statements (regardless of whether they are called hypotheses or not)? For different postpositivists have different views on certain issues. How shall one decide
which of them shall be accepted and which rejected? How shall one determine what are the results of their investigations? We will not get any
clear answer to this question from postpositivists. Postmodernists perhaps
would even be irritated by the mere question. According to them, the criteria restrict the freedom of the thinker and is a manifestation of the worst
of oppressions—intellectual oppression. The concept of "results of investigations," according to them, is also too positivistic, since the question
whether such results are achieved or not can only be judged on the basis
of strictly formulated criteria.

Readers familiar with postmodernist social theories would probably
agree with the claim that their authors have no clear and strict criteria for
accepting scientific statements on the basis of empirical observations. But
can one claim that *other* postpositivists do not formulate and employ any
such criteria?

My answer is as follows. First, the notion of postpositivism is sometimes given a broader meaning that in this book. If constructivists would
also be considered postpositivists (and that happens rather often), then one
would also have to acknowledge that they have the criteria for accepting
scientific statements and theories. But I don't consider constructivists (at
least moderate constructivists) to be postpositivists.

The next most important and, I think, most influential form of postpositivism after postmodernism is critical theory. Thus in order to prove that postpositivists have no clear and strict criteria of acceptance of statements and theories *on the basis* of *empirical data*, it remains to be proven that there are no such criteria in critical theory. I believe this is not a difficult task. Critical theorists reject the fact/value distinction, and thus the constructions that they call theories contain not only factual but also value or normative elements. Since facts are not derivable from valuations (as well as valuations from facts), it is simply impossible to submit the constructions of critical theorists to empirical tests.

The representatives of critical theory, Jürgen Habermas among them, understand this perfectly well. When explaining the difference of their methodological views from those of positivism or classical Marxism, they point out that both Auguste Comte and Karl Marx aimed to provide an empirically adequate description of social reality and its processes (although they did that in different ways), while critical theorists take their main task to be not a description or explanation, but critique of social reality. Thus the correspondence between their theories and reality would be (according to them) not an advantage, but a shortcoming of their theory, because the theory which provides an empirically adequate description (and explanation) of social reality justifies rather than criticizes *status quo* social reality.

Precisely the recognition or rejection of the fact/value distinction is the fundamental difference between positivism and postpositivism. Postpositivists look with arrogance at the thorough and meticulous examination of social phenomena on the basis of empirical data, just as most philosophers looked at the investigation of "insignificant" natural phenomena before the scientific revolution of the seventeenth century. They consider themselves to be above special (and thus *essentially* positivist) science. They look at humanity and society precisely as *philosophers*, not as the representatives of the *science* of anthropology or sociology. Although they call themselves *post*positivists (or even *post*modernists), regarding the fundamental issue of the relation between philosophy and (special) sciences their stance is characteristic of the *premodern* epoch, when the sciences that later became independent, were subordinated to the first philosophy, i.e., metaphysics. This relation was reversed by the seventeenth century scientific revolution, but they are loath to acknowledge the change. Postpositivists engage in many discussions and criticize incessantly. But their criticism is predominantly negative. The researchers who consider themselves postpositivists achieve little positive scientific results. They obviously respond to such reproach by saying that only the positive mind requires positive results, while they are not bound by the attitudes and criteria of such a mind. It seems that the discussion between positivists and radical postpositivists cannot be very productive. Nevertheless, it has produced at least one, not too expected, result. *Constructivism* was born. Quite often it is regarded as the middle position between positivism and postpositivism.

The term was first used by Giambattista Vico at the beginning of the eighteenth century. He regarded the process of knowledge as a process of construction, and claimed that human reason can only know what it itself creates. In the twentieth century the psychologist Jean Piaget began investigating how a child constructs his understanding of the world, and the term began to be increasingly used in psychology and education science, and later in other social sciences (see Glasersfeld 1989).

Constructivists emphasize that in ascribing certain meanings to words, people construct their understanding of the world. An individual constructs his worldview in social interactions with other individuals, thus this view is intersubjective in character. The main principles of constructivism may be formulated as follows:

> (1) [T]hat the structures of human association are determined primarily by shared ideas rather than material forces, and (2) that the identities and interests of purposive actors are constructed by these shared ideas rather than given by nature. The first represents an "idealist" approach to social life, and in its emphasis on the sharing of ideas it is also "social" in a way which the opposing "materialist" view's emphasis on biology, technology, or the environment, is not. The second is a "holist" or "structuralist" approach because of its emphasis on the emergent powers of social structures, which opposes the "individualist" view that social structures are reducible to individuals. Constructivism could therefore be seen as a kind of "structural idealism." (Wendt 1999: 1)

I am not willing to agree with Wendt that constructivism is a branch or a kind of idealism. Wendt himself notes that there are many forms of constructivism. The most radical of them are very close to postmodernist idealism. But Wendt himself takes a much more moderate position, for, according to him, an important role in social reality is played not only by ideal, but also by material factors, which he would not reduce only to purely physical or natural ones. Precisely the analysis of the interplay of ideal and material factors of social life, the combined application of approaches of methodological collectivism (holism) and methodological individualism, constitutes the strength of moderate constructivism. This version of constructivism, according to Wendt, "concedes important points to materialist and individualist perspectives and endorses scientific approach to social inquiry" (Wendt 1999: 1).

The spread of the constructivist approach in social sciences was facilitated by the book of Peter Berger and Thomas Luckmann, titled *The Social Construction of Reality* (1966). Both authors are representatives of the sociology of knowledge and they wrote their work as a treatise in the sociology of knowledge (as its subtitle says). The sociology of knowledge contains derivations of many ideas of previously mentioned and seminal thinkers in sociology, such as Émile Durkheim and Max Weber, as well as Alfred Schutz. Probably this is why Berger and Luckmann succeeded in taking a stance towards social reality, which appeared to be attractive to certain participants of the debate between positivists and postpositivists that broke out shortly after.

Berger and Luckmann (just like positivists) do not get involved in the philosophical debate about the ontological status of (social) reality and confine themselves to the remark that it is sufficient "to define 'reality' as a quality apertaining to phenomena that we recognize as having a being independent of our own volition (we cannot 'wish them away')" (Berger and Luckmann 1966: 13). According to them, society is objective in this sense, since, for good or ill, we really cannot wish it away. On the other hand, society could be *understood* as subjective reality, because our impressions and ideas are socially significant. In this sense, society is a constant process of externalization, objectivation, and internalization (see ibid., 149). In other words, one could say that in our social life we constantly realize ideas, turn them into objective reality, and at the same time become subjectively aware of objective reality.

It is absolutely clear that Berger and Luckmann do not think that the nature of social reality is the same as that of natural reality—for the process of externalization and objectivation does not take place in nature (without human participation). However, this does not imply that social reality cannot be known in the sense in which positivists understand knowledge. Both authors define knowledge as "the certainty that phenomena are real and that they possess specific characteristics" (Berger and Luckmann 1966: 13). Such certainty can be acquired by formulating a hypothesis and testing it.

The researchers adhering to constructivist views, not only within the sociology of knowledge, but in various social sciences, rely on the view, so convincingly presented but by no means invented by Berger and Luckmann, that, first, social reality is constructed, and, second, it can be known in the ordinary sense of the word. To this I could add that social reality is created, but creations can be known no less than the conditions of their creation. Even values and subjective human opinions can be investigated by statistical, and all too positive, methods. On the other hand, social reality is created by entire generations of people, and not by an arrogant radical postmodernist, who believes and tries to convince others that it is merely a text of his creation. (Let's recall that George Berkeley once claimed that reality is nothing but perception.) Any results of the knowledge of social reality may be and ought to be intersubjectively tested. And, of course, they must be relied upon in order to successfully realize the reforms which attempt to implement certain ideals and aims. Otherwise they will remain the visions of a dreamer. If the dreamer calls himself postmodernist, reforms will be met with no less resistance. And reforms always are resisted because any reform faces conflicting interests.

Hardly all social scientists who consider themselves constructivists have read the text of Berger and Luckmann, not to speak of the writings of Vico or Piaget. Some of them declared themselves being constructivists because constructivism is now in fashion. According to a recent study that meets the best positivist standards, approximately 15 percent of researchers of international relations in the United States consider themselves constructivists, although most of them still adhere to (neo)realist or (neo)liberal views (see Peterson 2005: 31).

Why is constructivism becoming a new fashion? Certain constructivists, for instance, Alexander Wendt, whom we've mentioned above, have accomplished works worthy of attention and have persuasively argued for the reconsideration of the issues of anarchy, security, and identity. Constructivism invites investigators to solve new *problems* related with the social construction of reality. They attract the attention of many inquisitive researchers, especially newcomers. Some of them believe that dealing with the new issues requires elaboration of a new constructivist methodology. I am convinced, however, that most or even all of those problems might be inquired into by making use of scientific, positive *methods*.

Positivists never claimed that social reality *is not* constructed. Methodological naturalism is no more than their *methodological* attitude. They never claimed that social reality is not different from natural reality. But the differences between those two realities are not so vast that would not allow both of them to be investigated by the same or similar methods. Constructed social reality can be investigated no less effectively than natural reality. According to Comte, the three stages of the development of society correspond to the three stages of the development of mind precisely because the development of society is determined by the development of the mind. The thinking mind is the creative mind, and among other things, it creates social reality. Berger and Luckmann, using the terms widely accepted in sociology, psychoanalysis, and psychology, could say that Comte clearly understands that our ideas are externalized and objectified. Hence Comte, and, by the way, Mill, have shared the same attitudes as contemporary (moderate) constructivists. The very idea of social construction of the world is, as I've mentioned, even older.

According to Wendt, constructivism demands the investigation of certain previously unexamined problems and to approach them in a certain way, first of all by revealing the interplay of the material and ideal factors of social life. Such investigation must be scientific, and its results must be intersubjectively testable, just as positivists required. Sometimes one may hear that constructivism relies on postpositivist ontology on the one hand, and positivist epistemology and methodology on the other hand, and hence it is a middle ground between positivism and postpositivism. However, for reasons given above, I cannot agree with the claim that, for instance, Wendt's ontology is postpositivist, because whatever meaning is given to the term "ontology," there is no unified positivistic ontology. Thus it is senseless to talk about postpositivist ontology.

We have another important point to make: the admission that the social world is constructed and that therefore social facts must be distinguished from raw natural facts *involves no methodological commitments*. Other fundamental constructivist claims, variously formulated in different social sciences, are also methodologically irrelevant (i.e., they don't require the change in methods of scientific work), although they may be interesting and valuable from the *theoretical point of view*. The second constructivist thesis in the theory of international relations could be given as follows: anarchy is what states make of it. It was formulated by Wendt (1992). Bearing in mind the recognition of the two theses, another well-known researcher of international relations Chris Brown and his younger colleague at the London School of Economics, Kirsten Ainley, who assisted him in the preparation of the fourth edition of his *Understanding International Relations*, write: "in the 1990s a number of essentially empirical IR scholars have proclaimed themselves to be 'constructivist' in so far as they accept the above points, but have not changed their working methods in any significant way, at least not in any way that outsiders can discern" (Brown and Ainley 2009: 49). Neither of them, evidently, is elated with the situation in their field.

Yet it seems that Brown and Ainley do not comprehend that *moderate* constructivists like Wendt, i.e., those who are interested in the scientific, positive investigation of how social reality is construed, did not need to substantially change their methodology. *Radical* constructivists, i.e., those who reject the scientific approach to inquiry, examine the social construc-tion of reality on the basis of methodological principles characteristic of postpositivists, i.e., critical theorists and postmodernists. Thus, from my point of view, constructivism is a *theoretical* research program concerned with the *issues* of the social construction of reality, rather than a new *meth-odological* approach, distinct from a positivist or postpositivist one.

The Positive Mind and Law

The relation of positivism and law is worth separate discussion, for the notion of law is clearly wider than that of legal science or legal theory. It also includes important procedures and norms regulating social, eco-nomic, political, and other human relations. Positivism influenced not only legal theory, but also legal praxis.

Two schools have competed heatedly in law and legal theory from the nineteenth century until the present, one of them being precisely legal positivism. The recent and authoritative *Companion to Philosophy of Law and Legal Theory* informs its readers that "[a]long with natural law the-ory, legal positivism is one of two great traditions in legal philosophy" (Coleman and Leiter 1996: 241). Both traditions have continued until today. Natural law theory, which prevailed in the field until the emergence of legal positivism, is the principal object of criticism of legal positivists.

Traditional natural law theory asserts that not every properly enacted law is just and valid. Above laws enacted by human beings is a higher, natural law which enables the appraisal of enacted law. This law of higher status is contained in sacred scriptures. It may also be discovered by way of inquiry into human nature, by undergoing religious revelation, or by relying on moral intuition. Cicero, who was not the first, but was a very famous proponent of natural law theory, held that "natural law is unchang-ing over time and does not differ in different societies; every person has access to standards of this higher law by use of reason; and only just laws [i.e., in accordance with natural law] 'really deserve [the] name' law" (cf. Bix 1996: 224).

In the middle ages, the best known proponent of natural law theory was Thomas Aquinas. He distinguished four kinds of law, but the most important *in retrospecto* are *natural* law and *positive* law, i.e., the law enacted by human beings. According to Aquinas, natural law is not only above positive law: positive law must be derived from natural. The most conspicuous theorist of natural law in the eighteenth century, William

Blackstone, followed Cicero's and Aquinas's views, although he allowed more freedom for the creators of positive laws, by claiming that they need not be directly derived from natural law. According to him, it is essential that no human laws *contradict* natural law.

Precisely this basic idea of natural law theory was attacked by John Austin, who is commonly regarded as the leading representative of legal positivism in the nineteenth century. By opposing this idea and the theory of natural law in general, he laid emphasis on the fact that courts of justice deliver their sentences on the basis of positive law. If an act is innocuous or even beneficial from the point of view of natural law but strictly prohibited by a positive law, a delinquent will be sentenced according to positive law. The defense's appeals to natural law will not help.

From this observation, it clearly follows that there is no direct link between law and morality. In his *The Province of Jurisprudence Determined*, Austin expressed this idea in a famous phrase: "the existence of law is one thing; its merit or demerit is another" (Austin 1954 [1832]: 184). All legal positivists agree that whatever is (without violation of accepted procedures) properly enacted by a lawmaking agency is law, disregarding its conforming to the natural law and moral norms. In the twentieth century, legal positivists modified this view, and claimed "that what counts as law in a particular society is fundamentally a matter of social convention" (Coleman and Leiter 1996: 241). The latter thesis is often called "the social thesis" and together with the "separability (of law and morality) thesis" it constitutes the crux of (the modern reading of) a doctrine of *legal positivism*.

Although Austin did not consider himself a representative of positivist *philosophy*, his links with it were rather close. Hume's positivist philosophy greatly influenced him and he, in turn, was John Stuart Mill's teacher. Austin's philosophical positivism is best revealed by his conviction that what "actually exists" must be clearly distinguished from what "ought to exist."

The positivist fact/value distinction takes a peculiar form in Austin's philosophy of law and in legal positivism in general. The actual laws exist, they oblige all members of a society, and sanctions are imposed on those members who breach them. Thus laws must be treated as *social facts*. They may be judged against some moral standards as good or bad, they may have merits or demerits, but they exist and are valid irrespective of our moral attitude towards them. Our *values* do not affect their status of binding laws—one must obey them even though divine revelation or human nature impels us to speak against them. The law as it *is* and the law as it *should be* are very different matters. Hence, law and morality must be carefully separated.

One of the most important representatives of legal positivism in the twentieth century is Hans Kelsen. Sometimes he is called simply "the jurist of our century" (Weinberger 1973: ix). There is no doubt that the influence of positivist philosophy on his legal theory is deep and profound. In the years when the Vienna Circle was active, Kelsen was in its midst, he also took active part in The Unity-of-Science Movement (as a member of a major Committee for International Congresses for the Unity of Science formed in 1935). He was a speaker at the fifth Congress at Harvard in 1939, the sixth Congress in Chicago in 1942, and a contributor to the "Library of Unified Science" in which he published his *Retribution and Causality* in 1940. Yet in his main works, *The Pure Theory of Law* (*Reine Rechtslehre*, 1st ed. 1934, 2nd ed. 1960; English transl. 1968) and *General Theory of Law and State* (1945), Mach's influence is felt as well as that of logical positivism. Although the purification of the theory of law is, of course, different from the purification of experience, Kelsen clearly indicated that his concept of the "pure" is derived from Mach's ideas. According to Kelsen, his theory "is called 'pure' because it seeks to preclude from the cognition of positive law all elements foreign thereto" (Kelsen 1957 [1941]: 266). By listing "elements foreign thereto," Kelsen sounds like a member of the Vienna Circle who is interested in a clear separation of law from other disciplines: "The limits of this subject and this cognition must be clearly fixed in two directions: the specific science of law, the discipline called jurisprudence, must be distinguished from the philosophy of justice, on the one hand, and from sociology, or cognition of social reality, on the other" (ibid.).

Kelsen also clearly favors the distinction between fact and values, and declares himself "incompetent to answer either the question whether a given law is just or not, or the more fundamental question of what constitutes justice" (Kelsen 1957 [1941]: 266–67), because this questions cannot be answered in a scientific way. Kelsen's separation of facts and norms is even stricter than Austin's.

The third major representative of legal positivism is Herbert L. A. Hart, whose book *The Concept of Law* (1961; 2nd ed. with an extensive "Post Scriptum" published two years after Hart's death, in 1994) is regarded as one of the most remarkable treatises in twentieth century legal theory. The ways of addressing philosophical problems in legal theory suggested by Hart "quickly gained such widespread acceptance that some legal philosophers take them as uncontroversial principles of legal philosophy" (Bayles 1992: 1). According to many, Hart's works have changed the understanding of the philosophy of law, at least in the English-speaking world.

Hart was most concerned with the question "What is law?" He formulates it as follows "How does law differ from and how is it related to orders backed by threats? How does legal obligation differ from, and how is it related to, moral obligation? What are rules and to what extent is law an affair of rules?" (Hart 1961: 13). All these questions are related to the opposition of legal positivism to natural law theory. We will focus on the second question, since it is of greatest importance from a philosophical point of view.

Hart's stance towards the distinction and opposition of law and morality is less categorical than Austin's or Kelsen's. He emphasizes that next to positive law there is positive morality—the totality of generally accepted moral norms. Both positive law and positive morality constitute social constraints on action, although their role and nature are different. According to traditional natural law theory, to be valid, laws must conform to moral principles. Hart is convinced that there is no logically necessary relation between legal and moral rules. Hart thinks that "reflection on some very obvious generalizations—indeed truisms—concerning human nature and the world in which men live, show that as long as these hold good, there are certain rules of conduct which any social organization must contain if it is to be viable. Such rules do in fact constitute a common element in the law and conventional morality of all societies which have progressed to the point where these are distinguished as different forms of social control" (Hart 1961: 188). To be more precise, such rules, although different in nature, partly coincide. "Such universally recognized principles of conduct which have a basis in elementary truths concerning human beings, their natural environment, and aims, may be considered the *minimum content* of Natural Law" (ibid., 189). These rules of conduct can be derived from human vulnerability, approximate (physical) equality, limited altruism, limited resources, and limited understanding and strength of will. However, Hart is eager to emphasize that specific laws which are incompatible with these fundamental rules are not thereby invalid. In Bayles's words, "[t]he minimum content does not exclude or rule out rules; rather it indicates types of rules that are incorporated. Moreover, the minimum content is not necessary for the existence of a legal system or rule, only for its viability. Thus if law or morality lacked such a necessary rule, for example, had no rule against murder or theft, the legal system might exist for a while, but it would not long endure" (Bayles 1992: 120).

In Hart's interpretation, a legal system is more open than in Kelsen's. Principles of a different kind—not exclusively of a legal nature—are used not only by judges when discharging their duties (especially when facing hard cases) but by ordinary citizens when appraising laws and legal systems in general. Since laws are incurably incomplete, we must decide the borderline cases rationally by reference to social aims. Hart is also eager to emphasize that the aura of authority which the official legal as well as political system may have, is acquired not through violence, but through satisfying certain moral criteria.

Legal positivism exerted an extremely strong influence not only on legal theory but also on legal practice. In fact, this practice, at least in democratic countries, is mainly based on the doctrine of legal positivism which was developed precisely with the aim to make law more transparent and better capable of serving the people. These aims also inspired legal positivists. Their intention when denying the *logically necessary* connection between law and morals was motivated not by distrust of morality, but rather by a wish to avoid anarchy and conservation of the existing legal system. They, especially Austin, were afraid of the undesirable social consequences of the natural law theory. Those who identify law and morals may ignore all laws which are incompatible with their moral views. This leads to anarchy. On the other hand, the lack of separation of law and morals can produce the conviction that all laws are moral and therefore unchangeable. When claiming that to insist on the logically necessary relation of law to morals is a mistake, all legal positivists were inspired by the progressivist spirit of the philosophical positivism: they were convinced that the doctrine of natural law might lead towards the ossification of the legal system.

At the same time it is worth noting that the initial radical opposition between natural law theory and legal positivism has become, in recent years, less acute as both doctrines gradually changed their shape and partially assimilated elements of each other. It was best evidenced by Hart's celebrated debates with two most famous representatives of the natural law school, Lon L. Fuller (in the late 1950s) and Ronald Dworkin (from the 1960s until the 1980s), which has shown that participants of the debates agree on many issues. Signs of convergence are also discernible in less important schools of legal science, such as the sociological and historical schools. However, the attempts to create a so-called integrative jurisprudence, which would combine all classical schools of legal theory (cf. Berman 1988), are not likely to succeed. It is more likely that legal positivism will remain an important and influential participant of the discussion regarding the question "what is law?"

Positivism and Politics

Those who realize that the strict opposition of facts and values is one of the fundamental positivist views might regard the examination of the relationship between positivism and politics as a rather eccentric or even exaggerated pursuit, since the opposition turns politics which is based on value judgments into an area of no interest for positivism. However, as we already know, various positivists had many interesting and sometimes even radical political ideas.

Comte's slogan was "Order and progress," but his progressivism was not an obstacle for writing the *Appeal to Conservatives* (1855), the text supporting Napoleon III, who initiated a *coup d'état* in 1851 and declared himself the emperor a year later. Comte hoped that the emperor would help to implement positivist ideas and views. Like the *philosophes*, he also believed that an enlightened monarch would collaborate in establishing the State of Reason. As a progressivist he was not an advocate of conservative political philosophy, but he certainly was not a liberal. On the other hand, Mill's *On Liberty* is widely considered as a classic treatise on liberalism. And Neurath's political views were undoubtedly Marxist.

The divergence of political opinions and attitudes of various positivists makes one doubt whether positivist methodology implies some definite kind of political philosophy. However, Friedrich August von Hayek, a leading representative of the Austrian school of economics, in his book *The Counter-Revolution of Science* (1952) argues that the founder of classical positivism was a socialist and an opponent of democracy not by coincidence, but precisely *because* of his positivist methodological position. *Methodological collectivism* (or holism) leads, Hayek claims, to *political collectivism*, hence to socialism and finally, to its most radical form—totalitarianism. Hayek's arguments deserve a closer analysis.

Positivists, as is well known, have always emphasized that scientific knowledge must be based on observation and be intersubjective, i.e., be publicly accessible, comprehensible, and testable, both in natural and social sciences. Hayek calls this view "positivist objectivism" and argues that methodological individualism, which calls to start any investigation from the subject's knowledge about himself, about his attitudes and reasons for his actions, is incompatible with such objectivism. Introspection, which forms the basis of methodological individualism, is not regarded by objectivists as a reliable source of scientific knowledge. In Hayek's opinion, positivists believe that only those thoughts, propositions, and theories can be objective, which correspond to the objects *independent* from a subject. This opinion, as I have mentioned, is unfounded, since positivists in principle, due to their antimetaphysical attitudes, avoided talking of objects independent from a subject.

According to Hayek, "[c]losely connected with the objectivism of the scientistic approach is its methodological collectivism, its tendency to treat wholes like society or the economy, capitalism […] or class or country as definitely given objects about which we can discover laws by observing their behavior as wholes" (Hayek 1952: 93). The basic error of such collectivist view is, according to Hayek, "that it mistakes for facts what are no more than provisional theories, models constructed by the popular mind to explain the connections between some of the individual phenomena which we observe" (ibid., 95). Facts must be given to us but neither society, nor its behavior is given. And Hayek has no doubt that we cannot observe what is not given, thus he takes methodological collectivists to be victims of the "fallacy of conceptual realism," who treat referents of concepts such as humanity, society, nation, or class as real entities although they are no more than theoretical constructs of individuals. He thinks that social science could achieve much more if it would rely on methodological individualism.

Nonetheless, according to Hayek, one of the most important defects of holist, collectivist methodology is that it inevitably leads to political collectivism: "though, logically, methodological collectivism and political collectivism are distinct, it is not difficult to see how the former leads to the latter and how, indeed, without methodological collectivism political collectivism would be deprived of its intellectual basis" (Hayek 1952: 162). In his opinion, socialists would have no convincing grounds for claiming that social goals are best achieved through "central" power, if they would not believe that human mind is able to directly apprehend "society" or "humanity" and their goals.

Most of the flaws of Comte's political collectivism were exposed already by Mill, and Hayek basically follows him. Two questions, however, loom large. Firstly, does methodological collectivism *directly* lead to political collectivism? Does, say, present-day methodological collectivism—the prevalent approach in the social sciences—constitute a threat to freedom and democracy? Secondly, is in fact methodological collectivism a characteristic feature of every positivist methodology as Hayek seems to suggest?

First, it must be noted that Hayek draws, without reservation, an equal sign between Comte's philosophy and positivism, thus failing to take into account the development of positivism before and after Comte. The groundless equating of Comte with the rest of positivism explains why

Hayek thinks that methodological collectivism and positivist objectivism leads to political collectivism and historicism. The majority of contemporary social scientists treat society as a unitary object of research, irreducible to separate individuals or their qualities. However, only a few of them think that there are any clearly defined historical stages of social development. And even those who accept such a view usually don't think that social processes should fall under total control of political institutions. The mere fact that this was Comte's view does not prove that there is a direct link between these ideas.

Hayek's claim that a positivist thinking leads to political collectivism is also refuted by the fact that far from every positivist is a methodological collectivist. Neither Hume, nor Mach, Mill, and Schlick were adherents of methodological collectivism. On the contrary, according to them the starting point of the social inquiry is the individual and his consciousness, not the social whole.

One more point here is essential. In *The Counter-Revolution of Science* Hayek initially claimed that social sciences "are concerned with man's actions, and their aim is to explain the unintended or undesigned results of the actions of many men" (Hayek 1952: 41). However, when examining approaches and methods characteristic of social sciences, he speaks of "the social sciences in the narrower sense, that is those which used to be described as the moral sciences [and which] are concerned with man's conscious or reflected action, actions where a person can be said to choose between various courses open to him" (ibid., 42–43).

It is doubtful, however, that contemporary economics (especially macroeconomics), sociology, or political science are moral sciences in Hayek's sense. These sciences deal mainly or at least quite often with those of men's actions which are not the result of rational choice. And even when human actions are conscious and reflected, an inquirer hardly has a right to claim as Hayek does that "the concepts and views held by individuals [...] are directly known to us" (Hayek 1952: 65). An inquirer may only *assume* that he knows those concepts and views or they are the same as his own. But this assumption can hardly be justified, especially when investigating alien cultures. In such a situation an inquirer cannot grasp them from inside; the only thing he may do is to observe the behavior of the persons from outside. He does not know directly their "concepts and views," he reconstructs them from their actions.

We must draw the conclusion that Hayek's attempt to demonstrate that Comte's political views are a direct consequence of positivist methodological orientation failed. He described positivist methodology in a rather distorted way. History shows that positivist methodology and epistemology is compatible with *various* political philosophies. It would be surprising if that was not the case. From the positivist point of view, we must rely on scientific knowledge if we wish ours social and political reforms to succeed, but the direction of those reforms are not imposed by science itself. In fact, representatives of classical positivism have contributed to spreading the ideals of altruism, of devotion to the progress of society, of elevating the good of mankind over the individual good, and other noble ideas. The general public of the nineteenth century (including, as we will soon see, quite a number of artists) associated such ideas with positivism. However, even the founders of classical positivism Comte and Mill disagreed about the nature of positivist political system.

Positivist ideas served as a basis for many political doctrines and various political movements. Positivist ideas concerning the reorganization of society, the necessity of advancing education and science and, especially, changing the character of the ruling elite suited relatively underdeveloped countries, because they facilitated their modernization. That explains why the credo of Comtean positivism[1]—*Order and Progress*—is emblazoned on the Brazilian flag. The ideas of social positivism had a considerable influence on the political and social development of Mexico and other states of Latin America (see Williamson 1992). They also had many supporters in Russia, Poland, and Lithuania.

In Western Europe positivist ideas of evolution and gradual, non-revolutionary social progress were adopted by such moderate social movements as the Fabian Society lead by Sidney and Beatrice Webb and George Bernard Shaw. The very idea of a welfare state, so popular in Europe nowadays, goes back to Turgot and Condorcet and also is of positivist origin. The idea of technocracy, that the world must be ruled by scientists, managers, and engineers, was most convincingly defended by Comte.

Based on the distinction between fact and values, positivism could not have a united political philosophy, for the latter is hardly conceivable without values. Sometimes it is claimed that by denying that value statements have any cognitive content, logical positivism caused the decline of political philosophy, which lasted for the greater part of the twentieth century and ended only in the 1970s with the beginning of the vehement discussions regarding the ideas proposed by John Rawls in his *A Theory of Justice* (1971).

---

[1] Auguste Comte's complete motto of positivism is as follows: love as a principle and order as the basis; progress as the goal.

I do not think that this accusation is justified, although if it was justified, that would testify even greater influence of positivism than I assume. The claim is not justified, because various representatives of the early and classical positivism had *their own* political philosophy. Some logical positivists also had their own, although not too elaborate, political ideas. Yet quite often they are also blamed for separating political science, whose propositions have cognitive content, from political philosophy, whose propositions lack such content, and thus contributing to the entrenchment of the purely instrumental understanding of politics, as a mere *technique* of acquisition, preservation, and execution of power. Treating politics this way, power is understood as an end in itself, and not as means of implementing certain values, such as justice, liberty, and equality.

The question whether those values are specifically (purely) political, or they are of a different, say moral, nature, is really worth attention (see, e.g., Jokubaitis 2008). However, I have my doubts about the thesis, defended by some authors, including the aforementioned one, that political science which does not posit values and does not rely on them, is a source of demoralization. One who adheres to such a view, should probably recognize that humans are also demoralized by psychology, especially neuropsychology (which today is not easily separated from neurophysiology), since, instead of instilling values, it explains, without even relying on values, the regularities of human behavior and is capable of revealing that not only mental states, but also characteristics and traits of behavior that often are evaluated in moral terms, such as faithfulness, depend on what chemical substances and in what quantities are produced by a particular human organism.

A more thorough consideration of those issues would take us far astray from the main theme of this book. In order not to go too far, I will confine myself to the observation that a purely instrumental construal of politics, understanding it as a mere technique of manipulation on humans, is much older than positivism, hence positivism cannot be held responsible for the proliferation of this view. Some politicians have treated politics like that from the earliest times, and this view was theoretically pondered by Niccolò Machiavelli in the beginning of the sixteenth century. However, none among the positivists we have discussed shared this view. They separated values from facts, but it absolutely does not mean that they failed to understand the significance of values, or had no values at all.

How could a positivist respond to the accusation that he, intentionally or not, turns politics into a technique which requires no values? First, he would probably note that no political values could be implemented without intersubjective knowledge of peculiarities and regularities of political life. In order to achieve any goals, including political, we have to rely on knowledge—political knowledge in this case. Of course, this knowledge can be acquired in various ways, and also through practice, but political theory is a more reliable ground for political activity. A politician who has *no* political knowledge will achieve *no* political goals; hence he will be a politician only in his own imagination. Politics is not a declaration, but an implementation of values.

On the other hand, a positivist would probably not miss an opportunity to notice that politicians usually treat power as *means* to achieve or acquire certain values. Wealth could be such a value, but the fact that some politicians are much more concerned with their *personal* wealth rather than material welfare of the *society*, is a different issue, reminding us of Joseph de Maistre's pronouncement that every nation has the government it deserves.

Of course, power may be an end in itself rather than a means for achieving other goals. But in itself it is also a *value* (just one among many). Moreover, it would most difficult to dispute its status as a purely political (rather than, for instance, moral) value.

Positivism's Impact upon Literature, the Visual Arts, and Architecture

The attempts to trace the impact of existentialism or *lebensphilosophie* on literature and arts raise no surprise. Most existentialist ideas reached the wide public mainly through literary fiction, and some existentialist philosophers—Jean-Paul Sartre or Albert Camus—were famous writers themselves. However, at least at the first sight, positivism seems quite distant from arts. Existentialists rejected almost everything positivists affirmed: they sought to separate philosophy from science, and they considered the methods, language, and the very form or style of science as absolutely inappropriate for philosophy. Looking for alternative means of expression, they often opted for fiction. The subjects they were interested in, such as the individual's concern with his own life, were especially suitable to belles lettres. Positivists, on the contrary, took science to be an example for philosophy and clearly separated scientific philosophy from literature. Nevertheless, positivism exerted considerable influence upon the development of literature as well as other arts.

Many philosophers influence the creative process in the arts through their aesthetic conceptions and their philosophy of art. Although positivist aesthetics lacks elaboration and systematization, at least in nascent form it exists. Some questions of aesthetics and art are discussed both in Comte's *Course* and in his later works. Comte saw art as a mediator between the intellect and feelings, and a factor playing an important role in motivating action. He was always interested in art as a tool to be used for the implementation of positivist ideas. As Arline Reilein Standley writes in her book on Comte, he was convinced that "art helps to integrate a personality; and, in making possible communication through image and sign, art thereby makes possible civilization itself" (Standley 1981: 128). In the *System* Comte emphasizes that the main function of art is the same as that of science: both serve social progress. Thus the best art form for this illustrious purpose, according to Comte, is poetry because it most strongly works on the feelings and by cultivating our sense of perfection best motivates action.

Comte's ideas, for example, that the history of art conforms to his law of the three stages, made impact on thinking about art in the second half of the nineteenth century. Hippolyte Taine's works on the philosophy of art show clear signs of Comte's influence. However, when speaking about Comte's influence upon literature and arts, it is important to emphasize that he exerted this influence mainly not as an author of original *aesthetic* ideas but as a philosopher of science and sociologist. A few elements of his thought were of special importance for writers and artists: his positivist empiricism, collectivism, altruism, and his vision of the future. It seems that Comte exerted more influence on literature, not on fine arts.

Of all literary movements of the nineteenth century, Comte's name most often is associated, of course, with naturalism and its leader Émile Zola. An important mediator between them was, it seems, Claude Bernard. The literary theory of Naturalism bears much similarity to positivist philosophy as it is presented in Comte's *Course*. "Both focus on the scientific methods of observation and experiment, and both stress the need for objectivity. Both deal with the relationship between the individual and the milieu; and furthermore, both regard the future as predictable (though perhaps modifiable), given the orderly progression of past and present" (Standley 1981: 144).

If Zola and the naturalists were impressed more by positivist methodology and tried to apply it with minor modifications to literature, George Bernard Shaw was more affected by positivist social vision. Yet, like naturalists, he also had no doubt that artist's main role is to serve mankind's progress by giving concreteness to ideas and motivating human actions.

Among other Victorian writers whose novels and essays bear traces of Comte's positivism, it is necessary to mention at least George Eliot (born Mary Ann Evans), who regarded herself a positivist. In her novels of the 1860s and 1870s, "emphasis is always on moral and social issues: on the need actively to cultivate altruism over egoism; [...] on the need for submission and discipline; on the concept of society as a living entity. [...] Comte's all-encompassed unity—the oneness of world, society, and self is reflected in George Eliot's work as well" (Standley 1981: 149–50).

This "all-encompassed unity" was something which Mill did not like. It seems that he himself did not influence Victorian literature as much as Comte. Yet it is worth mentioning Mill in the context of examining the issue of positivism and literature for at least one reason. His *Autobiography* is one of the select few of the famous philosophical autobiographies in the world and might be mentioned in the same line as the autobiographies of Augustine or Rousseau. There is no doubt that Mill has used the form of autobiography which belongs rather to literature than to philosophy with a similar intention as Shaw used his plays: to popularize some philosophical ideas and to teach some way of life.

It is not difficult to notice positivism's impact on the work of certain Russian writers of the nineteenth century, e.g., Nikolai Chernyshevsky's novel *What Is to Be Done?* (1863). Chernyshevsky can be called positivist without hesitation. And the most famous positivist literary character not only in Russia but in entire world probably is Yevgeny Bazarov, whom Ivan Turgenev in his *Fathers and Sons* (1862) introduces as a nihilist. Perhaps positivism had an even greater influence upon Polish literature. Its entire epoch from the January Uprising of 1863 until the beginning of the twentieth century is usually referred to as positivistic. After the uprising was crushed by Russians and the hope to regain statehood was lost, and after the ideals of romanticism deteriorated, special importance in the region (partly in Lithuania as well) was bestowed upon the ideas of organic work and work at the grassroots that already began to unfold earlier. Their adherents urged to develop domestic economy, education, and culture in spite of political distress, and to spare no efforts for the consolidation of the nation. The economic and political development was supposed to become the foundation on the basis of which the statehood could more easily be regained in the future, under more favorable circumstances. The conditions for expansion of positivist ideas in Poland were rather auspicious. And those ideas were spreading, also through literature, for quite a few Polish writers took it to be the most suitable means for dissemination of positivist views and attitudes. The most famous representatives of the positivist epoch in Polish literature are Bolesław Prus and Henryk Sienkiewicz, who in 1905 was awarded one of the first Nobel Prizes in literature. Positivism has also influenced the development of Lithuanian literature.

Not only classical positivism, but also empirio-criticism had a significant impact on the development of literature. Mach's influence on literature and the arts in the Central Europe at the dawn of the twentieth century and the following decades was comparable to that of Comte on literature in France and Britain in Victorian times. At the turn of the twentieth century, Vienna was a cultural center of world importance exceeding in many respects London and New York. Many outstanding new movements in psychology, philosophy, logic, politics, law, economics, and music started here: from psychoanalysis to the second Vienna School in music. The very line of names of intellectuals and artists of the highest rank acting in Vienna at that time sounds *today* when Vienna is a nice but rather unpretentious city almost incredible: Sigmund Freud, Alfred Adler, Kurt Gödel, Hans Kelsen, Ludwig von Mises, Friedrich von Hayek, Robert Musil, Ludwig Wittgenstein, Arnold Schönberg, Alban Berg.

Mach could be another name included in this list, although he spent most of his productive years in Prague, coming to Vienna only at the end of his academic career. It was mentioned already in Chapter 3 that Mach had been widely read by various people, including members of literary and artistic circles. Specifically two of his theses were revelations for these circles, although the second was much talked about in *philosophical* circles well before Mach. These theses were as follows: (1) the boundary lines between the physical and the mental world are only conventional, and (2) there is no stable "I": what we call "I" is only a bundle of sensations which changes permanently. And if we look for a single sentence of Mach which might be regarded as the quintessence of Mach's philosophy which impressed most men of letters and artists, the best candidate is, in all probability, his famous "The ego must be given up" ("*Das Ich is unrettbar*") (Mach 1996 [1886]: 24). Mach had in mind that the *ego* does not exist as self-sufficient entity.

In fact it was not a new idea, as from a purely philosophical standpoint this was only repeating what Hume had said much earlier, but the fine philosophical-psychological analysis on which Mach grounded and illustrated this thesis made a much greater impression on writers and painters than the simple and much more dry examination of the question of the self made by Hume. Mach wrote his philosophical text in simple and comprehensible language. He illustrates his claim that the strict opposition between subject and object is untenable and that it is always possible to analyze a so-called *objective* course of events in terms of *subjective* sensations by giving in *Knowledge and Error* an example of how the "objective" activity of going to university may be presented in such a way that the elements of the outer and inner worlds come together and interplay by producing plenty of associations. This fragment inspired many writers:

> 2. How various things I discover in myself, for example, on the way to a lecture: my legs move, one step triggers off the next, without my making any special effort, even if I have to turn an obstacle. I walk past the municipal park and recognize the town hall which reminds me of Gothic and Moorish buildings as well of the mediaeval spirit that pervades its chambers. In the hope of conditions more fit to be called civilized I am just imagining the future when in crossing the road I am grazed by a speeding cyclist and made involuntarily to jump sideways. My fantasies of the future are replaced by a silent grumbling against such reckless speedsters. The sight of the ramp of the university entrance finally reminds me again of my object, namely the task of the hour ahead, and speeds my pace.
>
> 3. Let us analyze this mental experience into its constituents. First we find those that are called *sensations* insofar as they depend on our bodies [...] but are *physical* qualities insofar as they depend on other physical features (presence of the sun, tangible bodies and so on): the green of the park, the greyness and shapes of the town hall, the resistance of the ground I tread, the grazing contact with the cyclist and so on. For the psychological analysis consider the term *sensation*. Toward sensations such as hot, cold, light, dark, vivid colors, the smell of ammonia or the scent of roses and the like, we are as a rule not indifferent. They are agreeable or disagreeable, that is to say our body reacts by means of more or less intense movements of approach or retreat, which in turn present themselves to introspection as complexes of sensations. [...] An otherwise indifferent look at the bottle that contained ammonia evokes the memory of smell and so ceases to be indifferent. The whole prior experience of sensations, insofar as it is preserved in memory, tinges every new experience. The town hall I passed would be no more than a spatial arrangement of colored patches, had I not previously seen many other buildings, walked their corridors and mounted their steps. Memories of the most varied sensations here become interwoven with the optical sensations to become a much richer complex, namely a perception, from which the more immediate sensation is difficult to detach as such. (Mach 1976 [1905]: 15–16)

This erasing of the lines dividing objective and subjective, physical and mental, fact and dream, appearance and reality, sensation and memory became in fact a feature which allows us to identify modern, twentieth-century literature by separating it from the older, especially nineteenth-century naturalist and realist literature. It is worth noting that in this respect the character of Mach's influence differs significantly, even dramatically from that of Comte. The latter was the representative of *physicalistic positivism* and thus was one of the most important sources of inspiration for naturalism which sought to describe *objective* facts as exactly as possible. Meanwhile the influence of Mach, the adherent of *phenomenalistic positivism*, was evidently antinaturalistic. Of course, he also invites writers to pay closest attention to facts, but the facts Mach has in mind have quite a different nature that those in which Comte was interested. The naturalist wouldn't call facts what Mach considers as facts. Yet, according to Mach, the facts cannot be blamed for that. He claims that "the question which is often asked, whether the world is real or whether we merely dream it, is devoid of all scientific meaning. Even the wildest dream is a fact as much as any other" (Mach 1996 [1886]: 11).

In order to demonstrate that the strict distinction between objective and subjective facts is unfounded, and to convince us that his conception of neutral elements is correct, Mach invites us to lie upon a sofa and look around the room with one eye closed. We will see the ridge of our nose, our body from chest to feet and its environment, but we will not see our head. Monocular vision will strengthen the impression, that whatever we see are not seemingly objective things, but mere elements of our field of vision.

In his *Analysis of Sensations* Mach also included a drawing, depicting a field of vision of a man lying on a sofa. One enormous mustache and the line of his nose and eyebrow forming a kind of an arch through which one can see a part of his body and a part of the room with a window. That is probably one of the most famous drawings ever produced by a philosopher. It visually deobjectifies what is seen: we perceive the elements of the drawing as visual sensations and not as objective things.

It is possible to see a greater or lesser, direct or indirect influence of Mach in the works of most classics of modernist literature of the twentieth century, be it James Joyce, Marcel Proust, Franz Kafka, William Faulkner, or Jorge Luis Borges. Joyce—probably the most famous writer of the twentieth century—is credited for being a founder of the stream of consciousness narrative mode. His *Finnegan's Wake* is a perfect loop of the stream of consciousness—its last sentence is the first half of. It is nothing less than a literary representation of Mach's ideas: the line between subject and object, between sensation and reality, is erased, and the reader is thrown into a surreal or half-surreal world of associations and dreams.

Mach's influence upon literature is clearly seen in Robert Musil's *The Man without Qualities* (1930–43)—one of the most remarkable masterpieces of twentieth-century literature. Musil studied engineering at the Technical University (Technische Hochschule) in Brünn (in the Czech Republic to which the city belongs now, it is called Brno), but later began studies in philosophy, logic, and experimental psychology at Berlin University. In 1908 he earned a Ph.D. in philosophy conferred to him for the dissertation *Contributions towards the Evaluation of Mach's Theories*. Yet Musil rejected the offer to pursue an academic career and decided to become a writer. His novel *The Man without Qualities* brought him international success, although his other works such as *The Confusions of Young Törless* or *Three Women* received very favorable reviews as well.

In his dissertation Musil does not try to conceal his sympathy for positivist philosophy. Moreover, he clearly states that Mach's works are full of fruitful stimuli, and not only philosophical. Later Musil writes in his diary: "It is not from Goethe, Hebbel, Hölderlin that we will learn, but from Mach, Lorentz, Einstein, Minkowski, Couturat, Russell, Peano" (Musil 1976 [1936]: 1251). It seems that he considered Mach to be his greatest teacher (see Diersch 1990: 38).

The title of Musil's main work refers to Mach's thesis that "the *ego* is not a definite, unalterable, sharply-bounded unity" (Mach 1996 [1886]: 24). Mach urged to give up the ego, and the protagonist of Musils book has accomplished that. Ulrich is the man without qualities, which have a solid foundation; he is the man who has given up his *ego*. Musil destroys the *egos* of the characters in his book by depersonalizing them, turning them into the streams of various sensations of all kinds without any underlying real substance. Values lose their absolute character as well, their systems are destroyed and the fragmented makeup of the novel unambiguously reminds of the splitting and collapsing world of the Austro-Hungarian Empire on the eve of the World War I (the novel takes place in 1913).

Almost all the features of Musil's literature, indicating his close connections with Mach, are also applicable, in one form or another, to the other famous writers of the twentieth century that we have mentioned above, especially James Joyce and Franz Kafka. These writers have shown that the traditional distinction between the objective and subjective worlds is not quite clear, as well as the difference between a dream and reality, which Kafka so ingeniously erased in his *Process*. Also in his *Castle*, K. is depersonalized even to a greater degree than Musil's Ulrich—he does not even have a name and is not able to see the faces of those whom he wants to meet. It would not be a great exaggeration to say that Mach belongs to a tiny group who, being nonwriters themselves, exerted an exceptional influence upon the development of literature of the twentieth century.

Mach also made a fundamental impact on the visual arts. In his main philosophical work, i.e., *The Analysis of Sensations*, he analyzed the space-sensations of the eye, the relations among various visual sensations,

the optical similarity as contrasted to geometrical similarity, the physiological significance of the direction of a given straight line or curve-
element, symmetry and asymmetry, repetition and regularity as a manifold
symmetry, the depth perception and other problems on the shared border-
lines of physics, physiology, and psychology (cf. Mahr 1988). These
analyses have directly influenced many artists. To everybody who is at
least a little interested in the development of visual arts in the twentieth
century, reading Mach's *Analysis of Sensations* leaves an impression that
many famous artists and authors of artistic manifestos have taken their
inspiration from this book. For instance, Kazimir Malevich's seminal abstract painting *Black Square* (1913) and his other squares correspond
rather precisely to what Mach said on the aesthetic significance of hori-
zontal and vertical lines.

Mach's fundamental philosophical idea concerning the desubstantiali-
zation of the world had, it seems, an even more universal impact on the
arts. In fact, all abstractionism is, to a degree, heir to Mach's rejection of a
strict opposition between the physical and the mental. Furthermore, many
of the graphic figures presented in the *Analysis of Sensations* (there are
more than thirty) served as a *direct* source of inspiration for many modern
artists, first of all for Paul Klee (see Mahr 1988: 406). It is possible to
trace Mach's influence on the oeuvre of Wassily Kandinsky and his sense
perception infused theories of color and line.

Unlike Mach, who was interested in various arts, logical positivists
were concerned primarily with one art—architecture, which started to play
an extremely important role in the twentieth century with its sweeping
urbanization (while Lenin was convinced that the most important art is
film, so logical positivists held the view that the most important art is
architecture). Progress in architecture had to be, according to them, a very
important aspect of social progress in which they were keenly interested.
Such also was the view of the many classical positivists. The successful
project of renovation of Paris carried out in the second half of the nineteenth century that brought fame to Georges-Eugène Haussmann is probably the most impressive objectification of the ideas of Comte and other
positivists about the bright new life. The guides of this city usually leave
out that a significant part of medieval Paris was destroyed while constructing wide avenues and boulevards.

In Austria during the era of rapid industrialization which began relatively late, there was an acute shortage, especially in Vienna, of housing,
particularly of housing for workers. Hence it is not surprising that the
most socially and politically active member of the Vienna Circle—Otto

Neurath—vigorously took part in planning of urban development also and that the only sentence in the manifesto of logical positivists, *The Scientific Conception of the World: The Vienna Circle*, in which any art is mentioned sounds as follows: "We witness the spirit of scientific world conception penetrating in growing measure the forms of personal and public life, in education, upbringing, architecture, and the shaping of economic and social life according to rational principles" (Hahn, Neurath, and Carnap 1973 [1929]: 317–18).

An architect, Josef Frank, the brother of Vienna Circle member Philipp Frank, delivered in 1929 a lecture in the series organized by the Ernst Mach Society (which, let's recall, was founded to popularize the ideas of Vienna Circle) in which he discussed the links between the scientific world conception and modern architecture. In this lecture Frank mostly spoke about the influence of the Vienna Circle on the leading school of architecture in Europe, namely Bauhaus. Bauhaus—Hochschule für Bau und Gestaltung—was founded in 1919 in Weimar. It aspired to combine crafts and the fine arts, but had no clear artistic profile for a while. In 1925 the school was moved to Dessau southwest of Berlin, and then became the most important school of industrial architecture in Europe. The teaching staff's "espousal of everything technical and scientific became ever more pronounced; art would act like science and serve as an initiator in the cycle of industrial production" (Galison 1990: 715).

Logical positivists regarded Bauhaus as an ally in the battle with the past. The most significant event of the past was the First World War. The legacy and wake of destruction it left in the material and spiritual sphere was overwhelming and this drive to overcome its legacy was much broader than a specific philosophical movement. The word *Aufbau* (reconstruction) was very popular in Germany and Austria in the twenties; scores of journals appeared with the word in their titles. Translating *Aufbau* as "rebuilding" overemphasizes the physical aspects of the construction. "On new, and for the first time firm foundations, they would erect a political, philosophical, and aesthetic world separate from everything that had come before" (Galison 1990: 715). Thus new life forms had to emerge from the ruins of loss and war. It seems that Carnap's choice for his book title *Logische Aufbau der Welt* (Logical reconstruction of the world) which appeared in 1928 was in the same mood of rebuilding. His friend Neurath was convinced that artists were leading the battle for spiritual liberation from the past and that in this battle Bauhaus's cultural role is decisive. Yet simultaneously Neurath criticized Bauhaus for being too preoccupied with style, instead of practical function (cf. ibid., 716).

Logical positivists opted to liberate Bauhaus from past "metaphysical" influences. The primary form of their organized action was lectures read at Bauhaus. The main ally of logical positivists in Bauhaus was architect Hannes Meyer, one of its leading figures. Like Neurath, he was convinced that aesthetics must be subordinated to engineering. In 1929 Meyer invited Herbert Feigl, a member of the Vienna Circle, to give lectures. At Bauhaus Feigl got to know Wassily Kandinsky, Paul Klee, and other artists who made Bauhaus world famous. A little later Carnap wrote to Neurath that Feigl's lectures provoked the interest of the Bauhaus staff in positivist ideas. The same year Carnap was also invited to lecture at Bauhaus. He emphasized that science and visible forms are two different sides of a single life and are drawn by the same impulses. Neurath lectured at Bauhaus as well.

There is enough evidence to claim that the ideas of the Vienna Circle had an impact on the Bauhaus style, which, in turn, influenced (and still influences) the development of modern architecture, as well as fine arts and design. Its characteristic constructions were more and more reminiscent of the Viennese logical constructions. From the late twenties, the Bauhaus style was becoming more geometrical, rational, constructive, and functional. It was a rather coldly clinical style, and it became known as "the new objectivity" (*neue Sachlichkeit*). The fact that logical positivists furthered its promotion is reflected by the circumstance that, as Feigl reports, Neurath and Carnap quite often referred to logical positivism as "an expression of the *neue Sachlichkeit*" (Feigl 1981 [1969]: 637).

Thus we may conclude that positivism played an important part in shaping the direction of the evolution of the arts. The character of this influence was different in different epochs and different fields of artistic creation. Yet, it seems, positivism generally succeeded in persuading many leading writers and artists that scientific thinking not only does not hinder artistic creation but, on the contrary, furthers artistic workmanship and proficiency. Quite a few artists could not resist such a lure.

The Positive Mind in Everyday Life. Positivism and Religion

Are we positivists? This question may seem odd or even naive, especially if we take it to be asking whether people (at least in Western civilization) adhere to principles of positivist philosophy. No theoretical philosophy in history was supported by the majority of the population.

Of course, some totalitarian systems, especially communism, required that people *show* support for a certain philosophy. Every educated person in the former Soviet Union knew that whenever officially asked, he or she must answer that, without doubt, he or she is a champion of Marxism. Such answer, however, was seldom sincere, and not because the person thought hermeneutics or phenomenological philosophy is preferable to Marxism, but because the person did not even know—in spite of all efforts to indoctrinate the entire population—what Marxism as a philosophy is really about.

Unlike in totalitarian societies, in democracy one cannot even expect an almost universal, even purely declarative, endorsement of a certain philosophy. A pluralistic society does not have at its disposal an enormous machinery of political indoctrination and different notions compete in the market of ideas, thus it is naive to hope that some particular philosophy will make a conquest of entire society, especially a philosophy like positivism. Nevertheless, in this section I will attempt to show that although the vast majority of people in Lithuania or Japan or the United States know nothing about Hume, Comte, Mach, or Carnap, patterns of our thinking are strongly influenced by positivism. Even without knowing what positivism is, we usually behave as people of positive minds.

Positive thinking and behavior is natural for a person living in a technological civilization whose main moving force was and remains science. Most of our environment, especially in urban areas, not to mention the tools and procedures we use in our work places, is a *reification and manifestation of the positive mind*, the positive mind materialized. Skyscrapers, cars, computers, mobile phones, and other devices and technologies are products of applying positive, i.e., scientific, exact, reliable, and useful knowledge. They are the embodiment of the positive, not of the metaphysical mind. Our appreciation of those things, the recognition that we need them, and our ability to use them presupposes that we are able to think positively. On the other hand, permanent contact with them strengthens our positive thinking and our orientation towards the technical (in the broadest sense) solution of various questions.

Most problems people face in the present-day materialistic and commercialized world require positive, not metaphysical or theological, thinking. Mundane problems clearly replace—at least during workdays—religious or metaphysical ones (reflections on faith, soul or being are reserved for the church on Sundays). A statistical person most often acts like a positive, practical individual. The general spirit of the present day is predominantly positive.

It is possible to point out some facts which may be interpreted as signs of the diminishing of cultural importance of positive thinking. Perhaps the most important among those is the strong criticism of science. Science has been a genuine cult object at the end of the nineteenth and the very beginning of the twentieth centuries, but today it is blamed for the worst calamities of the last and present centuries, and especially for both world wars because the mass scale of killing was made possible by wide use of scientific discoveries. Such accusations are unjustified. Every discovery, technology, or tool may be used for different purposes and political leaders or state officials rather than scientists are responsible for how these discoveries and technologies are used. It is true that in the eyes of the general public, science has lost its status as the savior, the solver of the world's problems. Some environmentalists would claim that it creates more problems than it solves. However, it is must be stated clearly that environmental problems are caused by economic development, industry, consumerism, and energy needs, rather than by science itself. And those problems cannot be solved without the help of ecologists and environmental engineers, i.e., without scientists and researchers. It is also worth noting that at least in the twentieth century positivists did not praise science as much as their predecessors used to, and they didn't think that science alone could solve all the problems humanity confronts. They were more sober, calm, and rational than their forebears of the nineteenth century. It is possible to call this attitude more positive as well, since, when talking about the positive mind in the context of everyday life and attempting to give its outline, one must undoubtedly mention that such a mind not only consciously or unconsciously follows the principles of positivist philosophy, but it also is *precise, clear, and keen*. Such a mind is rather skeptical and it does not credulously trust yet another remedy from each and every illness of humanity.

Hume was most certainly a skeptic. In this respect he is followed by d'Alembert, who claimed that the seventeenth century was excessively affirmative and that the eighteenth century was excessively repudiative. Logical positivists, who claimed that all our knowledge of the world, i.e., knowledge with factual content, is merely hypothetical, can hardly be considered dogmatic. They were notably self-critical and were repeatedly reconsidering their conceptions. One must admit that Comte was too enthusiastic about science and partly because of that at the end of the nineteenth century and the very beginning of the twentieth century a belief that science, like a messiah, will save humanity became quite popular. The educated minority got the impression that science defeated religion.

Hence, when some authors today see the (alleged) revival of religion, they think that science has lost its battle with religion and that Comte's law of the three stages must be modified: after the theological, metaphysical, and positive stages comes once more the theological stage. Some even claim that science itself is moving from the rational Athens towards the irrational Jerusalem. From my point of view that is an illusion.

First of all, according to many positivists, science and traditional religion are not opposed. Granted, Hume was skeptical about most dogmas of traditional religions and theological propositions. He considered arguments for the existence of God to be invalid. But do those arguments matter for the majority of believers? I want to say that Hume's skepticism about religion, which was also characteristic of many later positivists, was by far not equal to atheism. Atheists deny the existence of God, while skeptics merely doubt it. Moreover, Hume did not take the very idea of God to be meaningless. On the contrary, drawing on his theory of the origin of ideas from sensation and feelings, which forces one to abandon any metaphysics that has no connection with sensations, in his *Enquiry Concerning Human Understanding* Hume clearly says that the idea of God is connected with the source of all complex ideas—our sensations and feelings.

In fact, Comte was convinced that scientific truths are incompatible with religious truths, since, it seems, he failed to take into account that the grounds of those two sorts of truths are essentially different, and he did not entertain the possibility, which is nowadays often exploited by theologians, to interpret certain statements of the Bible rather metaphorically, as addressed to the people of those times with their knowledge of those times. However, by claiming that science is not compatible with *some* of the traditional religious dogmas, e.g., the creation of the world in seven days, Comte, more than any other positivist, emphasized the important role of both traditional religion in the theological society and his New Religion of Humanity in the forthcoming positive society.

The positivists of the twentieth century, like most famous scientists of our epoch, clearly understood that the questions raised by science and religion are quite distinct in character. Although the early Carnap argued that questions about God lack cognitive content, it could always be replied that they belong to the sphere of faith, and not to the sphere of science. The later Carnap did not even claim what he used to claim earlier—he only emphasized that that theological statements are not verifiable empirically, by means of external observation and experiment. But, after all, contemporary theologians do not say that their statements are testable on the basis of empirical methods or empirical data. They usually adopt the attitude that is worth emphasizing: *religious faith and scientific knowledge are distinct realms.* The propagation of this view was contributed not only by the latter Roman Catholic popes (Galileo's rehabilitation is probably the best example) and contemporary theologians, but also modern positivists. Their announced distinction between science and religion is a direct corollary of the distinction between facts and values. Science is the realm of facts and scientific knowledge, and religion is the realm of values and faith.

Hence it is absolutely natural that positivists, especially modern positivists, regarded religion with more respect than metaphysics. Religion talks about truth, but it is understood as the truth of Faith, and not the truth of Reason. The truth of Faith is revealed to human beings through acts of faith. Therefore a scientist who adopts positivistic attitudes cannot *qua* scientist say anything about it. At the same time, positivists are obviously inclined to reject the claims of metaphysicians, that their statements are truths of Reason.

Both science and religion talk not only about truth but also about experience. Yet again this word is treated differently. Experience in science means observation and experiment. Whereas religion talks about a different sort of experience: revelation is accepted by faith, not by reason. If these two notions of experience are kept separate, the conflict of science and religion is *logically* impossible, although quite a few such conflicts took place in *history*. But they had risen just because their participants, from the contemporary point of view, failed to grasp the differences in character and nature between the truths of Faith and truths of Reason. Debate raged in the Middle Ages (as well as later) regarding the question whether the truths of Faith can be derived from the truths of Reason, or vice versa, whether truths of Faith serve as the basis for the truths of Reason. However, from the positivist point of view, these disputes lacked much sense.

Therefore *the positive mind is not an antireligious mind.* Hence a religious person can think positively while thinking about earthly matters. However, religious (theological) and positive thinking are different modes of thought. Thus we should briefly discuss the social importance of positive and religious thinking.

Although one may sometimes hear about the revival of religion, we still find no convincing evidence that the role of religion in the world now is any greater than it used to be two hundred years ago. Neither scientists nor science-oriented philosophers claim that science can save humanity and solve all its problems. Critics of science claim that since scientists renounce their pretensions to give final answers to each and every question, it seems reasonable that this void must be filled by faith. Yet anyone who thinks positively understands that the questions that are formulated in scientific language and that cannot be answered today will with all probability be answered in the future. At the same time, they realize that not all questions may be formulated in scientific language and not all questions are scientific in character. Not only the question "What is God?" but also questions "What is Good?," "What is Justice?," or "What is Freedom?" are not questions of science, if science is understood in the way suggested by positivists. Religion may propose its own answers to them, but they can also be dealt with in philosophy. In fact, philosophy is dedicated to analyzing such fundamental but nonscientific issues.

Returning from the problem of the relation between positivism and religion to other aspects of the influence of positivism (and science) upon our everyday thinking and lifestyle, it must be said that the general public's interest in (and even more—the usage of) new technological achievements is not diminishing and, many observers would say, is growing exponentially. Probably nothing has made such a huge impact on everyday life in recent decades as the new information technologies. And whatever skeptics say about contemporary disillusionment in science, there is little doubt that the age of information (or knowledge) is just another form of the *age of science*. Nowadays most critics of science (including latter-day Luddites and some radical sectarian clerics) use the Internet, notwithstanding the inconsistency of such behavior. Moreover, on the Internet one easily finds many more indications that science and scientific education remains an important social value: it is sufficient to take a look at the sharply rising numbers of universities and students around the world. The exploration of outer space attracts less attention than it used to, but in the focus of public interest it is replaced by other areas of research, based on the attainments of molecular biology, such as genetic engineering and cloning. Robotics and nanotechnology also impress many people.

Of course, it is possible to retort that what is most fascinating about cloning or human stem cell research lies not in their purely scientific or technological aspects, but in the related ethical problems, and these problems have nothing to do with positive thinking. Yet it seems that when discussing ethical problems related to cloning—publicly or privately— most people accept without reservation the most fundamental principle of the positive mind: the dichotomy of facts and values. They are discussing values, the desirability to do or not to do certain things, but they don't dispute the scientific discoveries or the *technical possibility* of their application. At the same time they do not think that the scientific findings—present or future—will finally help to solve ethical problems. Thus they are thinking positively.

Belief in progress, prevailing in contemporary society, although withering at the time of crises, is also a legacy of positivist philosophy. Positivists were convinced that progress is possible in every area of human life and were persuaded that it is necessary to strive for it. This idea of progress inherited from positivist philosophy forms the ideological basis of most political and other reform programs. Today few would doubt that the development of education and science is the foundation of all social progress. Hence the positive spirit is still alive and it would be impossible to eradicate it fully without destroying the very foundations of current Western (and probably even global) civilization, although, of course, nothing prevents us from considering whether it could have been initially laid on a more solid or a more sustainable foundation.

References

Achinstein, Peter and Stephan F. Barker, eds. 1969. *The Legacy of Logical Positivism.* Baltimore, MD: Johns Hopkins University Press.

Adorno, Theodor W., Hans Albert et al. 1976 [1969]. *The Positivist Dispute in German Sociology.* London: Heinemann.

Austin, John. 1954 [1832]. *The Province of Jurisprudence Determined.* New York: Noonday Press.

Ayer, Alfred J. 1952 [1936]. *Language, Truth, and Logic.* New York: Dover.

Ayer, Alfred J. 1966 [1956]. *The Problem of Knowledge.* Harmondsworth: Penguin.

Ayer, Alfred J., ed. 1959. *Logical Positivism.* New York: The Free Press.

Baker, William J. 1992. "Positivism versus People: What Should Psychology Be About?" In *Positivism in Sociology,* ed. C. W. Tolman, 9–16. New York: Springer.

Baudrillard, J. 1988 [1976]. "Symbolic Exchange and Death." In *Selected Writings,* ed. M. Poster, 119–48. Stanford, CA: Stanford University Press.

Bayles, Michael D. 1992. *Hart's Legal Philosophy.* Dordrecht: Kluwer Academic Publishers.

Berger, Peter L. and Thomas Luckmann. 1966. *The Social Construction of Reality: A Treatise in the Sociology of Knowledge.* New York: Doubleday.

Berman, Harold J. 1988. "Toward an Integrative Jurisprudence: Politics, Morality, History." *California Law Review* 76 (4): 779–801.

Bix, Brian. 1996. "Natural Law Theory." In *A Companion to Philosophy of Law and Legal Theory,* ed. D. Patterson, 223–40. Cambridge, MA: Blackwell.

Blackburn, Simon. 1996 [1994]. *The Oxford Dictionary of Philosophy.* Oxford: Oxford University Press.

Blackmore, John T. 1972. *Ernst Mach: His Work, Life, and Influence.* Berkeley, CA: University of California Press.

Blaug, Mark. 1992 [1980]. *The Methodology of Economics, or How Economists Explain.* 2nd ed. Cambridge: Cambridge University Press.

Blumberg, Albert and Herbert Feigl. 1930. "Logical Positivism: A New Movement in European Philosophy." *Journal of Philosophy* 28: 281–82, 285–87, 292–93.

Blumenberg, Hans. 1982. *The Legitimacy of the Modern Age.* Cambridge, MA: MIT Press.

Borradori, Giovanna. 1994 [1991]. *The American Philosopher: Conversations with Quine, Davidson, Putnam, Nozick, Danto, Rorty, Cavell, MacIntyre, and Kuhn.* Chicago, IL: University of Chicago Press.

Brown, Chris and Kirsten Ainley. 2009. *Understanding International Relations.* 4th ed. New York: Palgrave Macmillan.

Burry, John Bagnell. 1960 [1932]. *The Idea of Progress*. New York: Dover Publications.

Caldwell, Bruce J. 1994 [1982]. *Beyond Positivism: Economic Methodology in the Twentieth Century*. London: Routledge.

Carnap, Rudolf. 1935. *Philosophy and Logical Syntax*. London: Kegan Paul.

Carnap, Rudolf. 1937 [1934]. *The Logical Syntax of Language*. New York: Harcourt & London: Kegan Paul.

Carnap, Rudolf. 1939. *Foundations of Logic and Mathematics* (*International Encyclopedia of Unified Science*, vol. 1, no. 3). Chicago, IL: University of Chicago Press.

Carnap, Rudolf. 1945. "The Two Concepts of Probability." *Philosophy and Phenomenological Research* 5: 513–32.

Carnap, Rudolf. 1949 [1936]. "Truth and Confirmation." In *Readings in Philosophical Analysis*, ed. H. Feigl and W. Sellars, 119–27. New York: Appleton-Century Crofts.

Carnap, Rudolf. 1950 [1936/37]. *Testability and Meaning*. New Haven, CT: Graduate Philosophy Club, Yale University.

Carnap, Rudolf. 1951 [1950]. *Logical Foundations of Probability*. London: Routledge and Kegan Paul.

Carnap, Rudolf. 1952. *Continuum of Inductive Methods*. Chicago, IL: University of Chicago Press.

Carnap, Rudolf. 1953 [1935]. "Formal and Factual Science." In *Readings in Philosophical Analysis*, ed. H. Feigl and W. Sellars, 123–28. New York: Appleton-Century Crofts.

Carnap, Rudolf. 1956. "The Methodological Character of Theoretical Concepts." In *Minnesota Studies in the Philosophy of Science*, vol. 1, ed. H. Feigl and M. Scriven, 38–76. Minneapolis, MN: University of Minnesota Press.

Carnap, Rudolf. 1956 [1947]. *Meaning and Necessity*. 2nd ed. Chicago, IL: University of Chicago Press.

Carnap, Rudolf. 1956 [1950]. "Empiricism, Semantics, and Ontology." In *Meaning and Necessity*, 2nd ed., 205–21. Chicago, IL: University of Chicago Press.

Carnap, Rudolf. 1956 [1952]. "Meaning Postulates." In *Meaning and Necessity*, 2nd ed., 222–29. Chicago, IL: University of Chicago Press.

Carnap, Rudolf. 1959 [1932]. "The Elimination of Metaphysics through Logical Analysis of Language." In *Logical Positivism*, ed. A. J. Ayer, 60–81. New York: The Free Press.

Carnap, Rudolf. 1959 [1932/33]. "Psychology in Physical Language." In *Logical Positivism*, ed. A. J. Ayer, 165–98. New York: The Free Press.

Carnap, Rudolf. 1963a. "Intellectual Autobiography." In *The Philosophy of Rudolf Carnap*, ed. P. A. Schilpp, 3–84. La Salle, IL: Open Court.

Carnap, Rudolf. 1963b. "Replies and Systematic Expositions." In *The Philosophy of Rudolf Carnap*, ed. P. A. Schilpp, 859–1016. La Salle, IL: Open Court.

Carnap, Rudolf. 1967 [1961]. "Preface to the Second Edition." In *The Logical Structure of the World & Pseudoproblems in Philosophy*, v–xi. Berkeley and Los Angeles, CA: University of California Press.

Carnap, Rudolf. 1968. "Inductive Logic and Inductive Intuition." In *The Problem of Inductive Logic*, ed. I. Lakatos, 258–67. Amsterdam: North-Holland.

Carnap, Rudolf. 1971. "A Basic System of Inductive Logic, Part I." In *Studies in Inductive Logic and Probability*, vol. 1, ed. R. Carnap and R. Jeffrey, 33–165. Berkeley, CA: University of California Press.

Carnap, Rudolf. 1975 [1958]. "Observation Language and Theoretical Language." In *Rudolf Carnap, Logical Empiricist*, ed. J. Hintikka. Dordrecht: Reidel.

Carnap, Rudolf. 1990. "Quine on Analyticity." In *Dear Carnap, Dear Van: The Quine–Carnap Correspondence and Related Works*, ed. R. Creath, 427–32. Berkeley, CA: University of California Press.

Carnap, Rudolf. 1995 [1966]. *An Introduction to the Philosophy of Science*. Ed. M. Gardner. Dover.

Carnap, Rudolf. 2003. [1928]. *The Logical Structure of the World*. Trans. R. George. Chicago: Open Court Classics.

Carnap, Rudolf. 2011 [1934]. *The Unity of Science*. New York: Routledge.

Cartwright, Nancy, Jordi Cat, Lola Fleck and Thomas Uebel. 1996. *Otto Neurath: Philosophy between Science and Politics*. Cambridge: Cambridge University Press.

Carus, A. W. 2007. *Carnap and Twentieth-Century Thought: Explication as Enlightenment*. New York: Cambridge University Press.

Chalmers, Alan F. 1999. *What Is This Thing Called Science?* 3rd ed. Brisbane: University of Queensland Press.

Cirera, Ramon. 1994. *Carnap and the Vienna Circle: Empiricism and Logical Syntax*. Amsterdam: Rodopi.

Cohen, Robert. S. 1963. "Dialectical Materialism and Carnap's Logical Empiricism." In *The Philosophy of Rudolf Carnap*, ed. P. A. Schilpp. La Salle, IL: Open Court.

Coleman, Jules L. and Brian Leiter. 1996. "Legal Positivism." In *A Companion to Philosophy of Law and Legal Theory*, ed. D. Patterson, 241–60. Cambridge, MA: Blackwell.

Comte, Auguste. 1975 [1822]. "Plan of the Scientific Operations Necessary for Reorganizing Society." In *Auguste Comte and Positivism: The Essential Writings*, ed. G. Lenzer, 9–67. New York: Harper Torchbooks.

Comte, Auguste. 1975 [1830/42]. "The Course of the Positive Philosophy." In *Auguste Comte and Positivism: The Essential Writings*, ed. G. Lenzer, 71–306. New York: Harper Torchbooks.

Comte, Auguste. 1975 [1851/54]. "The System of Positive Policy." In *Auguste Comte and Positivism: The Essential Writings*, ed. G. Lenzer, 309–458. New York: Harper Torchbooks.

Comte, Auguste. 2009 [1830/42]. *The Positive Philosophy of Auguste Comte*. Cambridge: Cambridge University Press.

Comte, Auguste. 2009 [1844]. *A General View of Positivism*. Cambridge: Cambridge University Press.

Condorcet, Marquis de. 1965 [1795]. "Sketch for a Historical Picture of the Progress of the Human Mind." In *The Idea of Progress since the Renaissance*, ed. W. W. Wagar. New York: Wiley.

Creath, Richard and Michael Friedman, eds. 2008. *The Cambridge Companion to Carnap*. New York: Cambridge University Press.

Daddow, Oliver. 2009. *International Relations Theory*. Los Angeles, CA: Sage.

D'Alembert, Jean le Rond. 1963 [1751]. *Preliminary Discourse to the Encyclopedia of Diderot*. Indianapolis, IN: Bobbs-Merrill.

Dahms, Hans-Joachim. 1994. *Positivismusstreit. Die Auseinandersetzungen der Frankfurter Schule mit dem logischen Positivismus, dem amerikanischen Pragmatismus und dem kritischen Rationalismus*. Frankfurt am Main: Suhrkamp.

Derrida, Jacques. 1978 [1967]. *Of Gramatology*. Baltimore, MD: Johns Hopkins University Press.

Dewey, John. 1938. *Logic: The Theory of Inquiry*. New York: Holt.

Dewey, John. 1969 [1939a]. "Unity of Science as a Social Problem." In *Foundations of the Unity of Science: Toward an International Encyclopedia of Unified Science*, ed. O. Neurath et al., 29–38. Chicago, IL: University of Chicago Press.

Dewey, John. 1969 [1939b]. "Theory of Valuation." In *Foundations of the Unity of Science: Toward an International Encyclopedia of Unified Science*, ed. O. Neurath et al., 1–84. Chicago, IL: University of Chicago Press.

Dewey, John. 1984 [1925]. "The Development of American Pragmatism." In *The Latter Works, 1925–1953. Vol. 2: 1925–1927*, 3–21. Carbondale, IL: Southern Illinois University Press.

Dewey, John. 1984 [1929]. *The Latter Works, 1925–1953. Vol. 4: 1929. The Quest for Certainty*. Carbondale, IL: Southern Illinois University Press.

Diersch, Manfred 1990. "Draussen, drinnen und Ich. Ernst Machs Spiegel der Erkenntniss als Anregung für österreichische Erzählkunst des 20. Jahrhunderts." In *Genauigkeit und Seele. Zur österreichischen Literatur seit dem Fin de siècle*, ed. J. Strutz and E. Kiss, 29–42. München: Wilhelm Fink Verlag.

Dilthey, Wilhelm 1989 [1883]. "Introduction to Human Sciences." In *Selected Works*, vol. 1, ed. R. A. Makreel and F. Rodi. Princeton, NJ: Princeton University Press.

Docherty, Thomas. 1993. "Postmodernism: An Introduction." In *Postmodernism: A Reader*, ed. T. Docherty, 1–31. New York: Columbia University Press.

Duhem, Pierre. 1954 [1906]. *The Aim and Structure of Physical Theory*. Princeton, NJ: Princeton University Press.

Earman, John. 1993. "Carnap, Kuhn, and the Philosophy of Scientific Methodology." In *World Changes*, ed. P. Horwich, 9–36. Cambridge, MA: MIT Press.

Feigl, Herbert. 1981 [1969]. "The Wiener Kreis in America." In *The Intellectual Migration: Europe and America, 1930–1960*, ed. D. Fleming, 630–73. Cambridge, MA: Belknap Press.

Feigl, Herbert. 1982 [1937/38]. "Moritz Schlick, a Memoir." In *Rationality and Science: A Memorial Volume for Moritz Schlick in Celebration of the Centennial of His Birth*, ed. Eugene T. Gadol, 55–82. Wien: Springer.

Fisch, Max H. 1964. "Was There a Metaphysical Club in Cambridge?" In *Studies in the Philosophy of Charles Sanders Peirce*, 2nd series, ed. E. C. Moore and R. S. Robin. Amherst, MA: University of Massachusetts Press.

Foucault, Michel. 1980. *Power/Knowledge*. Ed. C. Gordon. New York: Pantheon Books.

Foucault, Michel. 1984. *Foucault Reader: An Introduction to Foucault's Thought*. Ed. P. Rabinov. New York: Pantheon Books.

Frank, Philipp. 1907. "Kausalgesetz und Erfahrung." *Ostwald's Annalen der Naturphilosophie* 6: 445–50.

Frank, Philipp. 1949. *Modern Science and Its Philosophy*. Cambridge, MA: Harvard University Press.

Fraser, Nancy and Linda Nicholson. 1993 [1988]. "Social Criticism without Philosophy: An Encounter between Feminism and Postmodernism." In *Postmodernism: A Reader*, ed. T. Docherty, 415–32. New York: Columbia University Press.

Friedman, Michael. 1991. "The Re-evaluation of Logical Positivism." *Journal of Philosophy* 88: 505–23.

Friedman, Michael. 1992. "Philosophy and the Exact Sciences: Logical Positivism as a Case Study." In *Inference, Explanation, and Other Frustrations: Essays in the Philosophy of Science*, ed. J. Earman, 84–98. Los Angeles, CA: University of California Press.

Friedman, Michael. 1993. "Remarks on the History of Science and the History of Philosophy." In *World Changes*, ed. P. Horwich, 37–54. Cambridge, MA: MIT Press.

Friedman, Michael. 1999. *Reconsidering Logical Positivism*. Cambridge: Cambridge University Press.

Friedman, Milton. 1953. "The Methodology of Positive Economics." In *Essays in Positive Economics*, 3–43. Chicago, IL: University of Chicago Press.

Friedman, Milton. 1982 [1962]. *Capitalism and Freedom*. Chicago, IL: University of Chicago Press.

Galileo Galilei. 1890. *Opere*, vol. 4. Firenze.

Galison, Peter. 1990. "Aufbau/Bauhaus: Logical Positivism and Architectural Modernism." *Critical Inquiry* 16: 709–52.

Gellner, Ernest. 1975. "The Last Pragmatist or Behaviourist Platonist." *The Times Literary Supplement*, July 25, 843–53.

Georgen, Pedro. 1975. *Der Positivismus Auguste Comtes und seine Wirkung in Brasilien*. Unpublished doctoral dissertation. Ludwig-Maximillians-Universität, Munich.

Giddens, Anthony, ed. 1974. *Positivism and Sociology*. London: Heinemann.

Giere, Ronald N. and Alan W. Richardson, eds. 1996. *Origins of Logical Empiricism*. Minneapolis, MN: University of Minnesota Press.

Ginsburg, E. 1932. "On the Logical Positivism of the Viennese Circle." *Journal of Philosophy* 29: 121–29.

Glasersfeld, Ernst von. 1989. "An Exposition of Constructivism: Why Some Like It Radical." http://www.oikos.org/constructivism.htm (accessed January 7, 2015).

Golding, Martin P. 1975. *Philosophy of Law*. Englewood Cliffs, NJ: Prentice-Hall.

Goodman, Nelson. 1977 [1951]. *The Structure of Appearance*. Boston: Reidel.

Gouhier, Henri. 1965 [1933/41]. *La jeunesse d'Auguste Comte et le formation du positivisme*. 3 vols. Paris: J. Vrin.

Grice, Paul and Peter F. Strawson. 1956. "In Defense of a Dogma." *Philosophical Review* 65: 141–58.

Habermas, Jürgen. 1987 [1968]. *Knowledge and Human Interests*. Cambridge: Polity Press.

Habermas, Jürgen. 1994 [1985]. *The Philosophical Discourse of Modernity: Twelve Lectures*. Cambridge: Polity Press.

Hahn, Hans, Otto Neurath and Rudolf Carnap. 1973 [1929]. "Wissenschaftliche Weltauffassung: Der Wiener Kreis." In *Empiricism and Sociology*, ed. O. Neurath, 299–318. Dordrecht: Reidel.

Halfpenny, Peter 1992 [1982]. *Positivism and Sociology: Explaining Social Life*. Aldershot: Gregg Revivals.

Halfpenny, Peter and Peter McMylor. 1994. *Positivist Sociology and Its Critics*. 3 vols. Aldershot: Edward Elgar.

Hall, M. H. 1967. "An Interview with 'Mr. Behaviorist' B. F. Skinner." *Psychology Today* 1: 21–23, 68–71.

Haller, Rudolf. 1988. *Questions on Wittgenstein*. London: Routledge.

Haller, Rudolf. 1993. *Neopositivismus*. Darmstadt: Wissenschaftliche Buchgesellschaft.

Hanfling, Oswald. 1981. *Logical Positivism*. Oxford: Blackwell.

Hardcastle, Gary L. and Alan Richardson, eds. 2003. *Logical Empiricism in North America*. Minneapolis, MN: University of Minnesota Press.

Hart, Herbert L. A. 1961. *The Concept of Law*. Oxford: Clarendon Press.

Hart, Herbert L. A. 1977 [1958]. "Positivism and the Separation of Law and Morals." In *The Philosophy of Law*, ed. R. M. Dworkin, 17–37. Oxford: Oxford University Press.

Hart, Nibbrig. 1987. *Spiegelschrift*. Frankfurt am Main: Suhrkamp.

Hassan, J. 1993 [1987]. "Towards a Concept of Postmodernism." In *A Postmodern Reader*, ed. J. Natoli and L. Hutcheon, 273–86. Albany, NY: State University of New York Press.

Hayek, Friedrich August von. 1944. *The Road to Serfdom*. Chicago, IL: University of Chicago Press.

Hayek, Friedrich August von. 1952. *The Counter-Revolution in Science: Studies on the Abuse of Reason*. Indianapolis, IN: Liberty Press.

Heidegger, Martin. 1991 [1961]. *Nietzsche*. Vols. 1–2. San Francisco, CA: Harper.

Heidegger, Martin. 2000 [1953]. *Introduction to Metaphysics*. New Haven, CT: Yale University Press.

Heidegger, Martin. 2008 [1929]. "What is Metaphysics?" In *Basic Writings*, 89–110. New York: Harper Collins.

Heidegger, Martin. 2008 [1947]. "Letter on Humanism." In *Basic Writings*, 213–66. New York: Harper Collins.

Hempel, Carl G. 1959 [1950]. "The Empiricist Criterion of Meaning." In *Logical Positivism*, ed. A. J. Ayer, 108–29. New York: The Free Press.

Hempel, Carl G. 1966. *Philosophy of Natural Science*. Englewood Cliffs, NJ: Prentice-Hall.

Hintikka, Jakko. 1991. "Overcoming 'Overcoming Metaphysics through Logical Analysis of Language' through Logical Analysis of Language." *Dialectica* 45: 203–18.

Hintikka, Jakko. 1996. "Ludwig's Apple Tree: On the Philosophical Relations between Wittgenstein and the Vienna Circle." In *Ludwig Wittgenstein: Half-Truths and One-and-a-Half Truths*, 125–44. Dordrecht: Kluwer.

Hodges, Michael P. 1990. *Transcendence and Wittgenstein's Tractatus*. Philadelphia, PA: Temple University Press.

Hollis, Martin and Steve Smith. 1990. *Explaining and Understanding International Relations*. Oxford: Oxford University Press.

Holton, Gerard. 1993. "From the Vienna Circle to Harvard Square: The Americanization of a European World Conception." In *Scientific Philosophy: Origins and Developments*, ed. F. Stadler, 47–73. Dordrecht: Kluwer.

Horkheimer, Max. 1985 [1931]. "Die gegenwärtige Lage der Socialphilosophie und die Aufgaben eines Instituts für Socialforschung." In *Gesammelten Schriften*, 20–35. Bd. 3. Frankfurt am Main: Fischer.

Horkheimer, Max. 1985 [1932]. "Hegel und das Problem der Metaphysik." In *Gesammelten Schriften*, 295–308. Bd. 2. Frankfurt am Main: Fischer.

Horkheimer, Max. 1991 [1937]. "Die neueste Attacke gegen die Metaphysik." In *Critical Theory: Selected Essays*, ed. M. O'Connel, 132–87. New York: Continuum Press.

Horkheimer, Max and Theodor W. Adorno. 2002 [1947]. *Dialectics of Enlightenment*. Stanford, CA: Stanford University Press.

Hume, David. 1953. *David Hume's Political Essays*. Ed. C. W. Hendel. New York: Liberal Arts Press.

Hume, David. 1966 [1751]. "An Enquiry Concerning the Principles of Morals." In *Enquiries Concerning the Human Understanding and Concerning the Principles of Morals*, ed. L. A. Selby-Bigge. Oxford: Clarendon Press.

Hume, David. 1967 [1739/40]. *A Treatise of Human Nature*. Ed. L. A. Selby-Bigge. Oxford: Clarendon Press.

Hume, David. 1980 [1776]. "Hume's My Own Life." In *The Life of David Hume*, ed. E. S. Mossner. Oxford: Clarendon Press.

Hume, David. 1981 [1748]. "An Enquiry Concerning Human Understanding." In *Enquiries Concerning the Human Understanding and Concerning the Principles of Morals*, ed. L. A. Selby-Bigge. Oxford: Clarendon Press.

Hutchison, Terrence W. 1960 [1938]. *The Significance and Basic Postulates of Economic Theory*. New York: Augustus M. Kelley.

Irzik, Gürol and Teo Grünberg. 1995. "Carnap and Kuhn: Arch Enemies or Close Allies?" *British Journal for the Philosophy of Science* 46: 285–307.

Jacob, Pierre. 1980. *De Vienne a Cambridge. L'heritage du positivisme logique de 1950 a nos jours*. Paris: Galimard.

James, Henry, ed. 1920. *The Letters of William James*, 2 vols. Boston, MA: Atlantic Monthly Press.

James, William. 1978 [1909]. "The Meaning of Truth." In *Pragmatism, and The Meaning of Truth*. Cambridge, MA: Harvard University Press.

James, William. 1979 [1897]. "The Will to Believe." In *The Will to Believe and Other Essays in Popular Philosophy*. Cambridge, MA: Harvard University Press.

James, William. 1995 [1907]. *Pragmatism*. New York: Dover Publications.

Joergensen, Joergen. 1951. *The Development of Logical Empiricism*. Chicago, IL: University of Chicago Press.

Jokubaitis, Alvydas. 2008. *Politika be vertybių* [Politics without values]. Vilnius: Vilniaus universiteto leidykla.

Kamitz, Reinhardt. 1973. *Positivismus: Befreiung vom Dogma*. München: Müller.

Kelsen, Hans. 1957 [1941]. "The Pure Theory of Law and Analytical Jurisprudence." In *What Is Justice? Justice, Law, and Politics in the Mirror of Science*, 266–87. Berkeley, CA: University of California Press.

Kelsen, Hans. 1968 [1934]. *The Pure Theory of Law*. Berkeley, CA: University of California Press.

Keynes, John Maynard. 2010 [1921]. *A Treatise on Probability*. Kingston: FQ Legacy Books.

Keynes, John Neville. 1891. *The Scope and Method of Political Economy*. London: Macmillan.

Kołakowski, Leszek. 1972 [1966]. *Positivist Philosophy: From Hume to the Vienna Circle*. Harmondsworth: Penguin Books.

Kołakowski, Leszek. 2005 [1976]. *Main Currents of Marxism: The Founders, the Golden Age, the Breakdown*. New York: W. W. Norton.

Kraft, Viktor. 1953 [1950]. *The Vienna Circle: The Origin of Neo-Positivism*. New York: Philosophical Library.

Kuhn, Thomas. 1970 [1962]. *The Structure of Scientific Revolutions*. Chicago, IL: University of Chicago Press.

Kuhn, Thomas. 1977. *The Essential Tension*. Chicago: University of Chicago Press.

Kuklick, Bruce. 1977. *The Rise of American Philosophy: Cambridge; Massachusetts, 1860–1930.* New Haven, CT: Yale University Press.

Kyburg, Henry Ely, Jr. 1970. *Probability and Inductive Logic.* London: Macmillan.

Laclau, Ernest. 1993 [1988]. "Politics and the Limits of Modernity." In *Postmodernism: A Reader*, ed. T. Docherty, 329–43. New York: Columbia University Press.

Lakatos, Imre. 1968. "Changes in the Problem of Inductive Logic." In *The Problem of Inductive Logic.* Amsterdam: North-Holland.

Leahey, Thomas H. 1980. *History of Psychology: Main Currents in Psychological Thought.* Englewood Cliffs, NJ: Prentice-Hall.

Lenin, Vladimir Ilich. 1977 [1909]. *Materialism and Empirio-criticism.* Moscow: Progress Publishers.

Łukasiewicz, Jan. 1913. *Die Logischen Grundlagen der Wahscheinlichkeitsrechung.* Krakow: Akad. der Wiss.

Lyotard, Jean-François. 1993 [1979]. "The Postmodern Condition: A Report on Knowledge." In *A Postmodern Reader*, ed. J. Natoli and L. Hutcheon. Albany, NY: State University of New York Press.

MacCormick, Neil. 1981. *H. L. A. Hart.* Stanford, CA: Stanford University Press.

Mach, Ernst. 1895. *Popular Scientific Lectures.* Chicago, IL: Open Court.

Mach, Ernst. 1942 [1883]. *The Science of Mechanics.* La Salle, IL: Open Court.

Mach, Ernst. 1976 [1905]. *Knowledge and Error: Sketches on the Psychology of Inquiry.* Dordrecht: Reidel.

Mach, Ernst. 1996 [1886]. *The Analysis of Sensations.* London: Routledge/Thoemmes Press.

Mahr, Peter. 1988. "Ernst Mach, Gestaltwahrnehmung, Minimal Art." In *Ernst Mach— Werk und Wirkung*, ed. R. Haller und F. Stadler, 404–31. Wien: Hölder-Pichler-Tempsky.

Marcuse, Herbert. 1941. *Reason and Revolution.* London: Oxford University Press.

Marx, Karl. 1961 [1844]. *Economic and Philosophic Manuscripts of 1844.* Moscow: Foreign Languages Publishing House.

Marx, Karl. 1977 [1873]. "Postface to the Second Edition." In *Capital. Vol. 1: A Critique of Political Economy*, 94–103. New York: Vintage Books.

Marx, Karl. 1987 [1866]. "Letter to F. Engels, 7 July 1866." In K. Marx and F. Engels, *Collected Works*, vol. 42. Moscow: Progress Publishers.

Marx, Karl. 1998 [1888]. "Theses on Feuerbach." In K. Marx and F. Engels, *The German Ideology*, 569–71. Amherst, NY: Prometheus Books.

McDonald, Marvin J. and D. Vaden House. 1992. "Positivist Influence on 'Environment-Behavior' Studies: Egon Brunswick and Contemporary Social Ecology." In *Positivism in Sociology*, ed. C. W. Tolman, 155–84. New York: Springer.

Menand, Louis. 2001. *The Metaphysical Club: A Story of Ideas in America.* New York: Farrar, Straus, and Giroux.

Menger, Karl. 1982. "Memories of Moritz Schlick." In *Rationality and Science: A Memorial Volume for Moritz Schlick in Celebration of the Centenial of His Birth*, ed. Eugene T. Gadol, 83–103. Wien & New York: Springer-Verlag.

Mill, John Stuart. 1949 [1843]. *A System of Logic, Ratiocinative and Inductive: Being a Connected View of the Principles of Evidence and the Methods of Scientific Investigation.* London: Longman.

Mill, John Stuart. 1961 [1873]. "Autobiography." In *Essential Works of John Stuart Mill*, ed. M. Lerner. New York: Bantam Books.

Mill, John Stuart. 1972 [1859]. "On Liberty." In *Utilitarianism, On Liberty, Considerations on Representative Government*, ed. H. B. Acton, 69–185. London: J. M. Dent.

Mill, John Stuart. 1972 [1861]. "Utilitarianism." In *Utilitarianism, On Liberty, Considerations on Representative Government*, ed. H. B. Acton, 1–67. London: J. M. Dent.

Mill, John Stuart. 1991. *On Liberty and Other Essays*. Oxford: Oxford University Press.

Mill, John Stuart. 1993 [1865]. *Auguste Comte and Positivism*. Bristol: Thoemmes Press.

von Mises, Richard. 1968 [1939]. *Positivism*. New York: Dover.

Monk, Ray. 1991 [1990]. *Ludwig Wittgenstein: The Duty of Genius*. London: Vintage.

Morris, Charles Richard. 1952 [1931]. *Locke. Berkeley. Hume*. London: Oxford University Press.

Morris, Charles William. 1937. *Logical Positivism, Pragmatism, and Scientific Empiricism*. Paris: Hermann.

Mossner, Ernest Campbell. 1980 [1954]. *The Life of David Hume*. Oxford: Clarendon Press.

Musil, Robert. 1976 [1936]. "Nachlass zu Lebzeiten." In *Tagebücher*, ed. A. Frise. Bd. 2. Reinbek bei Hamburg: Rowohlt.

Musil, Robert. 1978 [1938]. "Curriculum Vitae." In *Gesammelte Werke in neun Bänden*, ed. A. Frise, 949–51. Bd. 7. Reinbek bei Hamburg: Rowohlt.

Musil, Robert. 1980 [1908]. *Beitrag zur Beurteilung der Lehren Machs und Studien zur Technik und Psychotechnik*. Reinbek bei Hamburg: Rowohlt.

Nagel, Ernest. 1936. "Impressions and Appraisals of Analytic Philosophy in Europe." *Journal of Philosophy* 33: 6–7, 9–10, 29–30.

Nekrašas, Evaldas. 1979. *Loginis empirizmas ir mokslo metodologija* [Logical empiricism and methodology of science]. Vilnius: Mintis.

Nekrašas, Evaldas. 1987. *Verojatnostnoe znanije. Stanovlenije i razvitije logiko-empiristićeskoj programmy verojatnostnoj ocenki naučnogo znanija* [Probable knowledge: The rise and development of the program for probabilistic appraisal of scientific knowledge in logical empiricism]. Vilnius: Mintis. (Polish translation: *Wiedza prawdopodobna. Powstanie i rozwój w empiryzmie logicznym programu probabilistycznej oceny wiedzy naukowej*. Warszawa: Wydawnictwo Naukowe PWN, 1992.)

Nekrašas, Evaldas. 2012. *Filosofijos įvadas* [Introduction to philosophy]. 3rd revised ed. Vilnius: Mokslo ir enciklopedijų leidybos centras.

Neurath, Otto. 1959 [1932/33]. "Protocol Sentences." In *Logical Positivism*, ed. A. J. Ayer, 199–208. New York: The Free Press.

Neurath, Otto. 1973 [1928]. "Personal Life and Class Struggle." In *Empiricism and Sociology*, ed. M. Neurath and R. S. Cohen, 249–98. Dordrecht & Boston: Reidel.

Neurath, Otto. 1973 [1931]. "Empirical Sociology: The Scientific Content of History and Political Economy." In *Empiricism and Sociology*, ed. M. Neurath and R. S. Cohen, 319–421. Dordrecht: Reidel.

Nevo, I. 1995. "Richard Rorty's Romantic Pragmatism." In *Pragmatism: From Progressivism to Postmodernism*, ed. R. Hollinger and D. Depew, 284–97. Westport, CO: Praeger.

Nietzsche, Friedrich. 1911 [1881]. *The Dawn of the Day*. New York: Macmillan.

Nietzsche, Friedrich. 1917 [1886]. *Beyond Good and Evil*. New York: Modern Library Publishers.

Nietzsche, Friedrich. 1968 [1886/87]. *The Will to Power*. Ed. F. Kaufmann. New York: Vintage Books.

Nietzsche, Friedrich. 1988 [1886/87]. "Nachgelassene Fragmente: Ende 1886–Frühjar 1887." In *Sämtliche Werke. Kritische Studienausgabe in 15 Einzelnbänden*, ed. G. Colli and M. Montinari. Bd. 12. München: Deutscher Taschenbuch Verlag; Berlin: De Gruyter.

Nietzsche, Friedrich. 2008 [1887]. *The Gay Science*. Ed. B. Williams. Cambridge: Cambridge University Press.

Oberdan, Thomas. 1990. "Positivism and the Pragmatic Theory of Observation." In *PSA 1990*, ed. A. Fine, M. Forbes, and L. Wessels, vol. 1, 25–37. East Lansing, MI: Philosophy of Science Association.

Outhwaite, William. 1987. *New Philosophies of Social Science*. Houndsmills: Macmillan.

Passmore, John. 1980 [1952]. *Hume's Intentions*. London: Duckworth.

Peale, Norman Vincent. 1996 [1959]. *The Power of Positive Thinking*. New York: Ballantine Books.

Peirce, Charles S. 1966 [1876]. "How to Make Our Ideas Clear." In *Selected Writings*, ed. P. P. Wiener, 113–36. New York: Dover Publications.

Peirce, Charles S. 1966 [1905]. "What Pragmatism Is." In *Selected Writings*, ed. P. P. Wiener, 180–202. New York: Dover Publications.

Peirce, Charles S. 1982. "Lecture on Kant." In *Writings of Charles S. Peirce: A Chronological Edition. Vol. 1: 1857–1866*, ed. E. C. Moore, 240–54. Bloomington, IN: Indiana University Press.

Peirce Charles S. 1984. "Critique of Positivism." In *Writings of Charles S. Peirce: A Chronological Edition. Vol. 2: 1867–1871*, ed. E. C. Moore, 122–30. Bloomington, IN: Indiana University Press.

Peirce, Charles S. 1986. "Educational Text-Books II." In *Writings of Charles S. Peirce: A Chronological Edition. Vol. 3: 1872–1878*, ed. E. C. Moore, 1–7. Bloomington, IN: Indiana University Press.

Peterson, Susan et al. 2005. "Teaching and Research Practices, Views on the Discipline, and Policy Attitudes of International Relations Faculty at U.S. Colleges and Universities." Williamsburg: College of William and Mary, 2005. https://www.wm.edu/offices/itpir/_documents/trip/trip_summary2005.pdf (accessed January 7, 2015).

Plant, Raymond. 1991. *Modern Political Thought*. Oxford: Blackwell.

Poincaré, Henry. 1963 [1913]. *Mathematics and Science: Last Essays*. Dernières pensées. New York: Dover.

Poincaré, Henry. 2001 [1902]. "Science and Hypothesis." In *The Value of Science: Essential Writings of Henri Poincaré*, 3–180. New York: Modern Library.

Poincaré, Henry. 2001 [1905]. "The Value of Science." In *The Value of Science: Essential Writings of Henri Poincaré*, 181–356. New York: Modern Library.

Poncaré, Henry. 2001 [1908]. "Science and Method." In *The Value of Science: Essential Writings of Henri Poincaré*, 357–572. New York: Modern Library.

Pollard, Sidney. 1968. *The Idea of Progress: History and Society*. London: Watts.

Popper, Karl. 1962. *Conjectures and Refutations: The Growth of Scientific Knowledge*. New York: Basic Books.

Popper, Karl. 1966 [1944/45]. *The Poverty of Historicism*. London: Routledge and Kegan Paul.

Popper, Karl. 1972 [1935]. *Logic of Scientific Discovery*. London: Hutchinson.

Popper, Karl. 1972 [1959]. "Preface to the First English Edition, 1959." In *Logic of Scientific Discovery*, 15–23. London: Hutchinson.

Popper, Karl. 1973. *Objective Knowledge: An Evolutionary Approach*. Oxford: Clarendon Press.

Putnam, Hilary. 1976. "'Two Dogmas' Revisited." In *Contemporary Aspects of Philosophy*, ed. G. Ryle, 202–13. Stocksfield: Oriel Press.

Putnam, Hilary. 1992 [1990]. "The Greatest Logical Positivist." In *Realism with a Human Face*, 268–77. Cambridge, MA: Harvard University Press.

Putnam, Hilary. 1994 [1981]. *Reason, Truth, and History*. Cambridge: Cambridge University Press.

Quine, Willard Van Orman. 1963 [1960]. "Carnap and Logical Truth." In *The Philosophy of Rudolf Carnap*, ed. P. A. Schilpp, 385–406. La Salle, IL: Open Court.

Quine, Willard Van Orman. 1969. "Epistemology Naturalized." In *Ontological Relativity and Other Essays*, 69–90. New York: Columbia University Press.

Quine, Willard Van Orman. 1973. *The Roots of Reference*. La Salle, IL: Open Court.

Quine, Willard Van Orman. 1980 [1951]. "Two Dogmas of Empiricism." In *From a Logical Point of View*, 20–46. Cambridge, MA: Harvard University Press.

Quine, Willard Van Ormen. 1986. "Reply to Herbert G. Bohnert." In *The Philosophy of W. V. Quine*, ed. L. E. Hahn and P. A. Schilpp, 93–95. La Salle, IL: Open Court.

Quine, Willlard Van Ormen and Rudolf Carnap. 1990. *Dear Carnap, Dear Van: The Quine–Carnap Correspondence and Related Work*. Ed. R. Creath. Berkeley, CA: University of California Press.

Rawls, John. 1971. *A Theory of Justice*. Cambridge, MA: Belknap Press of Harvard University Press.

Reichenbach, Hans. 1939. "Dewey's Theory of Science." In *The Philosophy of John Dewey*, ed. P. A. Schilpp, 157–92. Evanston & Chicago, IL: Northwestern University.

Reichenbach, Hans. 1961 [1951]. *The Rise of Scientific Philosophy*. Berkeley, CA: University of California Press.

Reisch, George A. 1991. "Did Kuhn Kill Logical Positivism?" *Philosophy of Science* 58: 264–77.

Richardson, Alan and Thomas Uebel. 2007. *The Cambridge Companion to Logical Positivism*. Cambridge: Cambridge University Press.

Rorty, Richard. 1995 [1989]. *Contingency, Irony, and Solidarity*. Cambridge: Cambridge University Press.

Russell, Bertrand. 1903. *The Principles of Mathematics*. Cambridge: Cambridge University Press.

Russell, Bertrand. 1911. "Knowledge by Acquaintance and Knowledge by Description." *Proceedings of the Aristotelian Society* 11: 108–28.

Russell, Bertrand. 1956 [1918]. "The Philosophy of Logical Atomism." In *Logic and Knowledge*. London: Allen and Unwin.

Russell, Bertrand. 1969 [1914]. *Our Knowledge of the External World as a Field for Scientific Method in Philosophy*. London: Allen and Unwin.

Russell, Bertrand. 1973 [1905]. "On Denoting." In *Essays in Analysis*, 103–19. London: Allen and Unwin.

Russell, Bertrand and Alfred N. Whitehead. 1910/13. *Principia Mathematica*. 3 vols. Cambridge: Cambridge University Press.

Ryan, Alan. 1990. *The Philosophy of John Stuart Mill.* 2nd ed. Atlantic Highlands, NJ: Humanities Press International.

Saint-Simon, Henri. 1821. *Du Système industriel.* Paris: Renouard.

Saint-Simon, Henri. 1825. *Nouveau Christianisme.* Paris: Bossange Père.

Sarkar, Sahotra, ed. 1996a. *Logical Empiricism at Its Peak: Schlick, Carnap, and Neurath.* New York: Garland.

Sarkar, Sahotra, ed. 1996b. *The Legacy of the Vienna Circle: Modern Appraisals.* New York: Garland.

Scharff, Robert C. 2002 [1995]. *Comte after Positivism.* Cambridge: Cambridge University Press.

Schilpp, Paul Arthur, ed. 1963. *The Philosophy of Rudolf Carnap.* La Salle, IL: Open Court.

Schlick, Moritz. 1908. *Lebensweisheit. Versuch einer Glückseligkeitslehre.* München.

Schlick, Moritz. 1919. *Raum und Zeit in der gegenwärtigen Physik. Zur Einführung in das Verständnis der allgemeinen Relativitätstheorie.* Berlin.

Schlick, Moritz. 1959 [1930]. "The Turning Point in Philosophy." In *Logical Positivism,* ed. A. J. Ayer, 53–59. New York: The Free Press.

Schlick, Moritz. 1959 [1932/33]. "Positivism and Realism." In *Logical Positivism,* ed. A. J. Ayer, 82–107. New York: Free Press.

Schlick, Moritz. 1959 [1934]. "The Foundation of Knowledge." In *Logical Positivism,* ed. A. J. Ayer, 209–27. New York: Free Press.

Schlick, Moritz. 1962 [1930]. *Problems of Ethics.* New York: Dover.

Schlick, Moritz. 1974 [1925]. *General Theory of Knowledge.* (Transl. from the 2nd German ed.) Wien: Springer.

Schlick, Moritz. 1979 [1927]. "On the Meaning of Life." In *Philosophical Papers. Vol. 2 (1925–1936),* ed. H. L. Mulder and B. F. B. Van de Velde-Schlick, 112–29. Dordrecht: Reidel.

Schlick, Moritz. 1979 [1931]. "The Future of Philosophy." In *Philosophical Papers, Vol. 2 (1925–1936),* ed. H. L. Mulder and B. F. B. Van de Velde-Schlick, 210–24. Dordrecht: Reidel.

Schlick, Moritz. 1979 [1936]. "Meaning and Verification." In *Philosophical Papers, Vol. 2 (1925–1936),* ed. H. L. Mulder and B. F. B. Van de Velde-Schlick, 456–81. Dordrecht: Reidel.

Schwab, Richard N. 1963. "Introduction." In *Preliminary Discourse to the Encyclopedia of Diderot,* J. L. R. d'Alembert, ix–lvi. Indianapolis, IN: Bobbs-Merrill.

Singer, Charles. 1959. *A History of Scientific Ideas from the Dawn of Man to the Twentieth Century.* New York: Barnes and Noble.

Singer, Michael. 2006. *The Legacy of Positivism.* Palgrave Macmillan.

Sjolander, Claire Turenne and Cox, Wayne S., eds. 1994. *Beyond Positivism: Critical Reflections on International Relations.* Boulder, Co: Lynne Rienner Publishers.

Skarga, Barbara. 1977. *Comte.* Warszawa: Wiedza Powszechna.

Skinner, Burrhus Frederic. 1938. *The Behavior of Organisms: An Experimental Analysis.* New York: Appleton-Century-Crofts.

Skinner, Burrhus Frederic. 1971. *Beyond Freedom and Dignity.* New York: Knopf.

Smith, Laurence D. 1986. *Behaviorism and Logical Positivism.* Stanford, CA: Stanford University Press.

Smith, Steve, Ken Booth and Marysia Zalewski. 1996. *International Theory: Positivism and Beyond.* Cambridge: Cambridge University Press.

Smythe, William E. 1992. "Positivism and the Prospects for Cognitive Science." In *Positivism in Psychology*, ed. C. W. Tolman, 103–18. New York: Springer.

Sommerville, J. 1936. "The Social Ideas of the Wiener Kreis's International Congress." *Journal of Philosophy* 33: 296–97, 300.

Stadler, Friedrich. 1997. *Studien zum Wiener Kreis. Ursprung, Entwicklung und Wirkung des Logischen Empirismus in Kontext.* Frankfurt am Main: Suhrkamp.

Stadler, Friedrich, ed. 2003. *The Vienna Circle and Logical Empiricism: Re-evaluation and Future Perspectives.* Dordrecht: Kluwer Academic Publishers.

Stadler, Friedrich. 2015. *The Vienna Circle. Studies in the Origins, Development, and Influence of Logical Empiricism.* 2nd Edition. Dordrecht: Springer.

Stadler, Friedrich and Christoph Limbeck-Lilienau, ed. 2015. *The Vienna Circle. Texts and Pictures of an Exhibition.* Münster-Berlin-London: LIT Verlag.

Standley, Arline Reilein. 1981. *Auguste Comte.* Boston: Twayne Publishers.

Szaniawski, Klemens. 1994. *O nauce, rozumowaniu i wartościach* [On science, inference, and values]. Warszawa: Wydawnictwo Naukowe PWN.

Tarnas, Richard. 1993 [1991]. *The Passion of the Western Mind.* New York: Balantine Books.

Tolman, Charles W., ed. 2009 [1991]. *Positivism in Psychology: Historical and Contemporary Problems.* Berlin: Springer.

Turgot, Anne Robert Jacques. 2007 [1750]. "Notes on Universal History." In *Life and Writings of Turgot*, ed. A. R. J. R. Turgot and W. W. Stephens. Kessinger Publishing.

Uebel, Thomas. 1992. *Overcoming Logical Empiricism from Within: The Emergence of Neurath's Naturalism in the Vienna Circle's Protocol Sentence Debate.* Amsterdam: Rodopi.

Uebel, Thomas. 2013. "'Logical Positivism'—'Logical Empiricism': What's in a Name?" *Perspectives on Science* 21(1): 58–99.

Waismann, Friedrich. 1930/31. "Logische Analyse des Wahrscheinlichkeitsbegriffs." *Erkenntniss* 1: 228–48.

Waismann, Friedrich. 1965. *The Principles of Linguistic Philosophy.* Ed. G. P. Baker, B. McGuinness and J. Schulte. London: Macmillan.

Waismann, Friedrich. 1976. *Logik, Sprache, Philosophie.* Stuttgart: Reclam.

Waismann, Friedrich. 1979 [1938]. "Foreword." In M. Schlick, *Philosophical Papers*: Vol. 2 (1925–1936), ed. H. L. Mulder and B. F. B. Van de Velde-Schlick. Dordrecht: Reidel.

Walsh, W. 1987. "Philosophy and Economics." In *The New Palgrave: A Dictionary of Economics*, vol. 3, ed. J. Eatwell, M. Milgte and P. Newman. London: Macmillan.

Walters, Alan. 1987. "Friedman, Milton." In *The New Palgrave: A Dictionary of Economics*, vol. 2, ed. J. Eatwell, M. Milgte and P. Newman. London: Macmillan.

Warner, Martin. 1984. "Philosophical Autobiography: St. Augustine and John Stuart Mill." *Philosophy and Literature: Supplement to "Philosophy."* Royal Institute of Philosophy Lecture Series 16: 189–210.

Wartoffsky, Max. 1996. "Positivism and Politics. The Vienna Circle as a Social Movement." In *The Legacy of the Vienna Circle: Modern Reappraisals*, vol. 6. ed. S. Sarkar, 79–101. New York and London: Garland Publishing.

Watson, John B. 1913. "Psychology as the Behaviorist Views It." *Psychological Review* 20: 158–77.

Watson, John B. 1967 [1924]. *Behavior: An Introduction to Comparative Psychology.* New York: Holt.

Weber, Max. 1949. "Objectivity in Social Science and Social Policy." In *Positivist Sociology and Its Critics*, vol. 1, ed. P. Halfpenny and P. McMylor, 448–61. Aldershot: Edward Elgar.

Weinberger, Ota. 1973. "Introduction: Hans Kelsen as Philosopher." In H. Kelsen, *Essays in Legal and Moral Philosophy*, ix–xxviii. Dordrecht: Reidel.

Wendt, Alexander. 1992. "Anarchy Is What States Make of It: The Social Construction of Power Politics." *International Organization* 46: 391–425.

Wendt, Alexander. 1999. *Social Theory of International Politics.* Cambridge: Cambridge University Press.

Westbrook, Robert B. 1991. *John Dewey and American Democracy.* Ithaca, NY: Cornell University Press.

White, M. 1969. *Pragmatism and the American Mind.* New York: Oxford University Press, 1973.

Williamson, Edwin. 1992. *The Penguin History of Latin America.* New York: Viking Press.

Wilson, Daniel J. 1995. "Pragmatism, Science, and Logical Positivism." In *Pragmatism: From Progressivism to Postmodernism*, ed. R. Hollinger and D. Depew, 122–41. Westport, CO: Praeger.

Winch, Peter. 1971 [1958]. *The Idea of Social Science and Its Relation to Philosophy.* London: Routledge and Kegan Paul.

Wittgenstein, Ludwig. 1953. *Philosophical Investigations.* Oxford: Blackwell.

Wittgenstein, Ludwig. 1961 [1921]. *Tractatus Logico-Philosophicus.* London: Routledge and Kegan Paul.

Wittgenstein, Ludwig. 1988 [1980]. *Remarks on the Philosophy of Psychology.* 2 vols. Ed. G. E. M. Anscombe and G. H. von Wright. Chicago, IL: University of Chicago Press.

Index